AF688661

Yang, Chen, Lu, Gao

**Injection Molding Process Control, Monitoring, and Optimization**


**eBook Bonus.**

With the purchase of this book, you also receive access to its electronic version. Use the following personal code to download the electronic pages:

nfvuc-q8nlu-k2gzc-sxda2-eyir2

Register at

**plus.hanser-fachbuch.de/ebookbonus**

You can use the e-book on your computer*, tablet PC and e-book reader.

* Requirements: Internet connection and Adobe® Reader®

# Polymer Processing Society
# Progress in Polymer Processing Series

Y. Yang / X. Chen / N. Lu / F. Gao
**Injection Molding Process Control, Monitoring, and Optimization**

T. Kanai / G. A. Campbell
**Film Processing Advances**

T. Kanai / G. A. Campbell
**Film Processing**

I. Manas-Zloczower
**Mixing and Compounding of Polymers, 2nd Edition**

M. R. Kamal / A. I. Isayev / S.-J. Liu
**Injection Molding**

J. L. White / H. Potente
**Screw Extrusion**

W. E. Baker / C. E. Scott / G.-H. Hu
**Reactive Polymer Blending**

I. M. Ward / P. D. Coates / M. M. Dumoulin
**Solid Phase Processing of Polymers**

R. S. Davé / A. C. Loos
**Processing of Composites**

Yi Yang
Xi Chen
Ningyun Lu
Furong Gao

# Injection Molding Process Control, Monitoring, and Optimization

HANSER

Hanser Publishers, Munich

Hanser Publications, Cincinnati

The Authors:

*Dr. Yi Yang,* Kunda Mould (Shenzhen) Co. Ltd., Shenzhen, Guangdong Province 518129, P.R. China

*Prof. Xi Chen,* Zhejiang University, Hangzhou, P.R. China

*Prof. Ningyun Lu,* Nanjing University of Aeronautics and Astronautics, Nanjing, P.R. China

*Prof. Furong Gao,* Hong Kong University of Science and Technology, Hong Kong, P.R. China

Distributed by:
Carl Hanser Verlag
Postfach 86 04 20, 81631 München, Germany
eMail: info@hanser.de
www.hanser-fachbuch.de

The use of general descriptive names, trademarks, etc., in this publication, even if the former are not especially identified, is not to be taken as a sign that such names, as understood by the Trade Marks and Merchandise Marks Act, may accordingly be used freely by anyone. While the advice and information in this book are believed to be true and accurate at the date of going to press, neither the authors nor the editors nor the publisher can accept any legal responsibility for any errors or omissions that may be made. The publisher makes no warranty, express or implied, with respect to the material contained herein.

The final determination of the suitability of any information for the use contemplated for a given application remains the sole responsibility of the user.

Cataloging-in-Publication Data is on file with the Library of Congress

All rights reserved. No part of this book may be reproduced or transmitted in any form or by any means, electronic or mechanical, including photocopying or by any information storage and retrieval system, without permission in writing from the publisher.

© Carl Hanser Verlag, Munich 2016
Editor: Cheryl Hamilton
Production Management: Jörg Strohbach
Typesetting: Manuela Treindl, Fürth
Coverconcept: Marc Müller-Bremer, www.rebranding.de, München
Coverdesign: Stephan Rönigk

ISBN: 978-1-56990-592-0
E-Book ISBN: 978-1-56990-593-7

**PROGRESS IN POLYMER PROCESSING (PPP) SERIES**

Musa R. Kamal, *Series Editor*
McGill University, Canada

## Editorial Advisory Board


M. Cengiz **Altan**
University of Oklahoma
U.S.A.

Patrick **Anderson**
Eindhoven University of Technology
NETHERLANDS

Satinath **Bhattacharya**
RMIT University
AUSTRALIA

Mosto **Bousmina**
Hassan II Academy of Science and Technology
MOROCCO

Shia-Shih **Chen**
Chung Yuan Christian University
TAIWAN

Phil **Coates**
University of Bradford
UNITED KINGDOM

Philippe **Cassagneau**
Université Claude Bernard
FRANCE

Frank **Henning**
Fraunhofer-Institut für Chemische Technologie (ICT)
GERMANY

Sadhan **Jana**
University of Akron
U.S.A.

Furong **Gao**
Hong Kong University of Science & Technology
HONG KONG

Andrew **Hrymak**
Western University
CANADA

Dilhan **Kalyon**
Stevens Institute of Technology
U.S.A.

Samuel **Kenig**
Shenkar College
ISRAEL

Takeshi **Kikutani**
Tokyo Institute of Technology
JAPAN

Jose **Covas**
University of Minho
PORTUGAL

Masami **Okamoto**
Toyota Technological Institute
JAPAN

Chul **Park**
University of Toronto
CANADA

Luiz A. **Pessan**
Universidade Federal de São Carlos
BRAZIL

Changyu **Shen**
Dalian University of Technology
CHINA

Mark **Smith**
Carl Hanser Verlag GmbH & Co. KG
GERMANY

Giuseppe **Titomanlio**
Università degli Studi di Salerno
ITALY

Lih-Sheng (Tom) **Turng**
University of Wisconsin
U.S.A.

John **Vlachopoulos**
McMaster University
CANADA

Ica **Manas-Zloczower**
Case Western Reserve University
U.S.A.

# Foreword

The polymer and plastics industries have had a profound techno-economic impact on society for almost a century. In fact, it has been suggested that the advent and use of polymers and plastics products have represented a revolutionary technological change. They are used in packaging, furniture, construction materials, automotive, aerospace, sporting goods, biomedical, electronics, communications, and so on. More importantly, they have adapted to the ever changing social and technological demands. Thus, many of the current popular products, such as smart phones, computers, and other technological innovations would be difficult to contemplate in the absence of polymers. It does not seem likely that the foreseeable future will see a reduction in the important role that polymers and plastics will play in future technological development.

Cognizant of the role that polymers played and will continue to play in our lives, a group of polymer scientists and engineers from various countries around the world founded the Polymer Processing Society (PPS) in March 1985 at the University of Akron, Akron, Ohio, USA. According to its constitution, the goal of the PPS is to foster scientific understanding and technical innovation in polymer processing by providing a discussion forum in the field for the worldwide community of engineers and scientists. Thus, PPS has attempted to achieve this goal using the following mechanisms:

1. Organization of annual and regional conferences rotating among the various regions of the world and the dissemination of technical content of the conferences in the form of proceedings.
2. The publication of the International Polymer Processing (IPP) Journal.
3. The publication of the Progress in Polymer Processing (PPP) Series.

So far, these activities have allowed the PPS and its members to exchange information and ideas about the evolution of the principles and methods of polymer science and engineering and their application to the generation of innovative products, processes and applications.

Since the formation of PPS, eleven PPP volumes have been published. Four distinguished leaders in the polymer processing field have served as series editors:

Leszek Utracki, Warren Baker, Kun Sup Hyun, and James L. White. Two years ago, in Nuremberg-Germany, I was asked by the Executive of PPS to serve as PPP series editor. At the time, I indicated that with the help of the Advisory Editorial Board, our colleagues in the polymer processing field, and Hanser Publications, we would aim to publish at the rate of about one book every year. So far, we are meeting this goal. Already, we have two books under preparation for publication during the next two years, in addition to discussion with other potential authors/editors for subsequent years. Of course, we would be happy to produce more than one excellent book per year, if the opportunity arises. I encourage prospective authors to contact me or any of the Advisory Board members with their ideas and suggestions.

Injection molding is the most versatile, flexible, and dynamic plastics production operation. It has been used to manufacture products from practically all thermoplastic polymers, blends, composites, and nanocomposites. The versatile injection molding process can be used to manufacture, repetitively at high rates, products with complex shapes, micro to large sizes, multilayers and colors, with or without inserts. The injection molded products must satisfy a multitude of specifications relating to shape, dimensions, dimensional and shape stability, strength, surface characteristics, and other specifications associated with functionality and the requirements of the intended application. The large number of products, molders, and machinery manufacturers has led to varying types and sizes of machines and to the development of various optimum strategies for manufacturing products meeting the required specifications.

A critical aspect for the success of the injection molding process depends on understanding and control of the various steps of the injection molding process, the thermo-mechanical history experienced by the polymer throughout the process, and the impact of this history on the characteristics of the final product. As many of these interactions and concepts are complex, it is very important to develop a monitoring strategy that permits the identification of the status and responses of the critical process variables. Overall, a successful injection molding process must be coupled to a successful process monitoring, optimization, and control strategy.

In view of the above, it is a pleasure to introduce this year the important book entitled *Injection Molding Process Control, Monitoring, and Optimization*. I am confident that the book will represent a major contribution to the science and practice of injection molding. It should satisfy some of the critical needs of injection molding machine manufacturers, mold and product designers, and molders. Moreover, the book should be helpful to researchers and teachers in the fields of injection molding and process control.

Finally, on behalf of the Polymer Processing Society and the PPP Editorial Advisory Board, I would like to express our sincerest thanks and appreciation to the authors for the intensive effort they made to prepare this valuable and important book.

We owe a lot of thanks to Dr. Mark Smith and Ms. Cheryl Hamilton and other Hanser staff for their efforts to ensure a timely completion of this project and for the organization of the copyediting and production of the book.

*Musa R. Kamal*

Series Editor

# Contents

Foreword .......................................................... VII

1 Injection Molding: Background .......................................... 1
1.1 Plastic Materials and Properties ...................................... 1
1.1.1 Plastics Classification ............................................ 3
1.1.1.1 Molecular Structure ............................................. 3
1.1.1.2 Processability .................................................. 5
1.1.1.3 Method of Synthesis ............................................. 5
1.1.1.4 Monomer(s) in Molecules ......................................... 7
1.1.2 Structural Characteristics of Plastic ............................. 7
1.1.2.1 Molecular Weight and Distribution ............................... 8
1.1.2.2 Degrees of Crystallinity ........................................ 10
1.1.2.3 Orientation ..................................................... 11
1.1.3 Basic Rheology Terminology ........................................ 12
1.1.4 Non-Newtonian Flow: Phenomenon and Constitutive Equations .......... 15
1.1.4.1 Normal Stress Differences in Shear Flows ........................ 15
1.1.4.2 Viscoelastic Behavior ........................................... 17
1.1.4.3 Viscoelastic Models ............................................. 19
1.1.4.4 Extensional (Elongation) Flow ................................... 21
1.1.4.5 Polymer Melt Constitutive Equations for Viscous Flow ............ 22
1.1.4.6 Power Law Constitutive Equation ................................. 23
1.1.4.7 Effects of Temperature and Pressure on Viscosity ................ 26
1.1.4.8 Effect of Temperature on Viscosity .............................. 26
1.2 Plastics Processing Technologies ...................................... 29
1.2.1 Extrusion ......................................................... 29
1.2.2 Blow Molding ...................................................... 31
1.2.3 Injection Molding Machine, Process, and Key Variables ............. 32
1.2.3.1 Injection Molding Machine and Process ........................... 32
1.2.3.2 Injection Molding Key Process Variables ......................... 35
References ............................................................. 37

2 Feedback Control Algorithms Developed for Continuous Processes .... 39
2.1 Introduction of Feedback Control Background .... 39
2.2 Traditional Feedback Control: PID .... 41
2.3 Adaptive Control .... 44
2.3.1 Model Estimation .... 45
2.3.2 Pole-Placement Controller Design .... 46
2.3.3 Solving the Diophantine Equation .... 47
2.3.4 Injection Velocity Adaptive Control Result .... 48
2.3.4.1 Antiwindup Estimation .... 50
2.3.4.2 Adaptive Feed-Forward Control .... 52
2.3.4.3 Cycle-to-Cycle Adaptation .... 57
2.3.4.4 Adaptive Control Results with Different Conditions .... 58
2.4 Model Predictive Control .... 61
2.4.1 Basic Principle of MPC and GPC .... 61
2.4.2 Model Order Determination .... 64
2.4.3 Comparison with Pole-Placement Control .... 65
2.4.4 GPC Control with Different Conditions .... 68
2.5 Fuzzy Systems in Injection Molding Control .... 70
2.5.1 Fuzzy Inference System Background .... 70
2.5.2 Fuzzy V/P Switch-Over .... 71
2.5.3 Fuzzy V/P System Experimental Test .... 76
2.5.4 Further Improvement .... 76
References .... 80

3 Learning Type Control for the Injection Molding Process .... 83
3.1 Learning Type Control Background .... 83
3.2 Basic Iterative Learning Control .... 85
3.2.1 PID-Type ILC .... 85
3.2.2 Time-Delay Consideration .... 86
3.2.3 P-Type ILC for Injection Velocity .... 87
3.2.4 P-Type ILC for Packing Pressure .... 88
3.3 Optimal Iterative Learning Control .... 90
3.3.1 Problem Formulation .... 91
3.3.2 Optimal Iterative Learning Controller .... 92
3.3.3 Robust and Convergence Analysis .... 94
3.3.4 Selection of the Weighting Matrices .... 96
3.3.5 Simulation Results .... 97
3.3.6 Experimental Results of Optimal ILC .... 101
References .... 106

4 Two-Dimensional Control Algorithms ..... 109
4.1 Two-Dimensional Control Background ..... 109
4.2 Two-Dimensional Generalized Predictive Iterative Learning Control ..... 112
4.2.1 2D-GPILC Control Algorithm ..... 112
4.2.2 Injection Velocity Control with 2D-GPILC ..... 116
4.3 Two-Dimensional Dynamic Matrix Control ..... 121
4.3.1 Problem Formulation ..... 121
4.3.2 Controller Design ..... 122
4.3.2.1 2D Equivalent Model with Repetitive Nature ..... 123
4.3.2.2 2D Prediction Model ..... 123
4.3.2.3 Cost Function and Control Law ..... 125
4.3.2.4 Analysis of Convergence and Robustness ..... 127
4.3.2.4.1 Model of the Closed-Loop Control System ..... 128
4.3.2.4.2 Tracking Error and Convergence Conditions ..... 130
4.3.2.4.3 Robustness Analysis ..... 133
4.3.3 Simulation Illustration ..... 136
4.3.3.1 Case 1: Convergence Test ..... 138
4.3.3.2 Case 2: Repetitive Disturbances ..... 140
4.3.3.3 Case 3: Nonrepetitive Disturbances ..... 142
4.3.4 Experimental Test of 2D-DMC ..... 144
References ..... 148

5 Statistical Process Monitoring (SPM) of Injection Molding: Basics ..... 149
5.1 Process Monitoring ..... 149
5.2 Statistical Process Monitoring (SPM) ..... 150
5.2.1 Data Collection and Preprocessing ..... 154
5.2.2 Construction of Nominal Statistical Model ..... 155
5.2.3 Application of Statistical Models ..... 157
5.3 Multivariate Statistical Analysis Methods for SPM ..... 158
5.3.1 Principal Component Analysis and Partial Least Squares ..... 158
5.3.2 PCA/PLS-Based Statistical Process Monitoring ..... 160
5.3.3 Multiway PCA/PLS ..... 162
5.3.4 Multiway PCA/PLS-Based Batch Process Monitoring ..... 165
5.4 Challenges in Monitoring Injection Molding Process ..... 166
5.4.1 Multiple Operation Phases ..... 166
5.4.2 Within-Batch and Batch-to-Batch Dynamics ..... 168
5.4.3 Unequal Batch Length ..... 169
References ..... 170

6 Phase-Based SPM Strategies . . . 173
6.1 Introduction . . . 173
6.2 Phase-Division-Based Sub-PCA Modeling and Monitoring . . . 175
6.2.1 Overview . . . 175
6.2.2 Data Normalization . . . 176
6.2.3 Phase Recognition and Division . . . 177
6.2.4 Phase PCA Modeling . . . 180
6.2.5 Statistics and Control Limits . . . 181
6.2.6 Online Process Monitoring . . . 182
6.2.7 Summary . . . 183
6.3 Application of Phase-Based SPM to Injection Molding . . . 184
6.3.1 Experimental Setup . . . 184
6.3.2 Result Analysis of Phase Division and Modeling . . . 185
6.3.3 Result Analysis of Process Monitoring and Fault Diagnosis . . . 187
6.4 Improved Phase-Based SPM for Unequal-Length Batch Processes . . . 193
6.4.1 Overview . . . 193
6.4.2 Data Normalization . . . 194
6.4.3 Phase Recognition and Division . . . 196
6.4.4 Sub-PCA Modeling Procedure . . . 198
6.4.5 Process Monitoring Procedure . . . 199
6.4.6 Summary . . . 200
6.5 Application of Improved Phase-Based SPM to Injection Molding . . . 202
6.5.1 Experimental Setup . . . 202
6.5.2 Result Analysis of Phase Division and Modeling . . . 203
6.5.3 Result Analysis of Process Monitoring and Fault Diagnosis . . . 205
6.5.3.1 Monitoring of a Normal Batch . . . 205
6.5.3.2 Monitoring of Faulty Batches . . . 207
References . . . 211

7 Phase-Based Quality Improvement Strategies . . . 213
7.1 Introduction . . . 213
7.2 Phase-Based Process Analysis and End-Product Quality Prediction (Method A) . . . 214
7.2.1 Phase-Based PLS Modeling . . . 214
7.2.2 Phase-Based Quality-Related Process Analysis . . . 217
7.2.3 Online Quality Prediction . . . 219
7.3 Application of Phase PLS Model (Method A) to Injection Molding . . . 220
7.3.1 Experimental Setup . . . 220
7.3.2 Illustration of Phase-Based Process Analysis . . . 222
7.3.2.1 Phase Division . . . 222
7.3.2.2 Process Analysis in the Critical-to-Surface Phase . . . 224

7.3.2.3 Process Analysis in Critical-to-Dimension Phases ......225
7.3.3 Illustration of Phase-Based Quality Prediction ................228
7.4 Phase-Based Process Analysis and End-Product Quality Prediction (Method B) ......................................232
7.4.1 Critical Phase Identification ..........................232
7.4.2 Key Variable Selection Based on Variable-Wise Unfolding........235
7.4.3 Phase-Based PLS Modeling Algorithm ......................239
7.4.4 Online Quality Prediction................................241
7.5 Application of Phase PLS Model (Method B) to Injection Molding........242
7.5.1 Illustration of Correlation Analysis.......................242
7.5.2 Results of Quality Prediction.............................246
References..............................................250

8 In-Mold Capacitive Transducer for Injection Molding Process......... 251
8.1 Fundamentals of Capacitive Transducers...........................252
8.2 Dielectric Properties of Polymers ...............................255
8.3 Principle and Preliminary Tests of Capacitive Transducer in Injection Mold ..............................................257
8.4 Design of In-Mold Capacitive Transducer..........................260
8.4.1 Mold Base Design ........................................260
8.4.2 Mold Insert Design ......................................263
8.4.3 Capacitance Measurement...................................264
8.5 Applications in Melt Flow Detection during Filling Stage ........266
8.5.1 Detection of Filling Start ...............................266
8.5.2 Detection of V/P Transfer.................................267
8.5.3 Detection of melt flow during filling.....................269
8.6 Applications for the Packing and Cooling Stages .................279
8.6.1 Guide to Packing Pressure Setting ........................279
8.6.2 Detection of Gate Freezing-Off Time.......................282
8.6.3 Solidification Rate Monitoring............................284
8.7 Online Part Weight Prediction Using the Capacitive Transducer ........287
References..............................................292

9 Profile Setting of Injection Velocity............................ 295
9.1 Constant Melt-Front-Velocity Strategy ...........................295
9.2 Scheme Based on Average-flow-length .............................301
9.3 Neural Network Model of Average-flow-length ......................302
9.3.1 Inputs and Output of the Neural Network Model ...............302
9.3.2 Architecture of the Neural Network Model .................303
9.3.3 Training Algorithm.......................................305
9.3.4 Data Collection of Training and Validation Samples.............306

9.3.5 Model Performance. . . . . 308
9.4 Profiling Strategy via Optimization . . . . . 317
9.5 Parabolic Velocity Profile . . . . . 319
9.6 Piece-Wise Ramp Velocity Profile . . . . . 323
9.7 Conclusions . . . . . 325
References. . . . . 326

10 Profile Setting of Packing Pressure . . . . . 329
10.1 Online Autodetection of Gate Freezing-Off Point . . . . . 329
10.1.1 Gate Freezing-Off Detection . . . . . 330
10.1.2 Development of Autodetection System . . . . . 333
10.1.3 Tests of Constant Packing Pressure Cases . . . . . 337
10.1.4 Tests of Varying Packing Pressure Profile Cases. . . . . 342
10.1.4.1 Online Detection Results of Step Pressure Profile. . . . . 342
10.1.4.2 Online Detection Results of Ramp Pressure Profile . . . . . 343
10.2 Influence of Packing Profile on Part Quality . . . . . 346
10.2.1 Constant Packing Profile . . . . . 348
10.2.2 Ramp Packing Profile . . . . . 351
10.2.3 Step-Change Packing Profile . . . . . 359
10.2.4 Summary . . . . . 363
10.3 Profiling of Packing Pressure. . . . . 364
10.3.1 Profiling Rules. . . . . 364
10.3.2 Online Profiling of Constant Packing Pressure . . . . . 365
10.3.3 Ramp Profile for Specific Thickness Distribution . . . . . 368
10.4 Conclusions . . . . . 370
References. . . . . 370

11 Parameter Setting for the Plastication Stage . . . . . 371
11.1 Visual Barrel System Development. . . . . 372
11.2 Plastication Behavior . . . . . 373
11.2.1 Melting Behavior. . . . . 373
11.2.2 Processing Condition Effects . . . . . 380
11.3 Neural Network Modeling of Melt Temperature. . . . . 384
11.4 Optimal Parameter Setting for the Plastication Stage . . . . . 385
References. . . . . 389

Subject Index. . . . . 391

# 1 Injection Molding: Background

## 1.1 Plastic Materials and Properties

Human history has been defined in terms of materials categories: the Stone Age, the Bronze Age, and the Iron Age. It is well accepted that we are now living in a polymer age. Since the 20th century, polymer materials, including plastics, fibers, elastomers, and proteins, have gradually appeared in almost every area of people's everyday life, and there are a variety of applications in agriculture, industry, and even the defense industry. In all of the polymer materials, plastic is a major class.

Plastics are ubiquitous in modern society, with applications ranging from toys to electronic components, interior or structural parts of automobiles, and different components in trains and airplanes. There is hardly an area that does not use plastic parts in modern industry. The main advantages of plastic materials compared to other commonly used materials such as metal and woods are obvious. First of all, they have good physical or chemical properties, such as low density (light weight), chemical resistance, and durability, and are thermostatically and electrically insulating. Second, they are economical in producing massive quantities of products. Third, plastic materials are normally easy to fabricate, especially compared to metal, and the energy cost accompanying plastics processing is also significantly reduced. Although plastics also have some disadvantages, such as not being biodegradable and promoting crude oil mining, these problems could be solved with recycling and the development of biodegradable plastics and other environmentally friendly enhancements.

The applications for plastics in modern industry and in people's everyday life are almost limitless. Plastic products can be found everywhere. The largest application of plastics worldwide is the packaging industry, including numerous products like containers, bottles, drums, trays, boxes, cups and vending packaging, baby products, and protection packaging. The typical materials used in this area are low-density polyethylene, high-density polyethylene, polypropylene, polystyrene, and polyethylene terephthalate.

The second largest consumer of plastic products is the building and construction industry. Since plastic materials have great versatility and a combination of excellent strength-to-weight ratio, durability, cost effectiveness, corrosion resistance, and low maintenance cost, they are an attractive choice throughout the construction sector. Pipes, siding sheets, insulators, roofing materials, and door and window frames all use plastics, and the materials used are not limited to only those used in the packing industry, but may also be some other types, like polyvinyl chlorides.

Another important application area is electrical and electronic devices. The wires of most electronic devices are encased in plastic. Most outer casings of electronic devices, such as telephones, mobile phones, lighting fixtures, fans, computers, and televisions, are plastic as well. Plastic materials like polycarbonate, polyamide, acrylonitrile butadiene styrene, and styrene acrylonitrile are widely used in this area.

A variety of plastics have been used in the manufacture of automobiles, trucks, trains for high-speed railways, and even airplanes. Nowadays many critical parts of the automobile are made of plastic materials, such as the steering wheels, air-intake manifolds, windshield wipers, bumpers, headlights, fuel pumps, and emission canisters. The train windows of China Railway High-speed (CRH) are made of six layers, where two of them are plastic materials [1]. This kind of multilayer structure can also be found in the aircraft windshield, for example, as illustrated in Fig. 1.1.

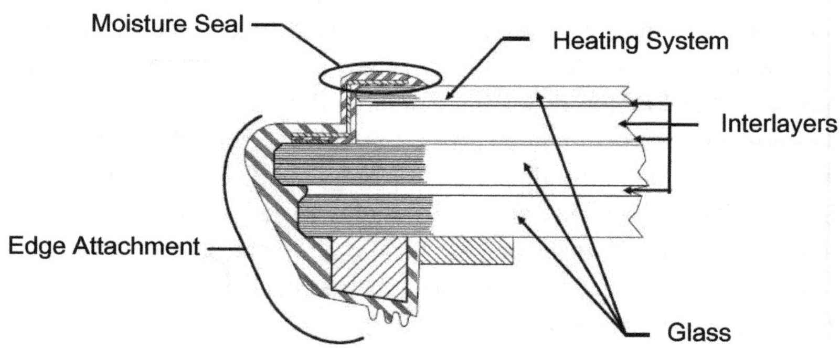


**Figure 1.1** Structure of Airbus SA windshield [2]

The transparent interlayers in this schematic are plastic materials used to bond the glass plies and other components together to form the aircraft windows. They also contribute to enhancing the windows' pressure fail-safe capability and bird-impact resistance.

Owing to the rapid development of plastic materials and their applications in recent decades, this material family is still growing. In recent years, materials scientists and engineers developed the shape-memory plastic materials, which are polymeric smart materials that have the ability to return from a deformed state induced by an external stimulus, such as temperature change [3], to their original or permanent shape. With the boom of nanotechnology, nanoplastic becomes a new engineering area. With the nanostructure units of nanocomposites added to the original plastic materials, the resulting nanoplastic can have some unique features, such as anti-bacterial, water resistance, high temperature tolerance, and high strength [4].

Plastic is a kind of organic polymer material with giant molecules, whose basic components are hundreds or thousands of monomers prepared synthetically or semisynthetically, mostly organic molecules such as carbon, hydrogen, oxygen, or nitrogen. Since plastics is a material group with vast variety, it can also include inorganic constituents, such as sulfur, chlorine, fluorine, or bromine.

Using polyethylene (PE) as an example, Fig. 1.2(a) shows the structure of PE. This structure can be shown in a more concise form as Fig. 1.2(b). It is clear that the molecular chain is constructed of many small molecules of $C_2H_2$, i.e., ethylene monomers. The notation "$n$" in Fig. 1.2(b) represents the number of repetitive units in the PE structure; its value could range from a few hundred to several thousand. Therefore, the molecular weight of PE may also be an undetermined value with a statistical distribution [5].

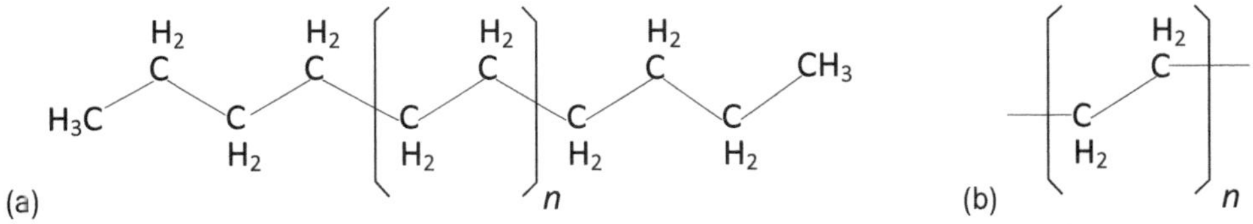


**Figure 1.2** Molecular structure of polyethylene (PE)

## 1.1.1 Plastics Classification

Plastics may be classified in many ways, based on the molecular structure, processability, method of synthesis, and so on.

### 1.1.1.1 Molecular Structure

In terms of the molecular structure, plastics can be roughly divided into linear, branched, and cross-linked structures. A schematic illustration of these typical structures can be found in Fig. 1.3, where the lines represent the polymer molecules, and the black dots indicate the monomers [6].

A linear structure [7] has repeating units linked end-to-end together in a continuous length. Each monomer in the linear plastic is thus linked to only two others, resulting in a long and narrow molecule. Poly(vinyl chloride) (PVC), high-density polyethylene (HDPE), and some polyamides (nylon) are typical linear plastics.

(a) Linear structure

(b) Branched structure

(c) Cross-linked structure

**Figure 1.3** Different structures of plastic materials

A branched structure refers to those plastic materials with branches distributed irregularly in the backbone molecular chain. These branches make it difficult to pack them in an array, so they are less dense and less crystalline and easier to process. The amount and type of the branches could affect some important physical properties of the material, such as viscosity and elasticity. Low-density polyethylene (LDPE) is a typical branched material. Although both HDPE and LDPE have ethylene as the basic unit, their structural difference leads to significantly different physical properties and applications.

A cross-link is a bond that links one polymer chain to another. The cross-linked structure contains short side chains or cross-links that connect different molecule chains into a network-fashion microstructure. The cross-linking structure makes the plastic material more elastic, meaning that when a force is exerted on the cross-linked material it could deform and then return to its original state after the force is removed. Polyacrylamide and cross-linked polyethylene are typical cross-linked plastics [5].

#### 1.1.1.2 Processability

Another classification based on processability divides the plastic materials into thermoplastic and thermoset. Briefly, the linear, branched, and slightly cross-linked plastic materials form the thermoplastics, while material that is heavily cross-linked after shaping of the products is known as a thermoset.

A thermoplastic is a plastic material that is solid and possesses significant elasticity at low temperature and can turn into a viscous liquid-like state at a high temperature, and this change is reversible. Since the thermoplastics can be softened and made to flow with the application of heat, they can be fabricated by processes like injection molding, extrusion, and blow molding. The molecular chains of a thermoplastic material are associated through intermolecular forces, hence allowing the thermoplastic to be remolded because the association force increases upon cooling and restores the bulk properties. This class of plastics includes PE, polypropylene (PP), polystyrene (PS), polycarbonate (PC), and acrylonitrile butadiene styrene (ABS).

Thermoset resin is a class of petrochemical in a soft solid or viscous state that changes irreversibly into an infusible, insoluble polymer network by curing. Curing can be induced by the action of heat or suitable radiation, or both. A cured thermosetting resin is called a thermoset. Prior to curing, the thermoset material is generally liquid or malleable with only partially polymerized states. The cure is actually a cross-linking process to form a network material. Generally, thermoset materials are stronger than thermoplastic materials because the molecular chain of the cured thermosets are connected by a three-dimensional network of bonds. However, the thermosets are more brittle, and their shapes are permanent, so they cannot be recycled to make new products. The fabrication of thermoset material must be done prior to the curing through, for example, reactive injection molding, transfer molding, extrusion, compression molding, or spin casting. Typical thermoset materials are polyurethanes (PU), polyimide (PI), and polyester resins.

#### 1.1.1.3 Method of Synthesis

All plastics are synthetic or manufactured materials, and polymerization is the most important process of converting monomer molecules through a chemical reaction into polymer chains or three-dimensional networks. Depending on the method of polymerization, plastics can also be classified into addition plastics and condensation plastics.

Addition plastics are formed by additional polymerizations or reactions, where many monomers bond together via rearrangement of bonds without the loss of any atom or molecule. For example, the following polymerization (Fig. 1.4) of polyethylene is an addition reaction [8]:

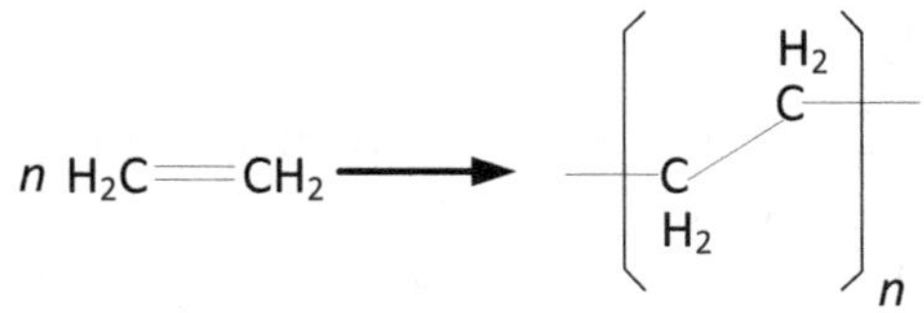

**Figure 1.4** Polymerization of polyethylene

The addition reaction normally involves unsaturated monomers, which in the previous example is ethylene.

Condensation plastics are those formed through condensation reactions, where monomer molecules join together, losing small molecules as byproducts, for example water or methanol. Using the polymerization of polyethylene terephthalate (PET) as an example, the whole reaction can be divided into two steps (Fig. 1.5). In the first step, terephthalic acid and excess ethylene glycol are reacted with a certain catalyst, with $H_2O$ as the byproduct. Excess ethylene glycol is distilled off at a high temperature with the aid of vacuum. The second step is the condensation polymerization with ethylene glycol as the byproduct. This step of the reaction proceeds at a temperature around 270–280°C and includes continuous distillation of ethylene glycol as well [9].

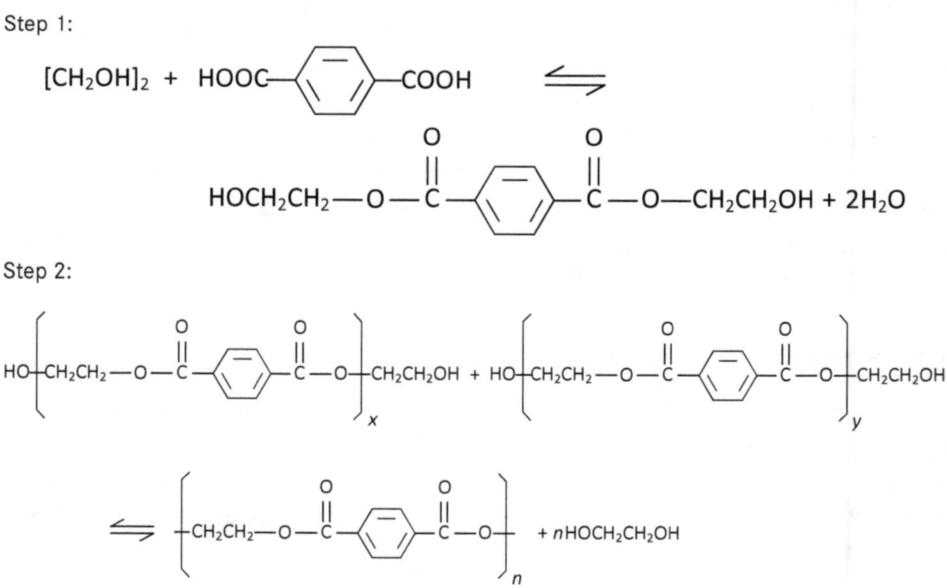


**Figure 1.5** Polymerization of polyethylene terephthalate (PET); step 1; terephthalic acid and excess ethylene glycol are reacted with a certain catalyst, with $H_2O$ as the byproduct. Excess ethylene glycol is distilled off at a high temperature with the aid of vacuum. Second step; condensation polymerization with ethylene glycol as the byproduct.

#### 1.1.1.4 Monomer(s) in Molecules

The monomer is the basic unit of plastics molecules: some plastics are constructed with a unique type of monomer, and others with more than one species. The plastics hence can be classified as homopolymer or copolymer. Typical homopolymer plastics include PE, PP, PS, and PC. The commercially relevant copolymers include ABS, styrene/butadiene copolymer (SBR), and ethylene-vinyl acetate.

Because the copolymer contains more than one type of basic unit, it can be further classified depending on how these units are arranged in the molecular chains.

- **Alternating copolymers** are those with regular alternating A and B units.

—A—B—A—B—A—B—A—B—A—B—

- **Statistical copolymers** are those with each unit distributed following a certain statistical rule. If the probability of finding a given type of monomer residue at a particular point in the chain is equal to the mole fraction of that monomer residue in the chain [5], the polymer can be referred to as a random copolymer.

—A—B—B—B—A—A—A—A—B—A—

- **Block copolymers** consists of two or more homopolymer subunits linked by covalent bonds. The union of the homopolymer subunits may need an intermediate nonrepeating subunit, known as the junction block.

—B—B—B—B—B—A—A—A—A—A—

- **Graft copolymers** are a special kind of branched copolymer in which the side chains are structurally distinct from the main chain. The individual chains of certain graft copolymers may be homopolymers or copolymers.

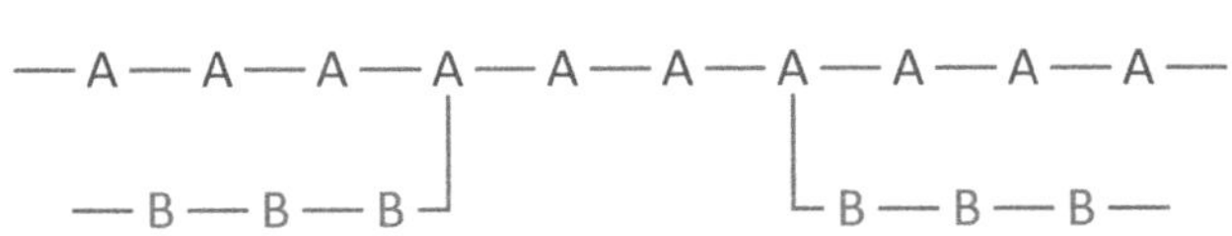


### 1.1.2 Structural Characteristics of Plastic

As a class of giant molecular polymer materials, plastics have a number of important structural characteristics that determine their processing and end-use mechanical, electrical, thermal, and optical properties. These properties are briefly introduced in this book; for a detailed introduction, the readers can refer to references [5–7].

#### 1.1.2.1 Molecular Weight and Distribution

Molecular weight is the mass of a molecule. It is calculated as the sum of the mass of each constituent atom multiplied by the number of atoms of that element in the molecular formula. For plastic materials, there is actually a group of molecular chains packed together, each with different length. They rarely have exactly the same degree of polymerization or molecular weight; there is always a distribution around an average value. A typical molecular weight distribution is shown in Fig. 1.6.

Some mean molecular weight indices can be defined to describe the relationship between the number of molecules of each chain length ($N_i$) and the molecular weight ($M_i$). The arithmetic mean, number-average, and weight-average molecular weights are the most commonly used.

- **The arithmetic mean** is simply the total mean value of the molecular weight distributions, as formulated in Eq. 1.1:

$$\bar{M} = f_1 M_1 + f_2 M_2 + \cdots + f_k M_k = \sum_i f_i M_i \tag{1.1}$$

where $f_i$ is the fraction of chains with molecular weight $M_i$. Almost all other mean values are related to this arithmetic mean.

- **The number-average molecular weight** is defined as the total molecular weight of all the polymer molecules in a given specimen divided by the total number of molecules:

$$M_n = \frac{\sum_i N_i M_i}{\sum_i N_i} \tag{1.2}$$

where $N_i$ is the $i$th kind of polymer molecule having molecular weight $M_i$.

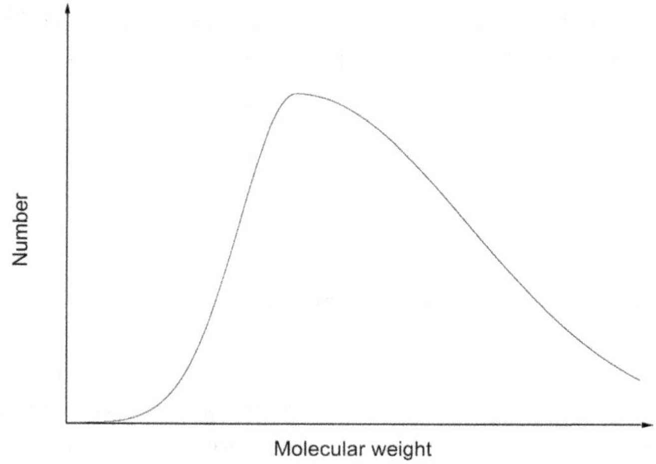


**Figure 1.6** An example of molecular weight vs. the corresponding number of chains

- **The weight-average molecular weight** is defined similar to the number average, except the number of molecules is replaced by the weight of the molecule. If $d_i$ is the total weight of the $i$th molecules, with molecular weight $M_i$, one can have

$$d_i = N_i M_i \tag{1.3}$$

The weight average $M_w$ can be obtained as

$$M_w = \frac{\sum_i d_i M_i}{\sum_i d_i} \tag{1.4}$$

Hence,

$$M_w = \frac{\sum_i N_i M_i^2}{\sum_i N_i M_i} \tag{1.5}$$

The molecular weight distribution can affect the bulk material properties directly. Using HDPE as an example, a smaller molecular weight distribution could improve the processability by reducing the viscosity of the material, while a much larger molecular weight distribution may improve the physical properties such as impact strength, toughness, and environmental stress cracking resistance. Traditional unimodal HDPE is produced using one reactor with one single distribution, so it cannot satisfy the processability and high physical properties requirements simultaneously. It has therefore appeared in industry to use the bimodal HDPE, whose molecular weight distribution has two distributions, as shown in Fig. 1.7.

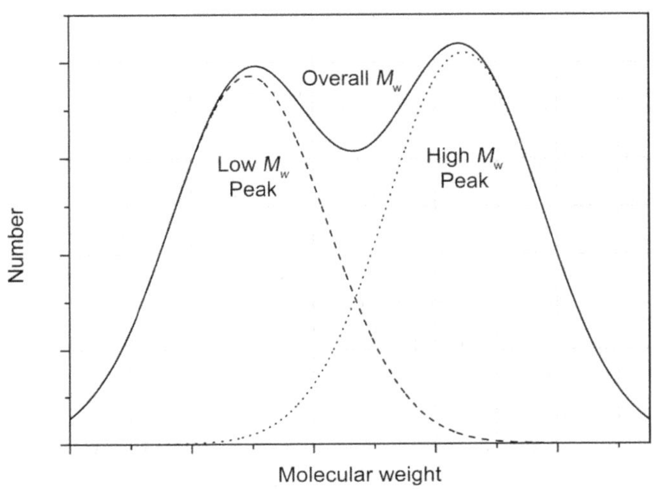


**Figure 1.7** Molecular weight distribution of a bimodal HDPE

Bimodal plastics, or in this particular example bimodal HDPE, have two molecular weight distributions, one in the low $M_w$ region and the other in the high $M_w$ region. The fabrication of this kind of plastics can be done through sequential or parallel polymerization. The material thus can have a performance balanced between the physical properties and the processability.

Bimodal HDPE is a clear illustration that structural characteristics like the molecular weight distribution can have strong impacts on the material physical properties.

#### 1.1.2.2 Degrees of Crystallinity

Crystallinity refers to the degree of structural order in the solid plastic materials. In the crystalline structure, the plastic molecules are aligned in a regular periodic manner. These chains fold together and form ordered regions called lamellae, which compose larger spheroidal structures known as spherulites [7]. The degree of crystallinity can be estimated by some analytical methodologies and typically range from 10 to 80% [7], as shown by Fig. 1.8. The crystallized plastics are often categorized as semicrystalline materials, while those with a randomly ordered molecular structure are called amorphous plastics.

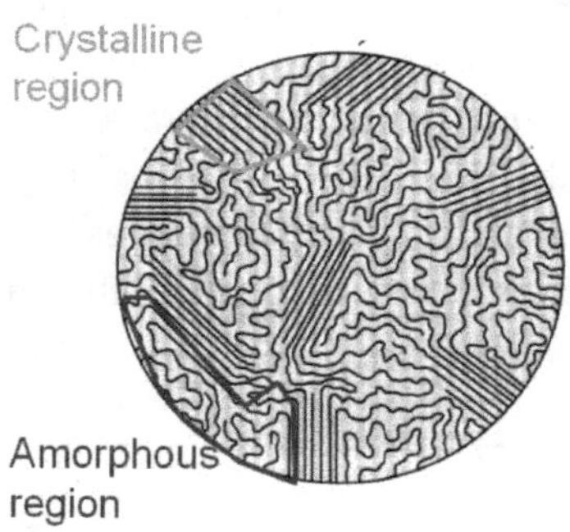


**Figure 1.8** Illustration of plastic material with amorphous and crystalline structure

For plastics molecules to crystallize, they must be first aligned and folded to form ordered regions. Another key factor that decides whether or not the crystallization can occur is the molecular structure: the straight chains with regularly spaced side groups favor crystallization. The crystallization process starts with nucleation, during which some seeds of parallel segments appear, resulting from heat motion. The detailed mechanism of nucleation is still not well discovered. The next step is the growth of the crystalline regions, which preferably occur in the direction of the largest temperature gradient. The growth is achieved by further addition of the folded polymer chain segments, and it happens only when the temperature is higher than the glass transition point $T_g$ and lower than the melting point $T_m$. To achieve a high degree of crystallinity, either the materials have small molecules, or the crystallization occurs in the material at temperatures just below the melting point for a long time. Therefore, the crystallization is actually a time-dependent process.

The degrees of crystallinity can be measured by different methods, such as density measurement, differential scanning calorimetry (DSC), X-ray diffraction (XRD), infrared spectroscopy, and nuclear magnetic resonance (NMR). Besides these analytical measurements, it is more important to understand the bulk properties of the crystalline structure. Amorphous plastics are normally hard and brittle below their glass transition temperatures, due to the low mobility of their molecules, and above $T_g$ the amorphous plastics begin to creep when a force is applied to the material. So amorphous plastics reduce their viscosity when heated, but seldom as quickly as semicrystalline materials. During processing, the amorphous materials normally exhibit lower mold shrinkage and less warpage than do semicrystalline materials. They lose their strength quickly above their glass transition temperatures.

Semicrystalline plastics have relatively stronger intermolecular forces and hence less tendency to be softened, even above the glass transition temperature. Higher crystallinity leads to a harder and more thermally stable, but also more brittle, material, while lower crystallinity helps on the elasticity and impact resistance. Another important feature of semicrystalline plastics is the strong anisotropy of their mechanical properties, along the direction of molecular alignment and perpendicular to it. Plastics nowadays are used for many optical applications, but the semicrystalline plastics are usually opaque because light scatters on the numerous boundaries between the crystalline and amorphous regions.

#### 1.1.2.3 Orientation

Orientation represents the alignment of the plastics molecules. Although usually associated with crystalline structure, orientations can and do occur in amorphous material. Under certain stress, either amorphous or semicrystalline materials will be oriented [7].

For amorphous materials, the molecular orientation can be categorized into three types, uniaxial, biaxial, and crossed orientations, and it has a strong influence on the mechanical performance [9]. The optimal pattern of orientation differs according to different application requirements. For example, a uniaxial orientation is preferable for fiber-reinforced amorphous plastics to ensure tensile strength and bending load resistance. However, for a shell structure subject to multiple stresses from different directions, the biaxial or crossed orientation can lead to maximum performance. The orientation in semicrystalline plastics can alter and reorganize the crystal morphology and meanwhile promote crystallization in the amorphous region to form an oriented crystalline structure. The physical performance of an oriented semicrystalline plastic material is determined by the orientations of both the amorphous regions and the crystalline regions.

### 1.1.3 Basic Rheology Terminology

Polymer rheology is the study of the flow of the material mainly in the liquid state, but also of deformation of the solid state in response to an applied force. Before further discussion of the major rheological behavior of the plastic materials, it is necessary to give some definitions used in the plastics melt-flow analysis [7].

- *Drag flow* is a flow caused by the relative motion of one or more boundaries with respect to the other boundaries that contain the flow. For example, Couette flow is a typical drag flow, as shown by Fig. 1.9. It is a laminar flow within two parallel plates with viscous fluid in between, where one plate is moving relative to another. The flow is driven by the viscous drag force on the fluid. In the plastics extrusion process, the relative motion of screw and barrel surfaces causes a drag flow.

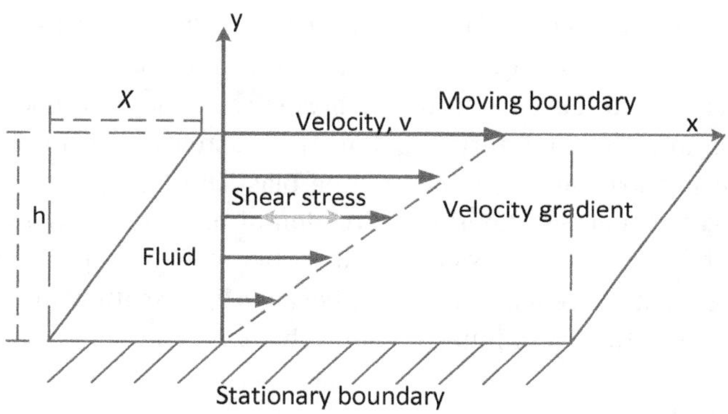


**Figure 1.9** A two-dimensional Couette flow between two infinite flat plates

- *Pressure flow* is a flow caused by the presence of pressure gradients or a local pressure difference in the fluid. Typical pressure flow includes the water flow out of the pipe, melt flow through the die in extrusion, and melt flow in an injection molding cavity. It must be noted that the pressure flow and drag flow may occur simultaneously during processing; see, for example, wire coating.
- *Shear* occurs when the velocity has a difference in any direction normal to the flow. An illustration can be seen in Fig. 1.9.
- *Elongation* occurs when there is a velocity difference in the direction of flow; an example is shown in Fig. 1.10.
- *Plug flow* is a flow where all fluid elements move at the same velocity; that is, there is flow without shear in the fluid, as illustrated in Fig. 1.11.
- *Shear rate* ($\dot{\gamma}$) is the difference in velocity per unit distance normal to the direction of flow. The shear rate, or rate of shear, is one of the most important parameters in plastics melt processing.

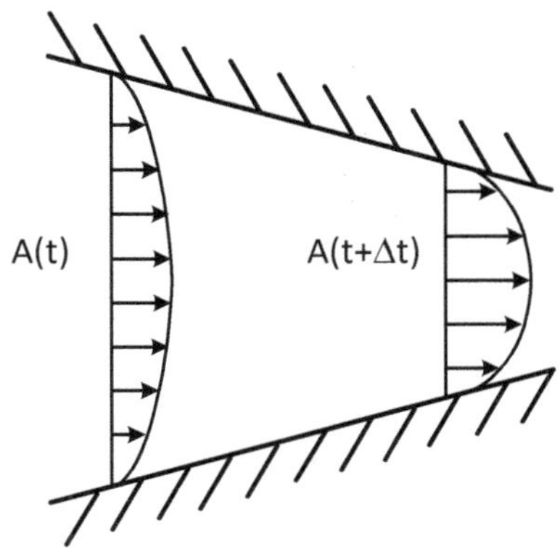


**Figure 1.10** Illustration of elongation

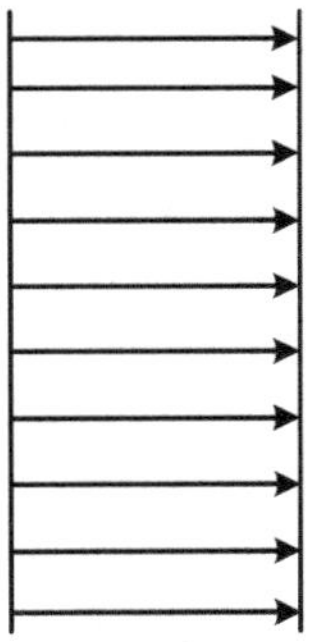

**Figure 1.11** Example of plug flow

- *Shear strain* ($\gamma$) is the deformation in the direction of flow per unit distance normal to the flow direction over a certain period of time.
- *Shear stress* ($\tau$) is the stress required to achieve a shearing type of deformation. When a fluid is sheared, a certain force will be required to bring about the deformation. The force divided by the area over which it works is the shear stress.
- *Shear viscosity* ($\eta$) is the resistance to shear flow. Quantitatively, the shear viscosity is determined as the ratio between shear stress and shear rate.

We can use the Couette flow shown in Fig. 1.9 to give the relationships between the shear rate, shear strain, shear stress, and shear viscosity. In this case,

$$\tau = \frac{F}{A}$$

where $F$ is the force applied to the moving plate, and $A$ is the cross-sectional area of the fluid parallel to the moving plate.

The shear strain is obtained by calculating the shear displacement per unit length:

$$\gamma = \frac{X}{h}$$

where $X$ is the deformation in the $x$ direction, and $h$ is the height of the flow channel. Because the shear rate $\dot{\gamma} \equiv \frac{d\gamma}{dt}$, it can be derived that

$$\dot{\gamma} = \frac{v}{h}$$

where $v$ is the velocity of the moving plate.

The shear viscosity can be written as

$$\eta = \frac{\tau}{\dot{\gamma}} = \frac{F/A}{dv/dy} = \frac{F/A}{v/h}$$

- *Newtonian fluid* is a fluid whose viscosity is independent of the shear rate.
- *Non-Newtonian fluid* is a fluid whose viscosity is dependent on the shear rate.

For a Newtonian fluid, the viscosity is only affected by pressure and temperature, so with the change of shear rate, viscosity is a constant value. For non-Newtonian fluids, although their viscosities vary with the change of shear rate, they can be further categorized into two types, shear-thickening fluid and shear-thinning fluid, as illustrated in Fig. 1.12. It is clearly shown that the viscosity of the shear-thickening fluid increases with increasing shear rate, while the viscosity of the shear-thinning fluid decreases with increasing shear rate. Wet sand is a typical shear-thickening fluid. Most plastic melts and solutions are shear-thinning.

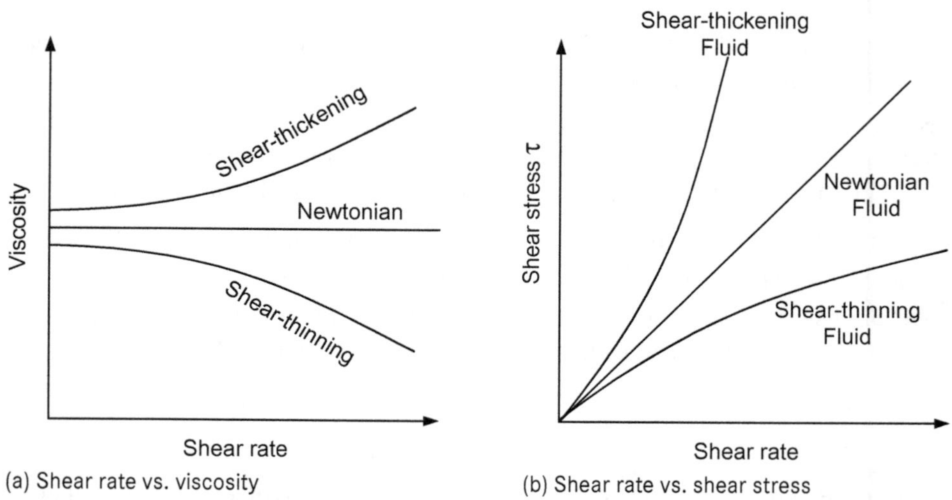


**Figure 1.12** Newtonian and non-Newtonian fluids [7]

### 1.1.4 Non-Newtonian Flow: Phenomenon and Constitutive Equations

Because the focus of rheology is the analysis of material flow in the liquid state, and the plastics melt is a non-Newtonian fluid, it is important to show how it is different from a Newtonian fluid.

#### 1.1.4.1 Normal Stress Differences in Shear Flows

Assume two identical stationary cups with a rotating rod arranged as shown in Fig. 1.13:

1. One cup contains a Newtonian fluid, such as water.
2. The other cup contains a polymer melt (or solvent).

When the rod is rotating, the fluids in these two cups behave differently.

1. In the cup with a Newtonian fluid, inertia would be the dominant factor, and the fluid is pushed away from the rod and to the edge of the cup by centrifugal forces.
2. In the cup with a polymer melt, the surface profile is the exact opposite of that in the first case. The fluid is drawn toward the rod and rises up around it. This phenomena is called the *Weissenberg effect*, named after Karl Weissenberg.

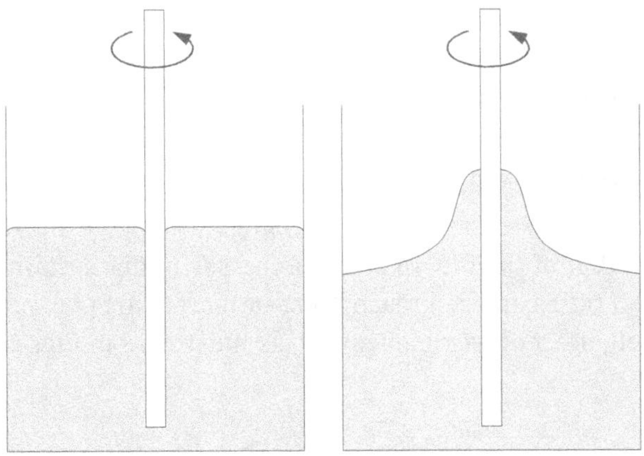

(a) Newtonian fluid (b) Plastics melt

**Figure 1.13** Illustration of Weissenberg effect [6]

The cause of the Weissenberg effect is attributed to the appearance of *strangulation stresses* exerted by the extended polymer chains. The plastics molecules become oriented with the annular flow, and in their tendency to return to the random coil state, they exert a hoop stress component on the layer of fluid next to them toward the rod.

The molecular-level explanation of this effect has its continuum mechanical counterpart. The annular tangential flow causes an extra tension in the tangential direction of flow compared to the perpendicular radial direction. That is, $\tau_{\theta\theta} - \tau_{rr} < 0$, and this normal stress difference is responsible for the tendency of the fluid to climb the rotational rod.

In addition to the Weissenberg phenomenon, many other rheological phenomena are attributed to the generation of normal stress differences. In simple shear flows, it is useful and necessary to establish a convention for labeling the stresses. Consider a shear flow that has only one nonvanishing velocity component that varies in only one direction, and define

Direction "1" as the velocity component direction,

Direction "2" as the direction along which the velocity changes, and

Direction "3" as the neutral direction.

Label the normal stress differences with the following convention:

$\tau_{11} - \tau_{22}$: the first (primary) normal stress difference, and

$\tau_{22} - \tau_{33}$: the second (secondary) normal stress difference.

The first normal stress difference is attributed to the following phenomena:

(a) In a Couette flow, the rod-climbing effect as introduced above.

(b) In a torsional flow between two parallel disks, a pressure is generated that increases with decreasing radius. In plastic materials processing, the normal stress extruder or Weissenberg extruder is designed based on this phenomenon.

(c) In a torsional flow between a flat disk and a cone, the pressure generated again increases with decreasing radius. A cone-plate viscometer exploits this phenomenon to measure the dynamic behavior of plastics melts.

(d) In the extrusion process, the extrudate passing through the die has a much larger diameter than that of the die. In some processes under certain conditions, the extrudate swell ratio can even reach 400% or more. This phenomenon is known as the die swell effect or extrusion swell, as illustrated in Fig. 1.14.

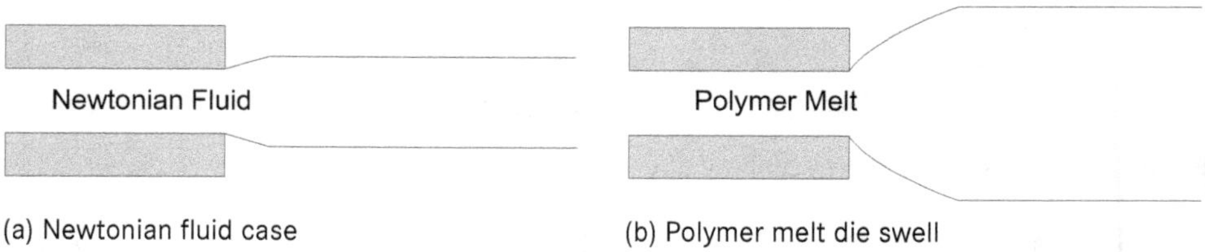


**Figure 1.14** Illustration of extrudate die swell

The secondary normal stress difference has the following properties:

1. It is an order of magnitude smaller than the primary normal stress difference.
2. Its precise measurement has not yet been achieved.

3. It is responsible for a small number of rheological phenomena, such as in wire coating: if the wire is eccentric to the die, there exist forces that tend to diminish this eccentricity. It has been shown by Tadmore and Bird [10] that such forces may be partly attributed to the secondary normal stress difference.

### 1.1.4.2 Viscoelastic Behavior

During plastics processing, the viscosity of the polymer melt is the most important variable. From a broader point of view, the materials viscosity can be categorized as linear or nonlinear, depending on the change of strain rate versus stress inside the materials. If the material exhibits a linear response, it can be categorized as a Newtonian material, where the stress is linearly proportional to the strain rate. The material can be categorized as a non-Newtonian fluid if its response to the strain rate is nonlinear. A plastics melt has the characteristics of both solid elastic and viscosity fluid behaviors. It is therefore called a viscoelastic fluid. The following analysis of strain rate and stress development clearly shows the viscoelastic behavior.

Assume that two infinitely large parallel plates at a distance $H$ are filled with fluid in between, as shown in Fig. 1.15. In an infinitesimally small interval of time, the upper plate is moved a small distance, $\delta x$, and then it is held there.

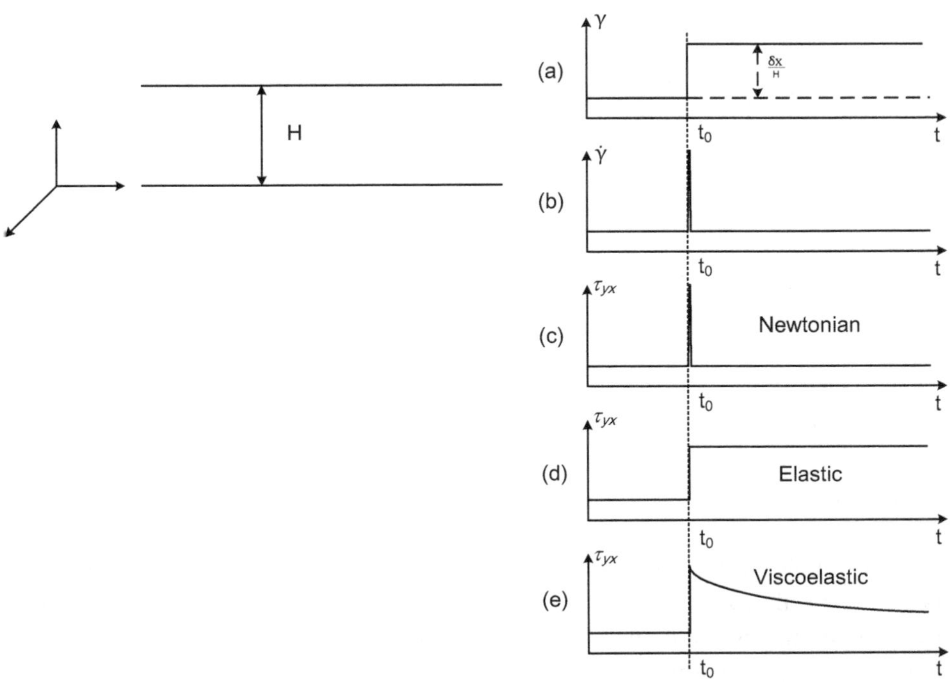


**Figure 1.15** Strain-stress relationship for Newtonian, elastic, and viscoelastic fluids

Figure 1.15(a) draws the strain in this case; the value of strain is changed from zero to $\frac{\delta x}{H}$ at time $t_0$, and the rate of strain is all zero except for the infinitesimally small interval during which the plate is moved, as shown by Fig. 1.15(b). The stresses $\tau_{yx}$ for Newtonian fluid, elastic fluid, and polymer melt are plotted in Fig. 1.15(c), (d), and (e), respectively. It is clearly shown that for the Newtonian fluid the viscosity is a constant value and there is no time dependency of the stress-strain relationship, so the stress is proportional to the rate of strain. For a pure elastic fluid, the stress is a constant value after time $t_0$ because the displacement $\delta x$ is kept constant, as shown by Fig. 1.15(d). For the plastics melt in Fig. 1.15(e), it shows characteristics of both the viscous and elastic fluids. The stress and strain relationship depends on time. In this particular example, the plastics melt exhibits as an elastic fluid at the very beginning of shear, and the stress slowly decreases until an equilibrium configuration is reached, and the fluid ultimately recovers fully after removal of the transient load.

Generally, a viscoelastic material under constant stress can accumulate strain; this phenomenon is also known as creep. If the strain is held constant, the stress decreases with time (relaxation), and the effective stiffness depends on the rate of strain, as indicated by Fig. 1.15(d). When a cyclic strain is applied, a phase lag occurs that results in a dissipation of mechanical energy. This phenomenon can be further illustrated as Fig. 1.16.

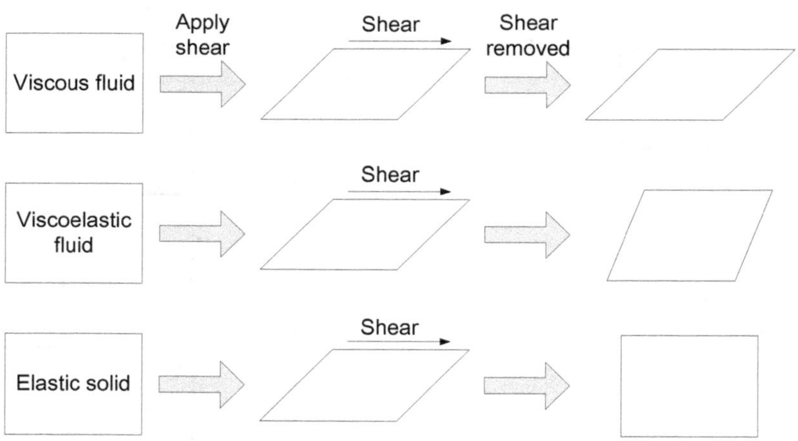


**Figure 1.16** Schematics of creep and recovery of different materials

In this example, when shear is applied to different materials like a viscous fluid, viscoelastic fluid (polymer melt), and elastic solid, they all show a certain degree of deformation. Once the shear force is removed, the viscous fluid will keep the deformation, whereas the elastic solid can recover the deformation fully and regain its original shape. For the viscoelastic fluid, it can regain its original form partly, and the remaining part of the deformation is kept permanently. The creep of the material gives the prefix visco-, and the recovery gives the suffix -elasticity.

#### 1.1.4.3 Viscoelastic Models

Viscoelastic response can be obtained, at least qualitatively, if the material is thought to have the "dual nature" of viscous fluid and Hookean elastic solid, with both aspects of this nature acting concurrently. This can be expressed in terms of simple mechanical models (Maxwell and Voigt).

**Maxwell Model (Relaxation Model)**

The Maxwell model [6, 7] can be represented by a purely viscous damper and a purely elastic spring connected in series, as shown in Fig. 1.17. The model can be represented by the following equations:

$$\gamma_{\mathrm{T}} = \gamma_{\mathrm{E}} + \gamma_{\mathrm{F}} \tag{1.6}$$

$$\gamma_{\mathrm{E}} = \frac{\tau}{G} \tag{1.7}$$

$$\frac{\mathrm{d}\gamma_{\mathrm{F}}}{\mathrm{d}t} = \frac{\tau}{M} \tag{1.8}$$

$$\frac{\mathrm{d}\gamma_{\mathrm{T}}}{\mathrm{d}t} = \frac{\mathrm{d}\gamma_{\mathrm{E}}}{\mathrm{d}t} + \frac{\mathrm{d}\gamma_{\mathrm{F}}}{\mathrm{d}t} = \left(\frac{1}{G}\right)\frac{\mathrm{d}\tau}{\mathrm{d}t} + \frac{\tau}{M} \tag{1.9}$$

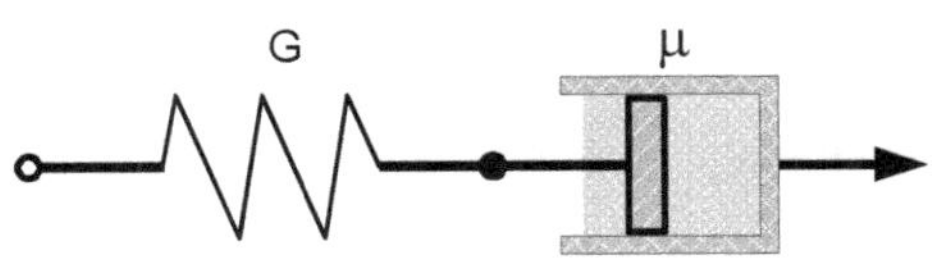


**Figure 1.17** Schematic representation of Maxwell model

Consider the situation where a substance is rapidly deformed and remains at a given deformation:

$$0 = \left(\frac{1}{G}\right)\frac{\mathrm{d}\tau}{\mathrm{d}t} + \frac{\tau}{\mu} \tag{1.10}$$

$$\tau(t) = \tau_0 \cdot \mathrm{e}^{-\frac{t}{\lambda}} \tag{1.11}$$

$$\tau(t) = \gamma_0 G \mathrm{e}^{-\frac{t}{\lambda}} \tag{1.12}$$

where $\lambda = \frac{\mu}{G}$ (relaxation time).

Equation 1.12 can be used to define the time-dependent (relaxation) modulus:

$$G(t)=\frac{\tau(t)}{\gamma_0}=G\,\mathrm{e}^{-\frac{t}{\lambda}} \tag{1.13}$$

The last expression, $G(t)=G\,\mathrm{e}^{-\frac{t}{\lambda}}$, is independent of the applied stress and strain (i.e., the experiment stimulus). This is due to the assumption of a linear Maxwell model.

**Voigt Model (Creep Recovery Model)**

The Voigt model consists of a Newtonian damper and a Hookean elastic spring connected in parallel, as shown in the picture. It is used to explain the creep and recovery behavior of polymers.

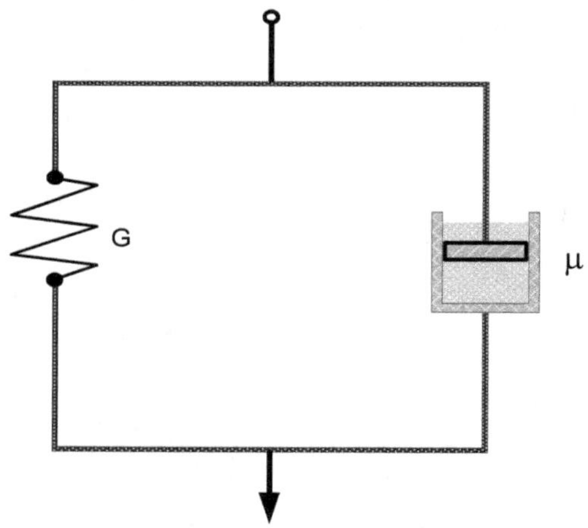


**Figure 1.18** Schematic representation of Voigt model

If a plastic melt is deformed by a stress, $\tau$, then

$$\mu\frac{\mathrm{d}\gamma}{\mathrm{d}t}+G\gamma=\tau \tag{1.14}$$

The stress $\tau$ is applied after time $t=0$, so $\tau=\tau_0$. The amount of strain is measured over the period of time of the experiment:

$$\mu\frac{\mathrm{d}\gamma}{\mathrm{d}t}+G\gamma=\tau_0 \tag{1.15}$$

Therefore,

$$\gamma(t)=\tau_0\frac{1}{G}\left[1-\mathrm{e}^{-\frac{t}{\lambda}}\right] \tag{1.16}$$

and again $\lambda = \frac{\mu}{G}$. Therefore, an experiment-independent quantity can be defined, the time dependent compliance:

$$J(t) = \frac{\gamma(t)}{\tau_0} = \frac{1}{G}\left[1 - e^{-\frac{t}{\lambda}}\right] \quad (1.17)$$

The actual stress relaxation and creep behavior of uncross-linked polymers is similar only qualitatively to the responses predicted by the Maxwell and Voigt models, even at small strain or stress levels, where such materials are linear viscoelastic.

#### 1.1.4.4 Extensional (Elongation) Flow

The most elementary type of extensional flow is uniaxial or simple extension, as illustrated by Fig. 1.19. One end of the cylinder is fixed, and force is applied to the other end to cause it to extend with a velocity $V$. The rheological question is then how the stress, the force on the cylinder end divided by the area, is related to the stretching rate.

The strain rate $(\dot{\varepsilon})$ is defined as

$$\dot{\varepsilon} = \frac{dL}{L\,dt} = \frac{V}{L} \quad (1.18)$$

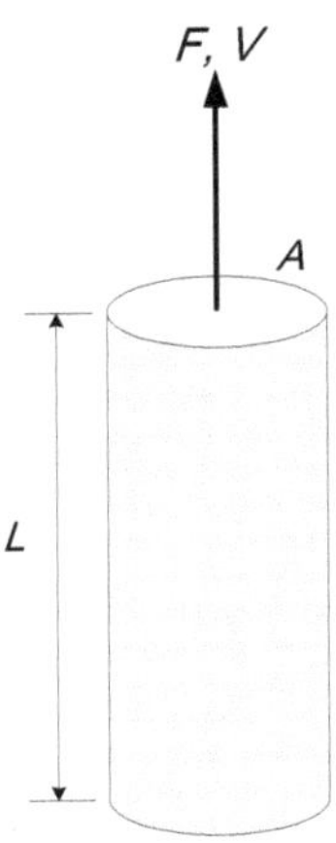


**Figure 1.19** Illustration of elongation flow

During a process, if the strain rate $(\dot{\varepsilon})$ is kept constant and if the tensile stress is constant with time, define an "elongation viscosity" as the ratio of the tensile stress to the extension rate, $(\dot{\varepsilon})$:

$$\eta_T = \frac{F/A}{\dot{\varepsilon}} \quad (1.19)$$

- For an inelastic fluid, the tensile stress rises to a constant value instantaneously after the extension begins.
- For a viscoelastic fluid, a steady state is reached only after the fluid has been stretched for a time longer than the longest relaxation time of the fluid. Because of this, $\eta_T$ is hard to get.

Extensional flow is fundamentally different from shear flow in its ability to impart high orientation to polymer molecules. This is not to say that orientation does not occur in shear flow, but extensional flow is much more effective in causing orientation to occur. Extension (or Trouton) viscosity is about three times the zero-shear viscosity. For a polymer melt, $\eta_T$ is approximately constant, but $\eta$ (shear viscosity) normally varies with $\dot{\gamma}$. Therefore, $\eta_T/\eta$ increases with high shear rate, and

$$\eta_T \approx 3\eta_0 \tag{1.20}$$

Elongation (extension) flows are important in analyzing film blow and bottle blow molding processes.

#### 1.1.4.5 Polymer Melt Constitutive Equations for Viscous Flow

From the previous discussions, it is clearly shown that polymer melts are pseudoplastic fluids. The fact that the plastic melt viscosity reduces with shear rate is of great importance in plastics processing: it allows for the realization of the extent of changes that will occur in a particular plastic material.

The typical shape of the viscosity-shear rate relationship for a plastic material can be seen in Fig. 1.20.

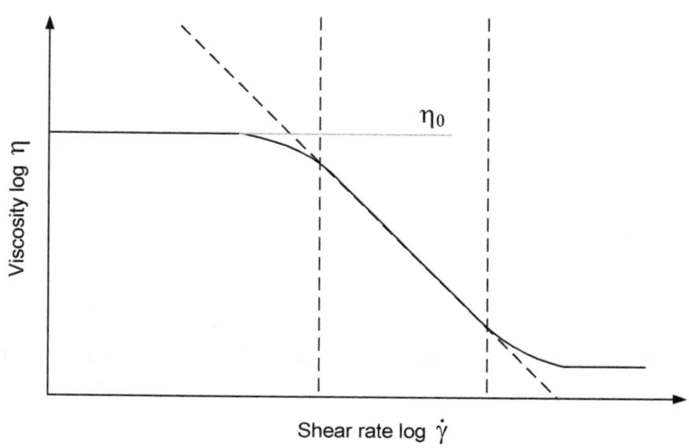


**Figure 1.20** Typical shape of viscosity-shear rate relationship for plastic material

It is clearly shown in the figure that:

- When $\dot{\gamma}$ (shear rate) is very low, viscosity $(\eta) \approx$ constant: the polymer behavior is similar to a Newtonian fluid when the shear rate is very low. The low shear rate plateau value $\eta_0$ is often referred to as the low shear limiting Newtonian viscosity.
- When $\dot{\gamma}$ (shear rate) is at a very high rate, viscosity $(\eta) \approx$ constant: the polymer behavior is similar to a Newtonian fluid when the shear rate is very high. The high shear rate plateau value $\eta_\infty$ is often referred to as the high shear limiting Newtonian viscosity. Note that $\eta_\infty$ is difficult to determine experimentally, due to the high viscous heating accompanying the high shear rate.
- The range of shear rate encountered in most polymer processing is approximately 1 to 10,000 $s^{-1}$. On the $\eta - \dot{\gamma}$ plot, it is within the range where the viscosity $(\log \eta)$ and shear rate $(\log \dot{\gamma})$ can be reasonably approximated by a straight-line relationship. This is true for most plastic materials, and this approximation is referred to as the *power law* relationship.

#### 1.1.4.6 Power Law Constitutive Equation

For a normal polymer processing operation, the shear rate may change from 1 to 10,000 $s^{-1}$. From the $\eta - \dot{\gamma}$ graph, it can be easily seen that $\dfrac{\log \eta}{\log \dot{\gamma}} \approx$ constant, so

$$\eta = m \cdot \dot{\gamma}^{n-1} \tag{1.21}$$

where $m$ is the consistency index, and $n$ is the dimensionless power law index. Parameters $m$ and $n$ can describe the log-log region of the $\eta - \dot{\gamma}$ relationship in the linear region. Equation 1.21 is the power law constitutive equation for plastics processing, also referred to as the power law of Ostwald.

The power index indicates how rapidly the viscosity reduces with the shear rate. For most thermoplastic fluids, the power law index ranges from 1 to 0.

#### Effects of Parameters in the Power Law Equation

It can be derived from the power law constitutive equation that when the power law index $n$ equals 1, the viscosity $\eta = m$ and it does not change with shear rate, and the fluid is Newtonian. When $n$ is close to 1, the fluid behavior is close to a Newtonian fluid; when $n$ decreases to 0, the non-Newtonian behavior gets stronger. For most plastic materials, $0.2 < n < 0.9$.

The parameter $m$ represents the plateau value of viscosity during a very low shear rate, so it is the upper limit of the viscosity value. This parameter has a close relationship with the molecular weight distribution, and it does change with temperature variation. In general, a higher molecular weight distribution leads to a larger value of $m$. The power law curve thus changes with different values of $m$, as illustrated in Fig. 1.22.

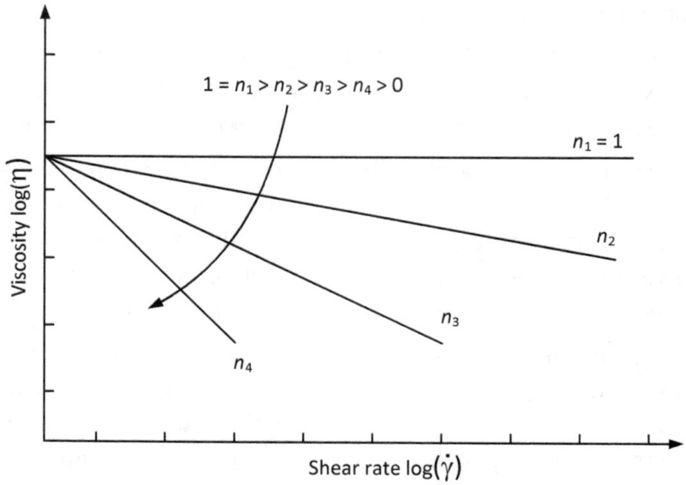


**Figure 1.21** Power law constitutive equation with different *n*

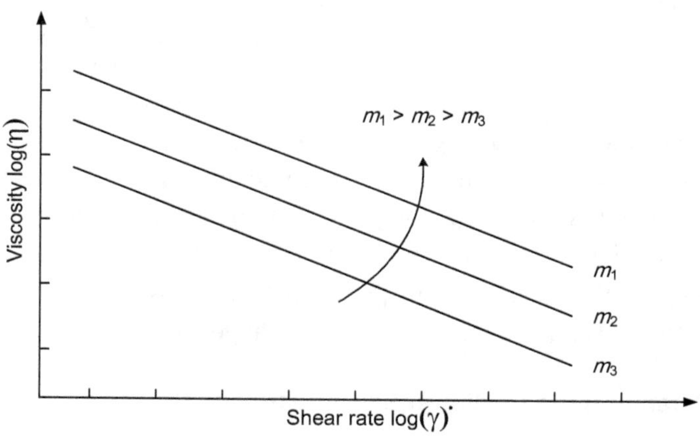


**Figure 1.22** Power law constitutive equation with different *m*

Since the shear stress $\tau = \eta\dot{\gamma}$, it is possible to write the power law equation in terms of the shear stress and shear rate $(\tau - \dot{\gamma})$ relationship as

$$\tau = m\dot{\gamma}^n \tag{1.22}$$

where *m* and *n* are the same as Eq. 1.21.

The $\tau - \dot{\gamma}$ relation for Newtonian and pseudoplastic fluids may be schematically shown in Fig. 1.23.

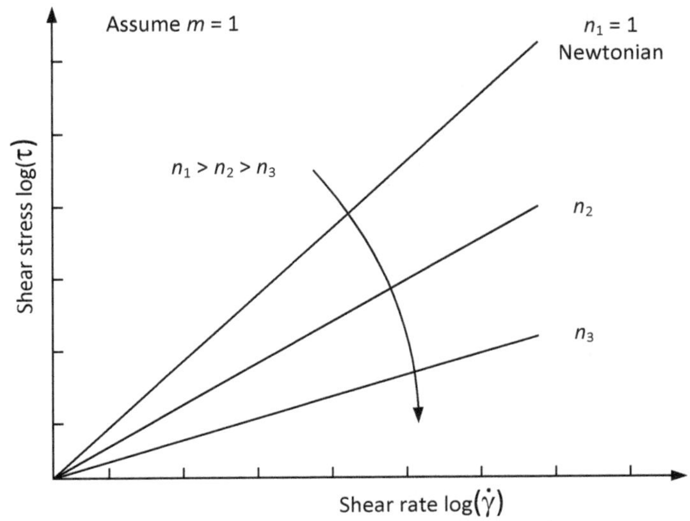


**Figure 1.23** $\tau - \dot{\gamma}$ relationship for Newtonian and pseudoplastic fluids

Equation 1.22 can be used if the shear rate is positive in the flow considered. If the shear rate changes sign at some points in the flow channel, a more general power law equation should be used:

$$\tau = m\left|\dot{\gamma}\right|^{n-1}\dot{\gamma} \tag{1.23}$$

There is another commonly used form of the power law equation to express shear rate ($\dot{\gamma}$) in terms of the shear stress:

$$\dot{\gamma} = \varphi\left|\tau\right|^{s-1}\tau \tag{1.24}$$

where $\varphi$ = the specific fluidity,

$s$ = the pseudoplasticity index, and

$n = \frac{1}{s}$, $\varphi = m^{-\frac{1}{n}} = m^{-s}$.

It is easy to understand that for Newtonian fluids, $n = s = 1$; for polymers, $1 < s < \infty$.

There are a few comments on the power law equation:

- It is not accurate over the entire range of shear rate. However, in most polymer processing problems, it serves as a good approximation.
- The major advantage of the power law equation is its simplicity. The stress and rate of deformation can be described with only two fluid properties: the consistency index $m$ and the power law index $n$.

- The power law equation does not allow for construction of a time constant from the constants $m$ and $n$; therefore, it cannot be used to analyze the transient flow phenomena where a characteristic time constant is necessary to describe the flow situation.

There are a number of other constitutive equations to describe the relationship between shear and viscosity. Since it is not the major focus of this book, they will not be listed here; please refer to references [5–7] for more details.

#### 1.1.4.7 Effects of Temperature and Pressure on Viscosity

The effect of shear rate on viscosity has been discussed in some detail in the previous sections. There are some other variables that also affect the viscosity, and these variables need to be carefully maintained or controlled during the process operations.

Two important variables that affect the viscosity are temperature and pressure. The effect of these variables is generally not as strong as the effect of shear rate, but in many cases, the effect of temperature or pressure on viscosity cannot be neglected.

#### 1.1.4.8 Effect of Temperature on Viscosity

The viscosity temperature-dependence effect is mainly reflected in the change of consistency index $m$. It has been introduced that $m$ is sensitive to the temperature variation, and the relationship may be described empirically by the following equation:

$$m = m_0 \, e^{-a(T-T_0)} \tag{1.25}$$

When the viscosity is shown against shear rate at several temperatures, the curve generally lowers with increasing temperature, as a result of the increased mobility of the polymer molecules. It must be noted that whenever experiments or processes are conducted at an elevated temperature, the possible effects of degradation have to be considered.

The power law equation including the temperature effect can then be written as

$$\tau = m_r a_T \dot{\gamma}^n \quad \text{or} \quad \eta = m_r a_T \dot{\gamma}^{n-1} \tag{1.26}$$

where $a_T$ is a shift factor, a function of temperature, and $m_r$ is the melt consistency at reference temperature $T$.

For semicrystalline and amorphous polymers above $T_g$ + 100°C, Andrade's law can be used to calculate $a_T$:

$$a_T = \exp\left[\frac{E}{R}\left(\frac{1}{T} - \frac{1}{T_r}\right)\right] \tag{1.27}$$

where $E$ is the activation energy, $R$ is the universal gas constant, and $T_r$ is the reference temperature in degrees K.

For amorphous polymers, the Williams-Landel-Ferry (WLF) equation is often used:

$$a_T = \exp\left[\frac{-c_1(T - T_r)}{c_2 + (T - T_r)}\right] \tag{1.28}$$

where $c_1$ and $c_2$ are material constants.

If the reference temperature $T_r$ is taken about 43 K above the glass transition point $T_g$, the constants $c_1$ and $c_2$ are essentially the same for a large number of amorphous polymers ($c_1$ = 8.86 and $c_2$ = 101.6). Therefore, the WLF equation becomes

$$a_T = \exp\left[\frac{-8.86(T - T_g - 43)}{101.6 + (T - T_g - 43)}\right] \tag{1.29}$$

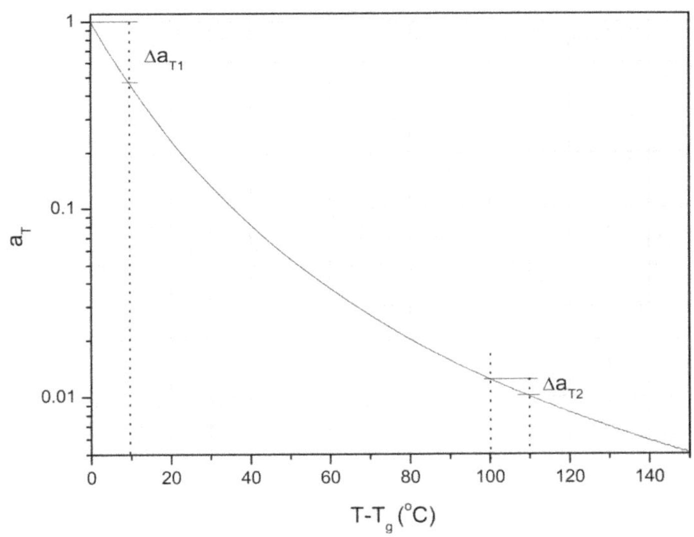


**Figure 1.24** Plot of Eq. 1.29

Figure 1.24, and from the plot, it can be noticed that a 10°C change around the reference temperature leads to a variation of $\Delta a_{T1} = 0.5321$, while a same 10°C change around 100°C above the reference temperature caused a variation of $\Delta a_{T2} = 0.0024$ only. It is clearly demonstrated in the above graphical analysis that the closer a plastic material is to its glass transition temperature, the larger the temperature sensitivity of the viscosity. The temperature sensitivity drops dramatically (several decades) with the increasing processing temperature. This explains why plastic materials whose normal process temperatures are close to their glass transition temperature exhibit high temperature sensitivities in processing. Normally, polymers are processed at a temperature considerably above their glass transition temperature (≈ 150°C > $T_g$).

The temperature sensitivity of viscosity varies widely for different plastics. As a general rule, amorphous plastic materials have higher temperature sensitivities, while semicrystalline materials have relatively lower temperature sensitivities.

**Table 1.1** Temperature Sensitivies of Viscosity for Various Plastics

| Materials | $T_g$ (°C) | $T_m$ (°C) | $\frac{1}{\eta}\frac{\partial \eta}{\partial T}$ [°C$^{-1}$] | $n$ |
|---|---|---|---|---|
| PS | 101 | - | 0.08 | 0.3 |
| PVC | 80 | - | 0.20 | 0.3 |
| PMMA | 105 | - | 0.20 | 0.25 |
| ABS | 115 | - | 0.20 | 0.25 |
| LDPE | -120/-90 | 120 | **0.03** | 0.35 |
| HDPE | -120/-90 | 130 | **0.02** | 0.5 |
| PP | -10 | 175 | **0.02** | 0.35 |
| PC | 150 | - | 0.05 | 0.70 |

**Effect of pressure on viscosity**

The effect of pressure on viscosity is relatively insignificant when pressure does not exceed 35 MPa (5000 psi). It has been found, however, that the effect of pressure on viscosity is quite significant at a pressure substantially above 5000 psi. The viscosity as a function of pressure is generally written as

$$\eta(p)=\eta(p_r)\exp\left[\alpha_p\left(p-p_r\right)\right] \tag{1.30}$$

where $p_r$ is the reference pressure, $\alpha_p$ is the pressure sensitivity coefficient, and $\alpha_p$ varies from one polymer to another and conditions.

**Table 1.2** Effect of Pressure on Viscosity

| Materials | Temp (°C) | Pressure changes (MPa) | Changes in Viscosity |
|---|---|---|---|
| **PS** | 150 | 0→100 | $\frac{\eta(p=100)}{\eta(p=0)}=200-1000$ |
| **PE** | 150 | 0→100 | $\frac{\eta(p=100)}{\eta(p=0)}=4-5$ |
| **PS** | 200 | 0→100 | $\frac{\eta(p=100)}{\eta(p=0)}=30-50$ |

Data on the pressure sensitivity of the viscosity is quite scarce.

It has been found empirically that the relative change in viscosity with pressure divided by the relative change in viscosity with temperature is approximately constant for many polymers:

$$\frac{\frac{1}{\eta}\left(\frac{\partial \eta}{\partial P}\right)}{\frac{1}{\eta}\left(\frac{\partial \eta}{\partial T}\right)} \approx -5\text{E}-7\,(°\text{C/Pa}) \tag{1.31}$$

where $\frac{1}{\eta}\left(\frac{\partial \eta}{\partial T}\right)$ can be determined by Andrade's law or the Williams-Landel-Ferry (WLF) equation.

From the above expression, it is clear that a polymer with a high temperature sensitivity of the viscosity will also have a high pressure sensitivity of the viscosity.

## 1.2 Plastics Processing Technologies

The purposes of all the plastics processing technologies are to transform the raw materials to products with commercial values. There are multiple books concerning the processing technologies [6, 7, 11], so only a brief overview is given in this book to explain the basic operations of some of the most widely used processes, including extrusion, injection molding, and blow molding. According to recent statistics, worldwide about 36% of all plastic materials go through extrusion, 32% through injection molding, 10% through blow molding, and 22% other technologies.

### 1.2.1 Extrusion

Extrusion is a continuous process during which the raw plastic material is melted and formed into continuous length of plastic products with a constant cross-sectional profile, and the products can then be cut into the desired length by certain post-die equipment. The extrusion products cover both industrial and domestic applications and typically include pipes and tubes, sheets, insulated wire, profile shapes such as window frames, cables, and even polymer granules for secondary processes. The major extrusion equipment is the extruder, which consists of four major components: (1) a feeding section, (2) a melting system, (3) a head with a die to form the desired profile, and (4) a controller to control key process variables [6].

There are different types of extruders depending on different melting mechanisms, for example, disk extruders, ram extruders, and screw extruders. The screw extruders are the most commonly used nowadays in the plastics industry, so the following description will focus on this type of extruder. Depending on the number of plasticizing screws, the screw extruders can be further divided into single-screw extruders or twin-screw extruders. The single-screw extruders have a relatively simpler structure and therefore much lower requirements for operational and maintenance efforts. The twin-screw extruders normally have higher plasticizing capability, so they are used in processing plastics with higher viscosities or those with a higher tendency to degrade. They are also used for specialty processing operations such as devolatilization and compounding.

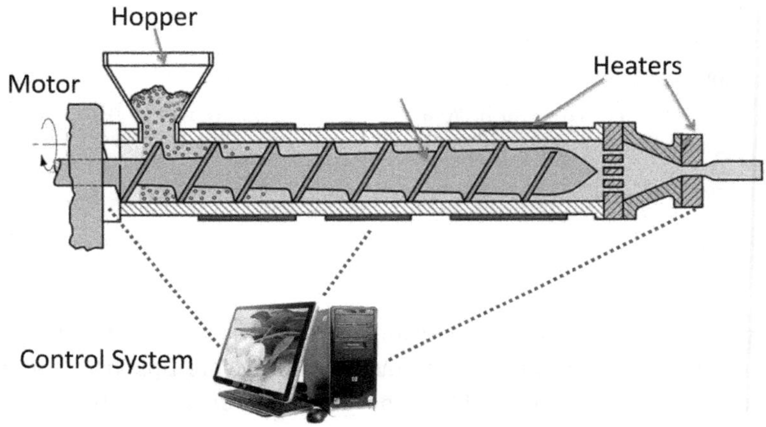


**Figure 1.25** Schematics of a single-screw extruder

The extrusion process starts with feeding the raw material, often plastic pellets and beads, into the hopper. Then the plastic granules are conveyed forward by a powerful rotating metal screw through the heated barrel. As the plastic material passes through the screw channel, it is gradually heated and melted by the shear heat generated by the rotating screw, as well as that conducted from the heaters around the barrel. The melting plastic is pumped to the tip of the screw and well mixed due to the design of the screw channel. Then the melt is forced through a screen pack to filter the contaminants and eventually pushed out of a die with properly regulated temperature and pressure to form the desired profiles. Finally, the continuous plastic product is pulled by a puller through a cooling system, typically cold water in industrial applications, and rolled or cut into pieces by the removal equipment or pelletizer.

Good product quality is accomplished under precise control of process variables [12]. During the extrusion process, there are many factors influencing the product quality, such as melt temperature, melt pressure, and viscosity. Among all these variables, melt pressure and barrel temperatures are the key process variables in the extrusion process. Extruder melt temperatures are well studied in extrusion research and applications. The fluctuations of melt temperatures influence the material viscosity, density of the polymer, and degradation kinetics and eventually affect the extrusion output rate, product quality, and process stability. Another key process variable is the melt pressure. It is as important as the melt temperature in product quality control. A melt pressure variation of 1% could lead to an equivalent extruder output variation of 1–3% depending on the rheological behavior of the polymer melt.

### 1.2.2 Blow Molding

Blow molding is a process for manufacturing hollow plastic parts, such as bottles, containers, and plastic bags. The blow molding industry started in the 1930s, and it has been one of the fastest growing industries in the past century; extensive innovations and technological developments have been made continuously. It is now the third largest plastic-processing technology for producing various products.

There are basically three types of blow molding process: extrusion blow molding, injection blow molding, and injection stretching blow molding. In extrusion blow molding, the plastic raw material is first melted and extruded into a hollow tube also known as a parison and captured by a metal mold with the desired shape. The air is then blown into the parison, inflating it into the shape of the hollow product. The mold is opened after the product is cooled enough, and the part is ejected, finishing the whole process, as schematically illustrated in Fig. 1.26.

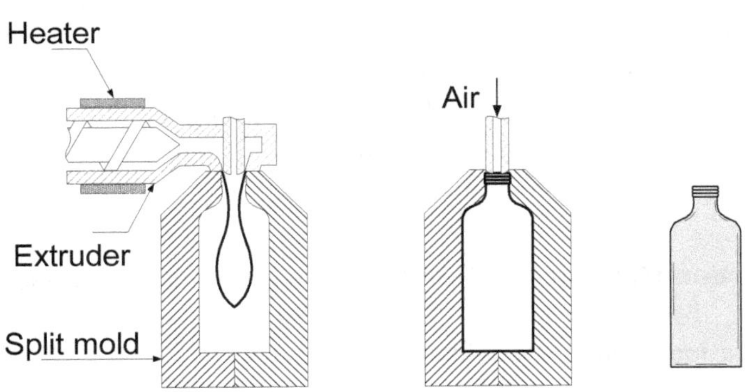


**Figure 1.26** Illustration of an extrusion blow molding process

Extrusion blow molding can also be adopted to produce plastic films. The machine is normally constructed vertically upward. An extruder is located on the bottom of the line to produce melt. The ring of molten plastic is drawn upward to form a generally hemispherical bubble on top of the die, and a pipe located inside the bubble with compressed air provides inflation and cooling for the film. As the bubble approaches the upper nip, it is converged by a roller or similar mechanism to change the shape from a hemispherical bubble to a flat and two-layer sheet. This sheet can be further cut and sealed to make bags or other products.

In the injection blow molding process, the plastic raw material is injection molded first onto a core pin, and the core pin is rotated to a blow molding subsystem to be inflated, cooled, and ejected. This kind of process can produce an injection-molded bottle neck with higher accuracy, but its capacity is limited and thus only used to produce small bottles. This kind of injection blow molding is thus the least used in the three blow molding processes.

In the injection stretching blow molding, a plastic preform with the necks of the bottles including the finishing threads is first made using the injection molding machine. These preforms can be packaged and stored for a later blow molding process, during which the preforms are reheated above their glass transition temperature, then blown, stretched, and inflated into the metal blow molds with high-pressure air. The injection stretching blow molding can have very high production speed and volume and little restriction on the bottle size and design, except the bottle must have a round shape. Preforms can be made with a separate injection molding factory other than the blow molding one. However, this process requires large floor space to facilitate the whole production line, and the capital cost is relatively high.

Like any other plastics processing technologies, precise controls of key process variables are the fundamentals for good product quality. The plastic temperature in the channel of a film blowing machine plays an important role in end-use product quality. Consequently, accurate control of parison heating temperature is needed in order to get good hollow products. It is recommended to set this temperature to be 20°C than the glass transition temperature ($T_g$). A poor control of the temperature may cause molding problems such as bubbles, nonuniform products, inproper swell, etc. Therefore, it is critical to control the temperature precisely.

### 1.2.3 Injection Molding Machine, Process, and Key Variables

#### 1.2.3.1 Injection Molding Machine and Process

Injection molding, a major polymer processing technology, transforms plastics into products of various shapes and types. It plays an important role in the plastics industry because of its high production rates, cost effectiveness, and ability to produce complex articles with high precision. Although, in terms of processing volume, injection molding is the second-largest among all the plastics processing technologies, it is the most value-added one. Both thermoplastics and thermoset materials can be processed by injection molding, but this book mainly focuses on the injection molding of thermoplastics.

Injection molding is accomplished by an injection molding machine, which has evolved from a single-stage plunger machine patterned by John Wesley Hyatt in 1872 to a modern reciprocating-screw machine developed in 1951 by William H. Willert. The process has been modified, extended, and refined over the last century. However, the basic concept remains similar to the one of Willert, with the exception of advances in the hydraulics and its control system. The reciprocating-screw injection molding machine is now the most widely used due to its reliable overall performance, such as improved melting rates, closer tolerances on shot size, and better control of temperatures. A reciprocating-screw injection molding machine comprises an injection unit, a clamping unit, and a hydraulic unit, as schematically shown in Fig. 1.27. It combines injection and plastication into a single unit, in which a reciprocating screw is used to both plasticize the material and inject the melt into the mold. The clamping unit holds, opens, and closes the mold automatically and ejects the molded product at the end of the cycle. The power unit generates and distributes either hydraulic or electrical power for every motion of the machine. The control system, as the brain of the whole machine, controls not only the machine sequencing but also some key processing variables to ensure the quality of the parts. It also provides a monitoring function to the process for consistently stable production. A modern injection molding control system also provides optimization of the process to automatically select the optimal operating point (or window) to meet the requirements of the manufacturer.

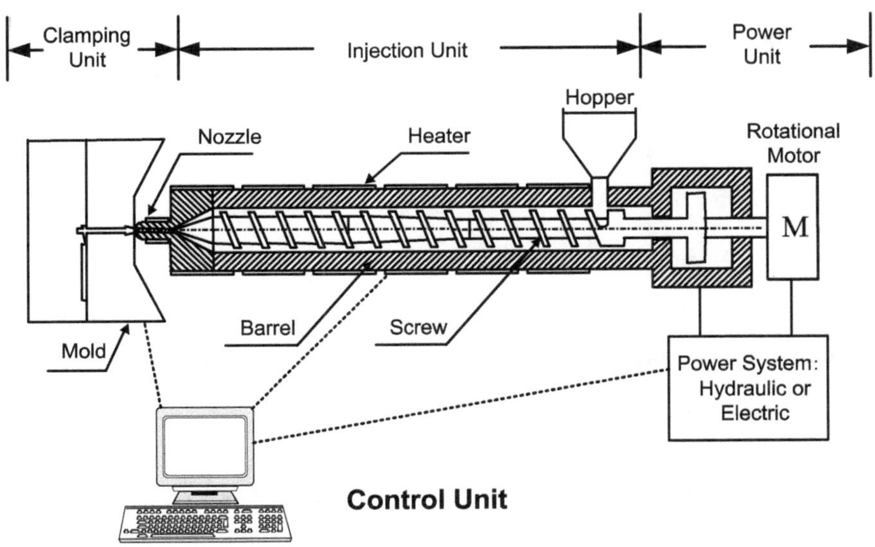


**Figure 1.27** A simplified schematic of a reciprocating-screw injection molding machine

As a cyclic process, injection molding can be divided into three main stages: filling, packing-holding, and cooling, as illustrated in Fig. 1.28 [13]. Filling is the first stage, starting after the mold is fully closed. During the filling stage, high pressure in the hydraulic cylinder forces the injection screw forward and pushes the plastic melt into the mold cavity until the mold is completely or nearly completely filled, as shown in Fig. 1.28(a). The process then switches to the packing-holding stage, during which additional material is "packed" into the mold cavity under pressure to compensate for the shrinkage associated with cooling and solidification, as shown in Fig. 1.28(b). The packing-holding stage continues until the material at the mold gate, a narrow entrance to the mold, is frozen and the material inside the mold is no longer influenced by that at the injection nozzle. The cooling stage commences, and the material is cooled inside the mold until it is rigid enough to be ejected. Concurrently with the material solidification in the cooling stage, plastication takes place inside the barrel, resulting in the melting and conveying of polymer melt to the screw tip by the screw rotation, as shown in Fig. 1.28(c). The screw rotation ceases after a sufficient amount of melt is generated in front of the screw. When the part in the mold becomes rigid enough, the mold opens, and the part is ejected, as shown in Fig. 1.28(d). The machine is then ready for the next cycle.

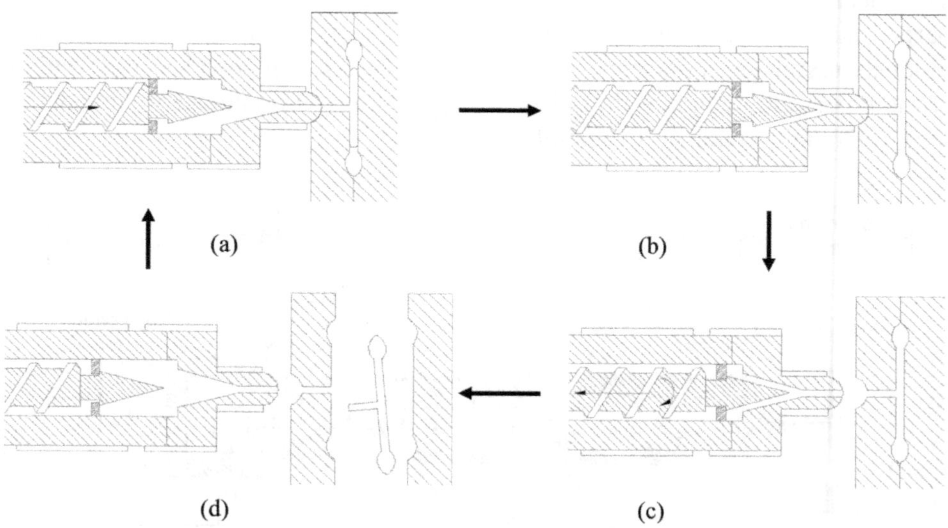


**Figure 1.28** Simplified illustration of injection molding process: (a) mold close and filling stage, (b) packing-holding stage, (c) cooling stage (plastication), and (d) mold open and part ejection

Though it appears to be simple, injection molding is rather a complex nonlinear dynamic process during which the machine parameters, material properties, and process variables interact with each other, as illustrated in Fig. 1.29 [13]. The qualities of the molded part, which may be characterized in terms of its dimensions, appearance, and mechanical properties, are strong functions of the processing conditions.

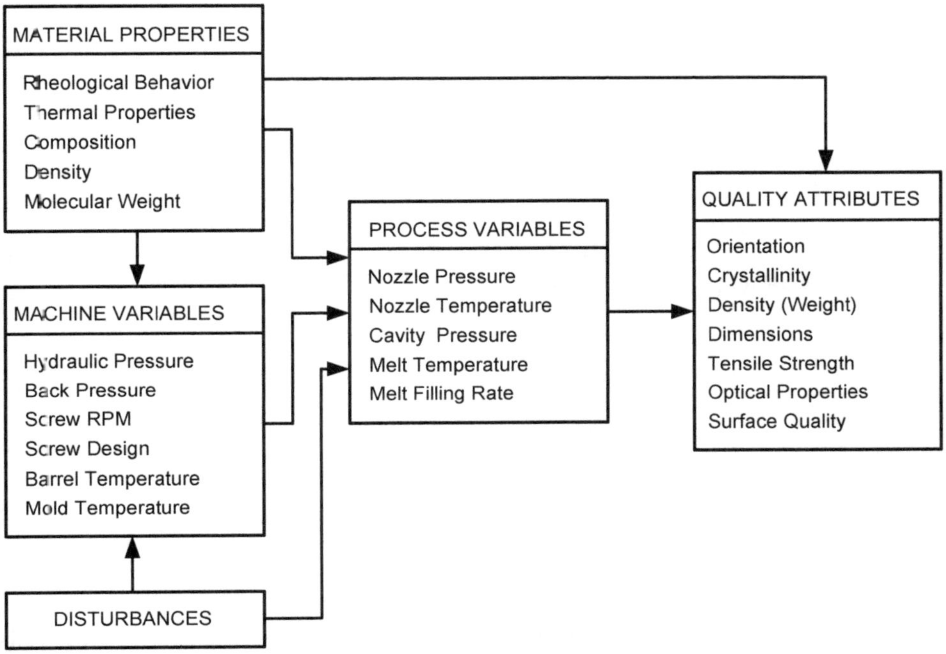


**Figure 1.29** Relationships among the machine, material, process variables, and part quality [13]

It is therefore necessary to seek precise control of these key process variables to ensure the consistency and accuracy of the injection molded product quality.

### 1.2.3.2 Injection Molding Key Process Variables

The injection molding process variables can be loosely categorized into two groups: stage-dependent variables and stage-independent variables. As a stage-based batch process, each injection molding substage needs to have its key variables be closed-loop controlled, see Fig. 1.30. During the filling stage, the nature of the polymer flow entering the cavity strongly influences the quality of the molded part, especially the mechanical properties, such as tensile strength, impact strength, heat distortion, and dimensional stability. The filling rate is approximated by the ram injection velocity (or simply injection velocity), which can be directly measured. In injection molding, a proper setting and good control of injection velocity are thus necessary. The packing stage commences after the filling. The packing pressure is a key variable in this stage because it determines how many materials will be packed into the mold. In the plastication stage, the melting of the polymer granules is mainly decided by shear heating generated from screw rotation and the back pressure. The screw rotation speed (unit: RPM) and the back pressure are therefore the key variables in plastication.

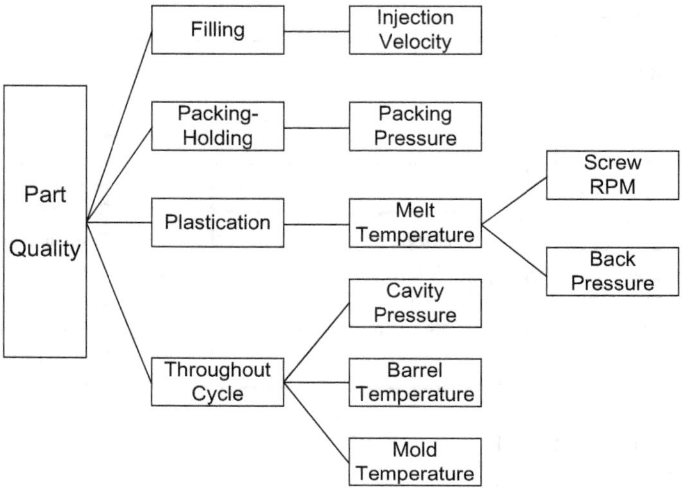


**Figure 1.30** Key process variables in the injection molding process

In the injection molding process, there are some variables that need to be controlled throughout the whole cycle, in particular the barrel temperatures, cavity pressure, and mold temperature. The barrel of the injection molding machine is a steel cylinder that contains the reciprocating screw. The outer surface of the barrel is fitted with heating media, such as electrical heater bands or an induction coil. The barrel heating system is often zoned such that a desired temperature profile can be achieved to meet the requirements of the injection molding operation. Prior to the normal operation, the barrel needs to be heated from room temperature to the required operating temperature. During the production, the barrel continuously provides the thermal energy to the plastic by means of heat conduction as the material travels through the barrel and screw channel. It is commonly accepted that about one-third of the heat contributed to the melting of plastic materials is provided by the barrel heating system. Closed-loop barrel temperature control is therefore critical to the consistent melt temperature and thermal history and hence needs to be executed throughout the molding cycle.

The cavity pressure, same as the nozzle pressure and hydraulic pressure, is directly manipulated by the hydraulic control valves. The change of the control valve opening first introduces a change in the hydraulic pressure, which then propagates to the nozzle pressure and finally arrives at the cavity pressure. The cavity pressure is the best choice as the controlled variable if the installation of the cavity pressure transducer is allowed in the product design. The cavity pressure is the clearest indicator of the status of the material in the cavity. Various studies [14–16] have confirmed that cavity pressure has the most direct effect on the microstructure and product quality.

## References


[1] H. Wu, R. Wang, S. Sang, J. Bao, L. Chen, G. Wang, C. Zhang, W. Liu, and X. Cheng, *Glass*, **3**, 34–36, 2012

[2] PPG Aerospace, Windshield Training Course (http://www.ppgprojects.com)

[3] A. Lendlein and S. Kelch, "Shape-Memory Polymers," *Angew. Chem., Int. Ed.*, **41**(12), 2034–2057, 2002

[4] S. Trimble, "Lockheed Martin reveals F-35 to feature nanocomposite structures," *Flight Int.*, 2011 (http://www.flightglobal.com/news/articles/lockheed-martin-reveals-f-35-to-feature-nanocomposite-357223/)

[5] J.R. Fried, *Polymer Science and Technology*, Prentice Hall, Professional Technical Reference (PTR), 1995

[6] Z. Tadmor and C.G. Gogos, *Principles of Polymer Processing*, John Wiley & Sons, Hoboken, 2006

[7] R.G. Griskey, *Polymer Processing Engineering*, Chapman & Hall, New York, 1995

[8] H. Köpnick, M. Schmidt, W. Brügging, J. Rüter, and W. Kaminsky, "Polyesters," *Ullmann's Encyclopedia of Industrial Chemistry A21*, Wiley-VCH, Weinheim, 233–238, 2005

[9] T. Alfrey, "Control of molecular orientation," *Polym. Eng. Sci.*, **15**(3), 234–234, 1975

[10] Z. Tadmor and R.B. Bird, "Rheological analysis of stabilizing forces in wire-coating dies," *Polym. Eng. Sci.*, **14**(2), 124–136, 1974

[11] C. Rauwendaal, *Polymer Extrusion* (3rd ed.), Hanser Publishers, 1994

[12] Z. Jiang, Y. Yang, S. Mo, K. Yao, and F. Gao, *Ind. Eng. Chem. Res.*, **51**, 14759–14770, 2012

[13] Y. Yang and F. Gao, *Control Eng. Pract.*, **8**, 1285–1296, 2000

[14] F. Gao, W.I. Patterson, and M.R. Kamal, *Polym. Eng. Sci.*, **36**, 1272–1285, 1996

[15] F. Gao, W.I. Patterson, and M.R. Kamal, *Polym. Eng. Sci.*, **36**, 2467–2476, 1996

[16] M.R. Kamal, W.I. Patterson, N. Conley, D. Abu Fara, and G. Lohfink, *Polym. Eng. Sci.*, **27**, 1403–1410, 1987

# 2 Feedback Control Algorithms Developed for Continuous Processes

## 2.1 Introduction of Feedback Control Background

Control of the injection molding process not only involves sequence manipulation, but also some key process variable regulation. The following discussion will mainly focus on the process variable control. Early injection molding controllers were mainly constructed using simple electrical components such as timing relays and switches. The control is thus naturally open-loop. As the molding process becomes more and more complex, manufacturers require greater accuracy and tighter tolerances. All of the variations and disturbances during the molding process must be properly dealt with, so closed-loop control becomes necessary, which also triggers the application of computer-based controllers to the injection molding machines. It is therefore essential to give a brief introduction to some basic control concepts.

A controlled process is the object or process of the control; if a certain variable is the control goal, it can also be referred to as a controlled variable. The controlled process or controlled variable is the major concern of the controller. The controlling variable or control variable is the one that can be changed or entered from outside that manipulates the controlled process or variable. The value of the control variable is actually the control decisions to be made by the controller. Disturbances are defined as those variables aside from the control variable that cause the controlled process or variable to deviate from the set point. Using the injection molding machine barrel temperature control as an example, the schematic of a control loop is illustrated in Fig. 2.1. In this case, the controlled variables are the different zones' barrel temperatures, the control variables are the powers for different heaters, and disturbances are the temperature of the environment, the air flow in the factory, and others.

The control system can be categorized as manual control and automatic control. Some slow processes can be adjusted manually by human operators, and this kind of control is known as manual control. In injection molding, the manual control mode is still useful in sequence control during machine setting or problem solving.

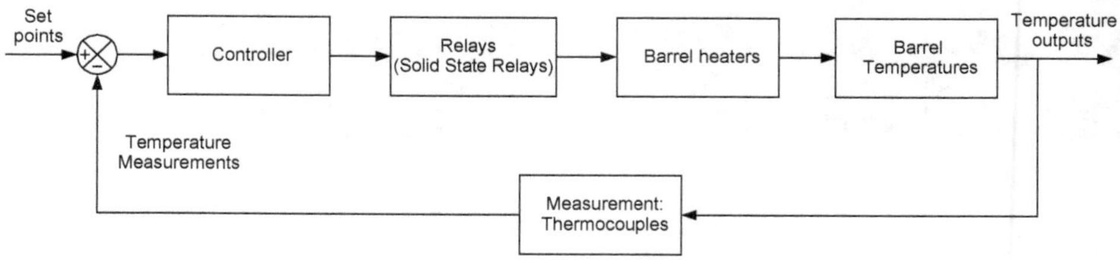


**Figure 2.1** Illustration of injection molding barrel temperature control

During normal production, since modern injection moldings are fast processes, some even with a total cycle time as short as several seconds, it is extremely difficult for the operator to maintain the machine sequence and perform key variable control. The machine and process must be controlled automatically with a mechanical or electrical controller. The objective of the automatic controller is to manipulate the controlled variable to maintain the controlled variable at its set point in spite of various disturbances.

There are some basic elements for the automatic control system to function properly: a measurement unit, a controller, and a control actuator. The measurement unit often includes a sensor that converts the controlled variable to an electrical or some other measurable signal, and a transmitter that takes the output of the sensor and transforms it into a signal strong enough to send to the controller. These two devices are also known as the primary and secondary elements, respectively. The controllers, with a certain control algorithm or strategy, decide what to do to maintain the controlled variable at the desired value. Based on this decision, the control signal is sent to the final control actuator to manipulate the process. Depending on how the controller uses the information, the automatic process control can be further divided into three types, open-loop, closed-loop, and combined controls, as shown in Fig. 2.2.

The open-loop system does not use the measurement of the controlled variable or any disturbances, and this kind of control is only applicable in cases with very good prior knowledge of the controlled process and a lack of disturbances; both cannot be satisfied in injection molding process control. In the closed-loop control, the controller receives the signal of the measurement unit and compares it with the desired values to make the control decision; for example, the $u$ in Fig. 2.2(b) is based on the observations of process output $y$. This kind of closed-loop control is also known as feedback control: the current control decisions are made based on the observations of the effects of former decisions.

It is necessary to note that in the closed-loop control scheme, there is no need for a complete and precise knowledge of the controlled process. The additional information of the disturbance loop of how $d$ affects $u$ can be obtained via the measurement of process output effects. Furthermore, all of the disturbances will eventually have

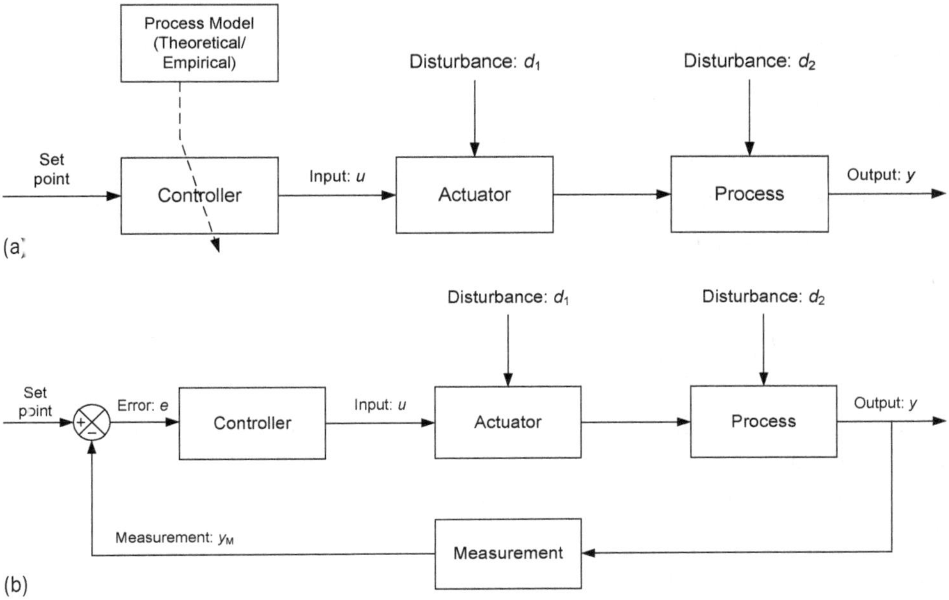


**Figure 2.2** Illustration of open-loop (a) and closed-loop (b) controls

certain influences on the output, and these influences can be compensated for by the feedback mechanism. However, in the closed-loop system, the control action counteracting the disturbances always lags behind the disturbance loop, and an open-loop system with measurement of the disturbance could be helpful in this kind of situation. Later in this book the combined control structure to improve control performance is presented.

## 2.2 Traditional Feedback Control: PID

The proportional-integral-derivative (PID) controller was first developed for automatic ship steering, and it is the most standard feedback control algorithm that measures the controlled variable, calculates the error between the output and set point, and generates the controller output based on the proportional, the integral, and the derivative of the errors. The controller uses not only the current error, but also the past error and the current rate of change. The original form of the continuous PID control algorithm can be written as follows:

$$u(t) = K_P\, e(t) + \tau_I \int_0^t e(\tau)\, d\tau + \tau_D \frac{d}{dt} e(t) \tag{2.1}$$

where $e(t)=y_{set}(t)-y_m(t)$ is the control error at time $t$, $t$ is the instantaneous time, $\tau$ is the variable of integration, and $K_P$, $\tau_I$, and $\tau_D$ are the proportional gain, integral gain, and derivative gain, respectively. In Eq. 2.1, $K_P\, e(t)$, $\tau_I \int_0^t e(\tau)\,d\tau$, and $\tau_D \frac{d}{dt} e(t)$ are the proportional term, integral term, and derivative term, respectively. The early PID controller was mostly implemented using analog circuits, but with the rapid development of computer-based control technology, the digital or discretized PID control has become widely used in modern control engineering:

$$u(k)=K_P\, e(k)+K_I \sum_{i=1}^{k} e_i + K_D\left[e(k)-e(k-1)\right] \tag{2.2}$$

The original positon form of PID control as formulated in Eqs. 2.1 and 2.2 could cause some problems, such as integration windup and difficulty in doing the integration and derivative. A more general expression is often used in industry, known as the velocity form:

$$\begin{aligned} u(k) &= u(k)-u(k-1) \\ &= K_P\left[e(k)-e(k-1)\right]+K_I\, e(k)+K_D\left[e(k)-2e(k-1)+e(k-2)\right] \end{aligned} \tag{2.3}$$

The velocity form of PID control algorithm has some advantages. In particular, it inherently removes the integration windup problem because the integration term is eliminated in Eq. 2.3.

As a simple and general-purpose control algorithm, the PID may be the most successful automatic controller in industry. For the injection molding process, PID control is commonly used in the barrel and mold temperature control. In the early practice, it was also used to control some key process variables, such as injection velocity and packing pressure. However, it also has some significant limitations. For example, it only works for linear and time-invariant processes and is not suitable for complex and nonlinear processes like injection molding, and the parameters of a PID controller are fixed, so it cannot be used as the core of an advanced control system. The PID control is suitable for continuous processes because, for a continuous process normally working around a certain operating point, the process dynamics can be linearized in a small range, and PID control can be effective under such a circumstance. As a typical batch process, the injection molding is stage-based and often operating over a wide range of conditions, and the traditional fixed parameter controller cannot ensure a satisfactory performance.

To illustrate the performance of a PID controller in an injection molding process application, a PID injection velocity control is presented below. The control experiments were conducted on an industrial injection molding machine with clamping tonnage of 88 tons, and the hydraulic power system was controlled by a fast-response servo-valve. For easy description and without loss of generality, a PI controller was designed and tuned for the injection velocity using the Ziegler-Nichols (Z-N) method [17] near the working point of 50 mm/s. The first step of the Z-N method is to experimentally determine the ultimate gain $K_{cu}$. The period of the resulting sustained oscillation is referred to as the ultimate period $P_u$. Then $K_P$ and $K_I$ can be calculated using the relations $K_P = 0.45K_{cu}$ and $K_I = K_P \frac{\Delta t}{P_u/1.2}$, where $\Delta t$ is the sampling period. For this specific problem, the parameters were determined to be $\Delta t = 5$ ms, $K_P = 50$, and $K_I = 4.167$.

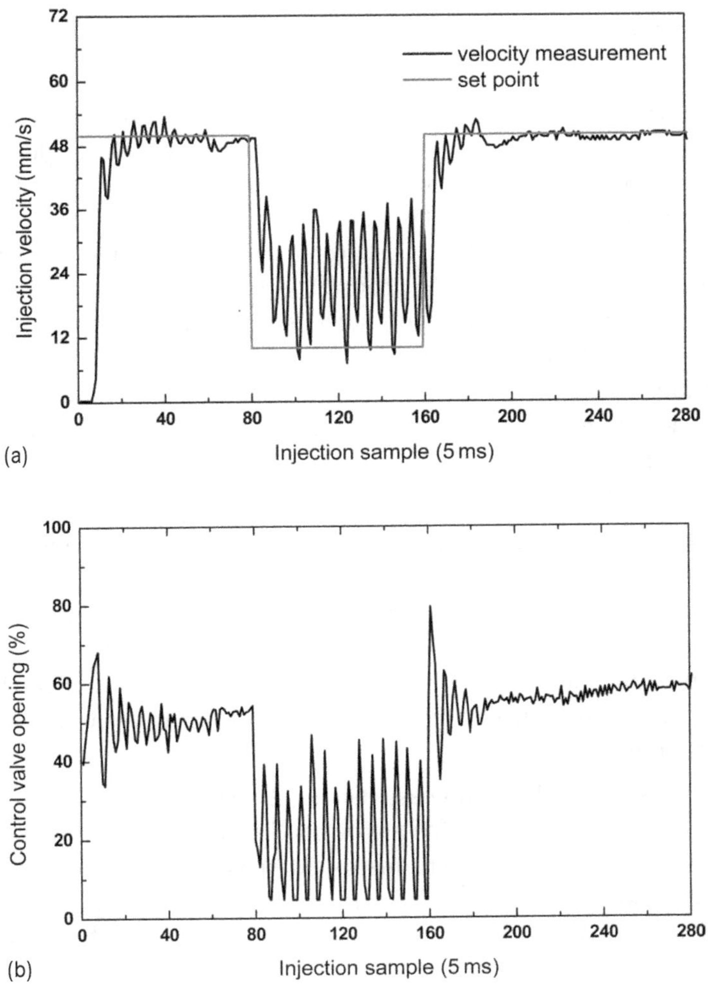


**Figure 2.3** PI controller response to a two-step change set-point profile: (a) the injection velocity response, and (b) the servo-valve opening

Figure 2.3 shows the PI controller response to a step-change velocity profile. The controller works well near the tuning point. However, when the set point changes from 50 to 10 mm/s, the PI control response oscillated strongly, due to the process nonlinearity. The oscillation of the response may be reduced by slowing down the response time of the PI controller, but a slow response is obviously undesirable in injection velocity control. The poor performance of the PI controller not only confirmed the nonlinear characteristic of the injection velocity, but also proved the necessity of using advanced control algorithms.

## 2.3 Adaptive Control

It was shown in the previous section that the traditional fixed-parameter control algorithm such as PID control cannot work well for the injection molding process, due to the batch operation nature and the nonlinear and time-varying characteristics of the process. Adaptive control is a good alternative for this kind of control. Its parameters are adapted in a certain way to conform to the nonlinear or time-varying process dynamics and provide a good control performance. There are many different types of adaptive control schemes, such as gain scheduling, model reference adaptive control, dual adaptive control, and self-tuning regulators (STR). The STR, as an important scheme of adaptive control, is used for illustration in this book to control some key process variables in injection molding. The basic principle of STR is briefly described in the following sections, and detailed discussions can be found in references [18–20].

A self-tuning system is graphically shown in Fig. 2.4 [14]. The system is composed of two loops: an ordinary feedback-control loop, as shown with the dashed line, and a controller parameter-adjusting loop, as shown with the dotted-line block. The latter, consisting of a parametric model estimator and a controller design calculator, gives an online adjustment of the parameters of the feedback controller. The process model parameters and controller design are updated during each sampling period, with a specified model structure.

There are several methods for process model parameter estimation, for example, least mean squares (LMS), projection algorithm (PA), and stochastic approximation (SA). In this book, a recursive least-squares (RLS) estimator is used because of its good sensitivity and superior convergence property [19]. A simple but effective pole-placement control design is adopted for the controller design first to demonstrate the working procedure of the STR.

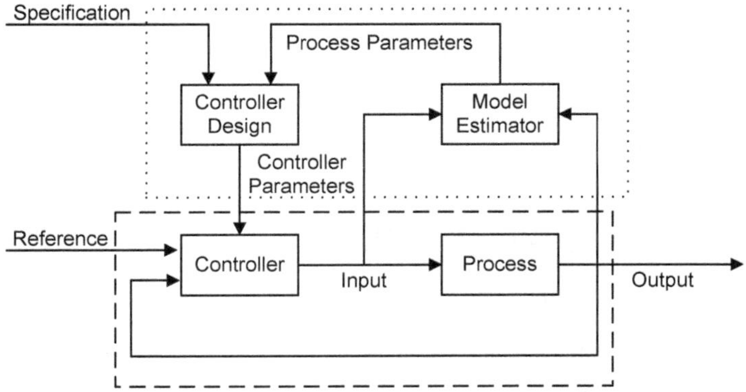


**Figure 2.4** Block diagram of an adaptive self-tuning regulator

## 2.3.1 Model Estimation

Assuming that the process dynamics may be modeled by a discrete time autoregressive with external input (ARX) model, we have

$$A(z)\,y(t) = B(z)\,u(t-n_\mathrm{d}) + e(t) \tag{2.4}$$

where

$$A(z) = 1 + a_1 z^{-1} + \ldots + a_{n_\mathrm{a}} z^{-n_\mathrm{a}}$$

$$B(z) = \left(b_0 + b_1 z^{-1} + \ldots + b_{n_\mathrm{b}-1} z^{-n_\mathrm{b}+1}\right) \cdot z^{-n_\mathrm{d}}$$

$u$: inputs to the process

$y$: corresponding observed process outputs

$z$: $z$-transform (time-shift) operator

$n_\mathrm{a}$, $n_\mathrm{b}$, and $n_\mathrm{d}$: the orders of $A$, $B$ polynomials and process delay, respectively.

Introduce the process model parameter vector:

$$\theta^\mathrm{T} = \left[a_1 \ldots a_{n_\mathrm{a}} \quad b_0 \ldots b_{n_\mathrm{b}-1}\right] \tag{2.5}$$

the regression vector:

$$\varphi^\mathrm{T}(t) = \left[-y(t-1) \ldots -y(t-n) \quad u(t-n_\mathrm{d}) \ldots u(t-n_\mathrm{d}-m+1)\right] \tag{2.6}$$

and the loss function:

$$V(\theta,t) = \frac{1}{2}\sum_{i=1}^{t}\left[y(i) - \varphi^\mathrm{T}(i)\theta\right]^2 \tag{2.7}$$

The model parameter $\theta$, which minimizes $V(\theta,t)$, the differences between the output observation, $y(i)$, and its prediction, $\varphi^{\mathrm{T}}(i)\theta$, in the least-squares sense, is given recursively by

$$\theta(t)=\theta(t-1)+K(t)\left[y(t)-\varphi(t)^{\mathrm{T}}\theta(t-1)\right] \tag{2.8}$$

$$K(t)=P(t-1)\varphi(t)\left[\lambda I+\varphi(t)^{\mathrm{T}}P(t-1)\varphi(t)\right]^{-1} \tag{2.9}$$

$$P(t)=\left[I-K(t)\varphi^{\mathrm{T}}(t)\right]\frac{P(t-1)}{\lambda} \tag{2.10}$$

The variable $\lambda$ in Eq. 2.9 is a forgetting factor that dictates how fast the model is updated. The value of $\lambda$ is $0<\lambda\leq 1$. The smaller $\lambda$ is, the faster the estimator can track the model changing, and a small $\lambda$ will also make the estimation more sensitive to measurement noises. In this project, $\lambda$ is set to be 0.98 for injection velocity control and 0.99 for packing pressure control because the selections produce good estimates. As a rule of thumb, the estimate is based on the last *N*-step results, and *N* can be calculated as below [19]:

$$N=\frac{2}{1-\lambda} \tag{2.11}$$

### 2.3.2 Pole-Placement Controller Design

A pole-placement design is adopted here as the feedback controller in STR for demonstration purposes, as shown in Fig. 2.5 [14]. The design is to find a controller that gives a closed-loop system response specified by the desired closed-loop pole locations. With the process described by Eq. 2.4, the controller has one output, $u$, and two inputs, the command signal, $u_{\mathrm{C}}$, and the measured process output, $y$. A linear controller relating the output to its inputs may be expressed by

$$R(z)u(k)=T(z)u_{\mathrm{C}}(k)-S(z)y(k) \tag{2.12}$$

where $R(z)$, $S(z)$, and $T(z)$ are controller parameter polynomials.

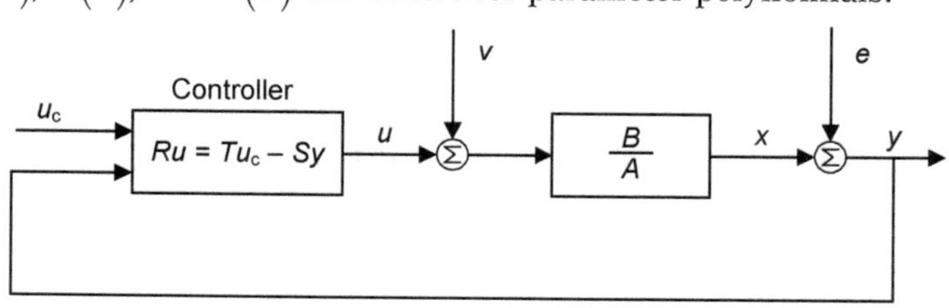


**Figure 2.5** Structure of pole-placement feedback controller

The goal of the controller design is to find these three polynomials for the given closed-loop poles. It can be derived that the closed-loop characteristic polynomial is [19]

$$AR + BS = A_C. \tag{2.13}$$

The desired closed-loop characteristic polynomial $A_C$ may be specified as

$$A_C = A_1 z^d = \left(z^2 + p_1 z + p_2\right) z^d \tag{2.14}$$

where $p_1 = -2e^{-\zeta\omega_0 h}\cos\left(\omega_0 h\sqrt{1-\zeta^2}\right)$, and $p_2 = e^{-2\zeta\omega_0 h}$.

The parameter$A_1$ is equivalent to the second-order system characteristic equation in the $s$ domain:

$$A_1(s) = s^2 + 2\zeta\omega_0 s + \omega_0^2 \tag{2.15}$$

Here $z^d$ is a deadbeat observer with all eigenvalues equal to zero. The damping factor, $\zeta$, and natural frequency, $\omega_0$, decide the closed-loop response of the system. A value of $\zeta$ larger than 1.0 yields a sluggish response, while $0 < \zeta < 1.0$ makes the responses exhibit oscillation and overshoot. The fastest response without overshoot is obtained for the critical damping factor, $\zeta = 1.0$. In terms of pole locations, a bigger $\omega_0$ makes the desired poles closer to the origin in the $z$ plane, which leads the response of the system to be faster but more sensitive to the model error. A smaller $\omega_0$, however, yields a slower response that is less sensitive to the model error and signal noise. Once the desired closed-loop characteristic equation is determined, the controller polynomials $R(z)$, $S(z)$, and $T(z)$ can be obtained by solving Eq. 2.9, which is known as the polynomial Diophantine equation.

### 2.3.3 Solving the Diophantine Equation

The most important part of the controller design is to solve the Diophantine Eq 2.12. There are several alternative methods to solve this equation [19]. The first is to equate the coefficients of the right and left sides of the equation. This method works only for low-order systems, and when the model order is changed, the equation has to be rearranged and solved again. The second alternative, called the Sylvester matrix method, is to use a matrix calculation, which is computationally inefficient and not suitable for online implementation. The Euclidean method, a polynomial method based on the extended Euclidean algorithm, is a good selection in this application as it is the most efficient method to solve the equation. This method works for a wide range of model orders since it exploits the structure of the problem [19]. The Diophantine equation is solved using the extended Euclidean algorithm through two steps [23]. The first step is to find four polynomials $X$, $Y$, $U$, and $V$ that satisfy the following equations:

$$AX + BY = G \tag{2.16}$$

$$AU + BV = 0 \tag{2.17}$$

where $G$ is the greatest common divisor of $A$ and $B$. If $A$ and $B$ were relatively prime, $G$ equals one. The generalized Euclid's algorithm is used to find polynomials $X$, $Y$, $U$, and $V$, as detailed in reference [21]. With the solutions of Eq. 2.20 and 2.21, the method proceeds to step two to find the minimum-degree solution for Eq. 2.12. The particular solution is given by Eqs. 2.17 and 2.18:

$$R_0 = XA_C \operatorname{div} G \tag{2.18}$$

$$S_0 = YA_C \operatorname{div} G \tag{2.19}$$

where div is the polynomial division operator.

The general solution is given by Eqs. 2.19 and 2.20:

$$R = R_0 + QU \tag{2.20}$$

$$S = S_0 + QV \tag{2.21}$$

where $Q$ is an arbitrary polynomial. It has been proved that there exists a unique solution to Eq. 2.9 such that $\deg X < \deg B$ or $\deg Y < \deg A$. This solution with $\deg Y < \deg A$ is obtained from Eq. 2.20 by choosing $Q = -S_0 divV$. Thus, the solutions to the Diophantine Eq. 2.12 are given by Eqs. 2.22–2.24:

$$Q = -S_0 \operatorname{div} V \tag{2.22}$$

$$R = R_0 + QU \tag{2.23}$$

$$S = S_0 \operatorname{mod} V \tag{2.24}$$

The above procedure is theoretically suitable for all model orders. In industrial implementation, however, the model order is always limited to reduce computation resource requirements, and the process dynamics of injection molding are unlikely to be too high.

### 2.3.4 Injection Velocity Adaptive Control Result

The adaptive controller described previously is applied to the injection velocity control again on the 88-ton molding machine with a hydraulic power system controlled by fast-response servo-valves [13]. A fixed-velocity set-point profile is used to test the adaptive controller. It is found that the adaptive controller is able to provide good control. However, during some cycles, sharp spikes appear in the response curve, as illustrated in Fig. 2.6. Another experiment with a velocity set point step-change profile is also carried out. The result is shown in Fig. 2.7. The injection velocity controlled by the adaptive controller follows the set point well, compared to that of the PI controller, but the response time is still not fast enough with respect to the requirements. These problems need to be solved to ensure the robustness and response time of the adaptive controller.

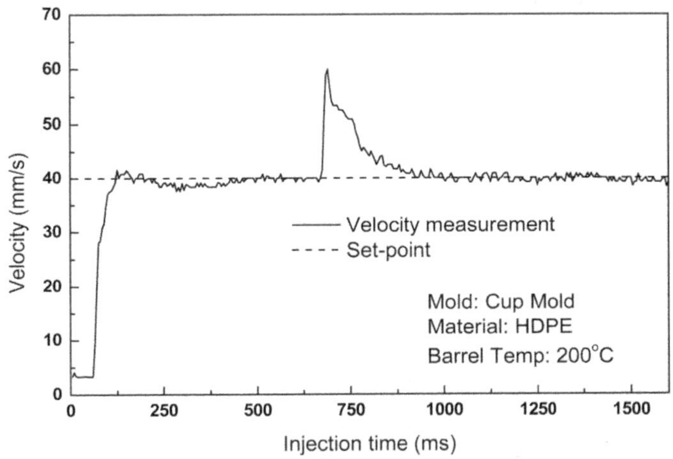


**Figure 2.6** Adaptive controller response to constant injection velocity set-point profile

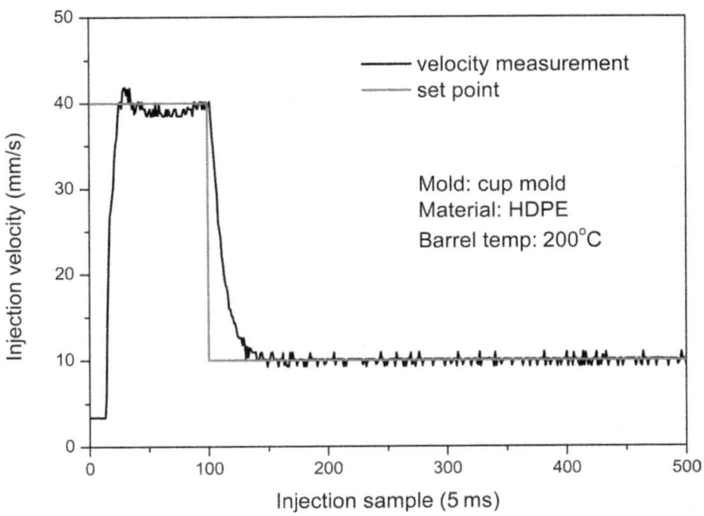


**Figure 2.7** Adaptive controller response to the injection velocity set-point step change

#### 2.3.4.1 Antiwindup Estimation

The spikes in the injection velocity response curve of Fig. 2.6 are found to be caused by the estimator in the STR. Figure 2.8 shows the corresponding model parameters during this cycle. It is observed that the parameters suddenly changed around the 130th samples, where the spike occurred. The spike is caused by the model estimation burst. In the adaptive controller design, an exponential forgetting RLS estimator is used to track the time-varying process model parameters, so recent data are given greater weighting than the earlier data. The estimator works well only if the process is properly excited. However, the burst problem happens when the control becomes perfect; correspondingly, the excitation to the estimator approaches zero.

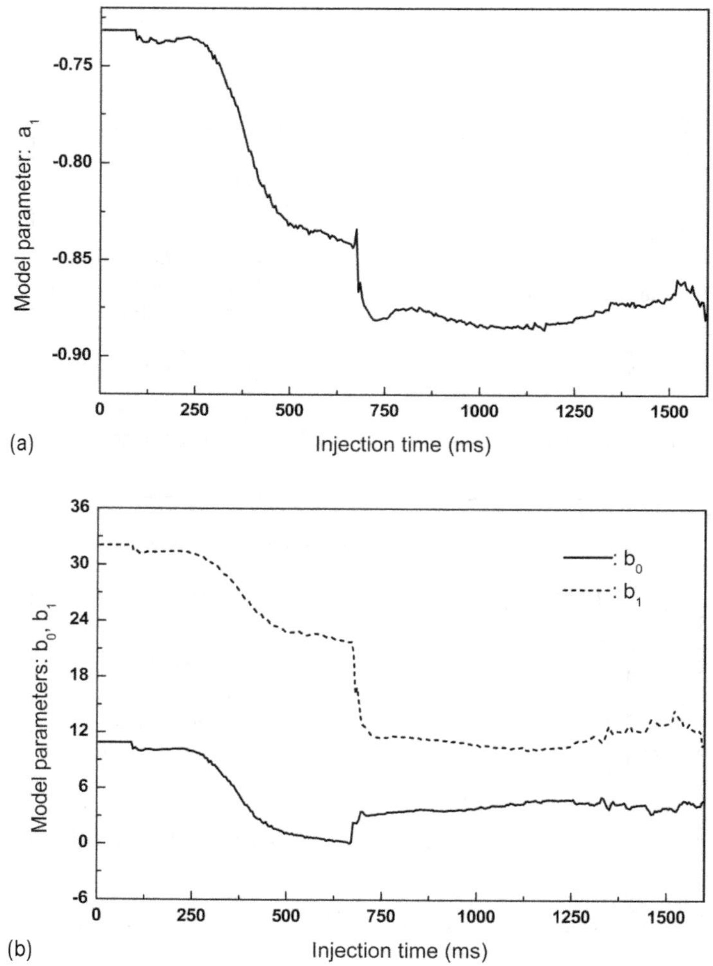


**Figure 2.8** Model parameters during the test cycle when windup happens: (a) model parameter $a_1$, and (b) model parameters $b_0$ and $b_1$

Considering an extreme case in which there is no excitation, that is, matrix $\varphi = 0$, the recursive estimate Eq. 2.9 becomes

$$P(t+1) = \frac{1}{\lambda} P(t) \tag{2.25}$$

Since $\lambda < 1$, the $P$ matrix will grow exponentially with time. The excessively large estimator gain, $P\varphi$, causes the model estimation to change drastically when $\varphi$ becomes nonzero, due to measurement noise, for example. This problem is referred to as the estimator windup.

The estimation windup problem has been reported and studied; see, for example [24–27]. One possible way to avoid windup in the estimator is to update the estimate and covariance only when there is excitation. This is conditional updating or estimation with dead zones. A correct detection of excitation should be based on the calculation of covariance or spectra. In practice, however, simpler conditions are often used to reduce the computation requirement. Another strategy is to keep the $P$ matrix bounded, by scaling the matrix at each iteration in such a way that the trace of the matrix is constant. This leads to the regularized constant-trace algorithm. The estimation Eq. 2.6 thus becomes

$$\vartheta(t) = \theta(t-1) + K(t)\left[y(t) - \varphi(t)^{\mathrm{T}} \theta(t-1)\right] \tag{2.26}$$

$$K(t) = P(t-1)\varphi(t)\left[\lambda + \varphi(t)^{\mathrm{T}} P(t-1)\varphi(t)\right]^{-1} \tag{2.27}$$

$$\bar{P}(t) = \frac{P(t-1) - \dfrac{P(t-1)\varphi(t)^{\mathrm{T}} P(t-1)}{1+\varphi(t)^{\mathrm{T}} P(t-1)\varphi(t)}}{\lambda} \tag{2.28}$$

$$P(t) = c_1 \frac{\bar{P}(t)}{tr\left[\bar{P}(t)\right]} + c_2 I \tag{2.29}$$

where $c_1 > 0$ and $c_2 \geq 0$, $c_1/c_2 \approx 10^6$, and $\varphi^{\mathrm{T}}\varphi c_1 \gg 1$.

Other methods to avoid the estimation windup, such as directional forgetting, leakage, and robust estimation design, are too complicated for the online implementation. The conditional update method is used for antimodel-burst for the injection velocity control. The estimation error, as shown below, is used as the excitation detection condition:

$$ee = y - \hat{y} \tag{2.30}$$

where $ee$ is the estimation error, $y$ is the output measurement, or the injection velocity, and $\hat{y}$ is the calculated output from the previous estimation.

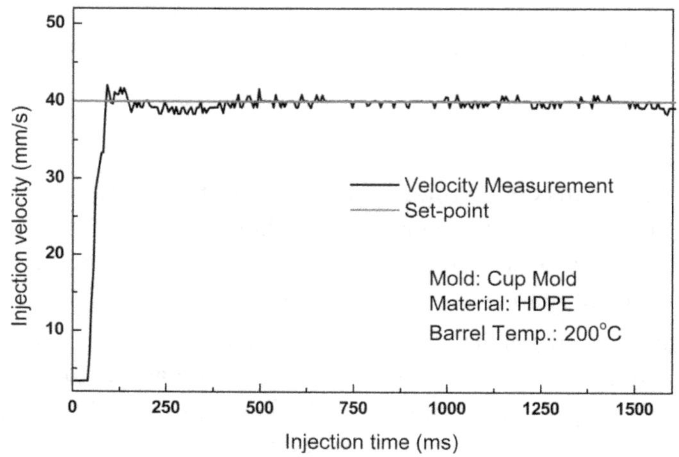


**Figure 2.9** Adaptive controller with estimation antiwindup (constant set point profile)

The control result after the introduction of anti-model-burst is shown in Fig. 2.9. The velocity response follows the constant set point, 40 mm/s, very well, suggesting that the windup problem can be solved with the conditional estimation.

#### 2.3.4.2 Adaptive Feed-Forward Control

The adaptive controlled injection velocity response shown in Fig. 2.7 is replotted in Fig. 2.10 with a zoomed-in plot around the time when the step change is introduced, to clearly illustrate the response time. The velocity is able to track the step change, but the settling time is relatively long. It takes about 30 steps, or 150 ms, for the velocity to respond to a step change of 30 mm/s. Compared to the time constant of the servo-valve, 10 ms, the response time is relatively slow. This slow response is improved by introducing an adaptive feed-forward control.

A feed-forward control is normally used to compensate for a measurable disturbance to the process. For injection molding, the set-point profile is determined prior to the cycle start, and the profile is normally not changed during the cycle. The preset set-point profile therefore may be treated as a measurable variable. The change in the set point can be used as the input to the feed-forward controller. Figure 2.11 gives a schematic diagram of an adaptive system combining a feedback regulator with a feed-forward controller. The adaptive feed-forward controller is designed to be proportional as

$$u_{\mathrm{ff}} = \frac{1+a_1}{b_0+b_1}\left(1-z^{-1}\right)U_{\mathrm{C}}\left(z\right) \tag{2.31}$$

where $U_{\mathrm{C}}\left(z\right)$ is the the set-point signal in the $z$ domain, and $u_{\mathrm{ff}}$ is the feed-forward controller output.

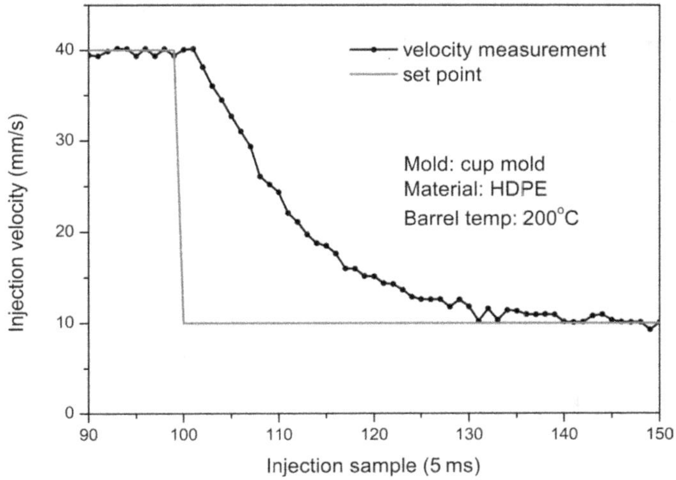


**Figure 2.10** Zoomed-in plot of the adaptive controller response to a set-point step change

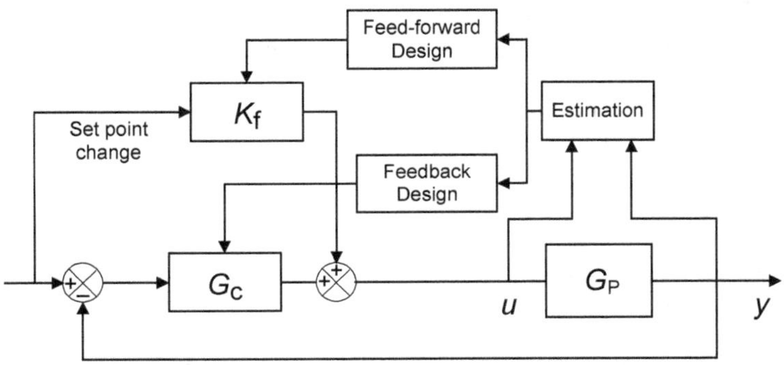


**Figure 2.11** Block diagram of the self-tuning feedback controller combined with an adaptive feed-forward controller

The feed-forward proportional gain, $K_f$, is set to be the reciprocal of the process gain estimated online by the recursive estimator, that is, $\frac{1+a_1}{b_0+b_1}$. The overall controller output to the servo-valve is a summation of the feedback controller output, $u_{fb}$, and the feed-forward controller output, $u_{ff}$. The injection velocity response using the adaptive feed-forward and feedback control is shown in Fig. 2.12(a); Fig. 2.12(b) shows the zoomed-in plot around the time when the step change is imposed. In comparison to the response of Figure 2.10 with the feedback control only, the response of Figure 2.12 is much faster in the set-point tracking: the response time is significantly reduced to six to seven steps, or 30 to 35 milliseconds. A careful inspection of the response indicates that there is a delay of $n_d$ steps before it responds to the set-point changes due to the process delay. The above feed-forward law may be further improved by shifting the injection velocity profile $n_d$ steps forward. Then the feed-forward control law Eq. 2.30 becomes

$$u_{ff} = \frac{1 + a_1}{b_0 + b_1}\left(z^{n_d} - z^{n_d - 1}\right)U_C(z) \tag{2.32}$$

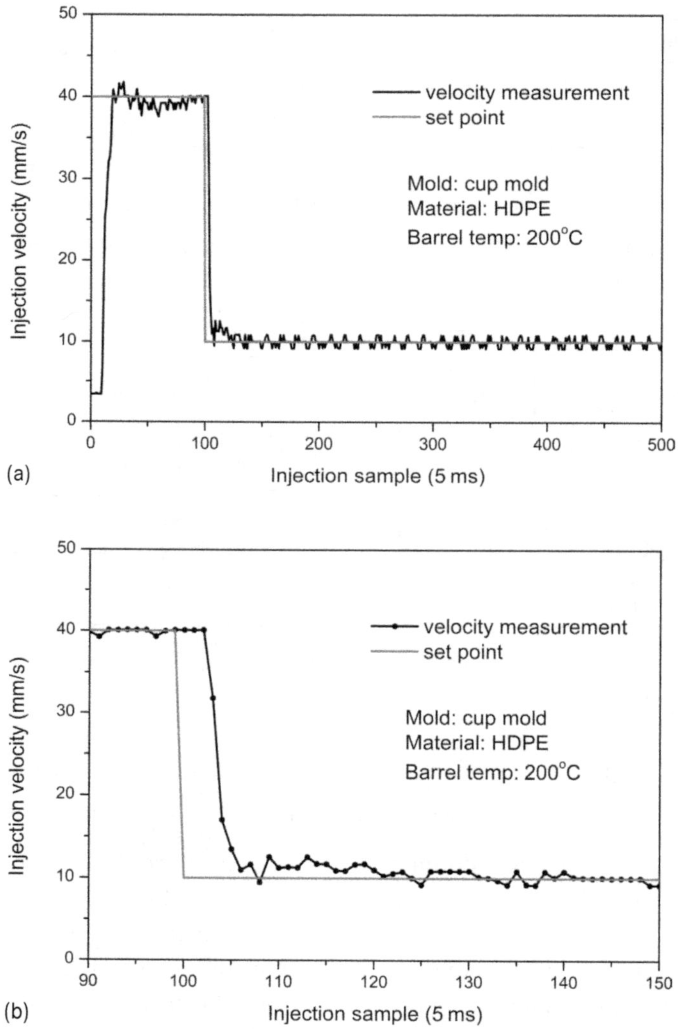


**Figure 2.12** Adaptive feed-forward/feedback injection velocity controller response to a set-point step change:
(a) injection velocity response throughout a complete injection phase, and
(b) the zoomed-in velocity response around the set-point step change

In a normal control situation, it is physically impossible to implement $z^{n_d}$ because this is an $n_d$ steps prediction. This can be implemented for injection molding control because the set-point profile is determined prior to the start of the cycle. The performance after this modification is shown in Fig. 2.13. Again, Fig. 2.13(a) shows the velocity response during a complete injection phase, and Fig. 2.13(b) shows the zoomed-in plot around the set-point step change. The tracking response time is further reduced to four steps, or 20 ms.

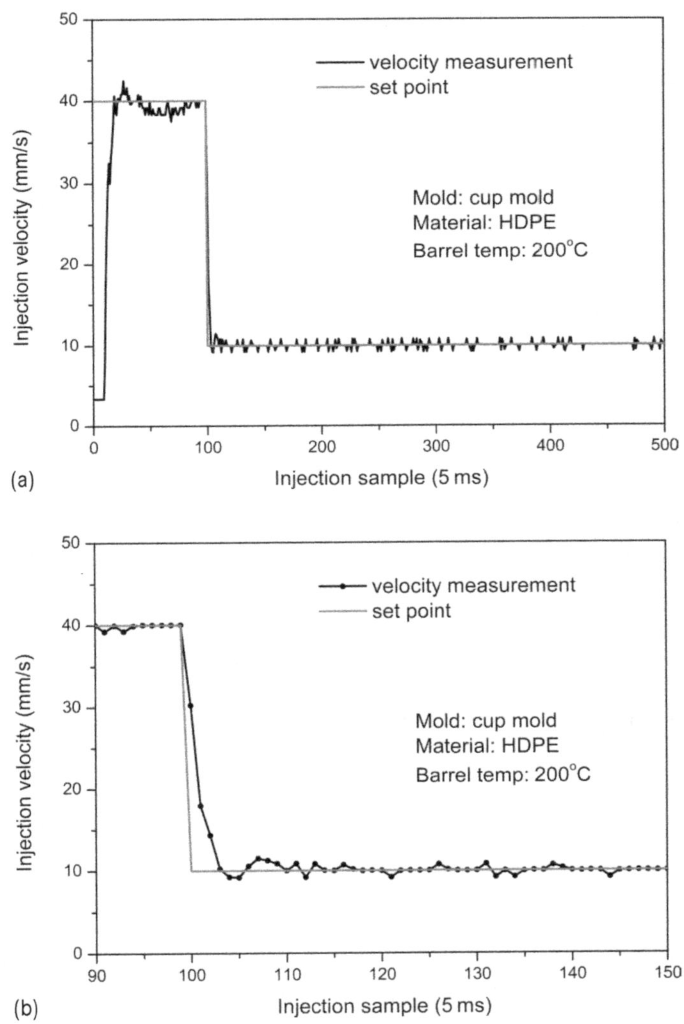


**Figure 2.13** Adaptive feed-forward/feedback injection velocity controller response to a set-point step change, with profile shift:
(a) injection velocity response throughout a complete injection phase, and
(b) the zoomed-in velocity response around the set-point step change

Figure 2.14 presents a comparison among these three control strategies. It is obvious that the adaptive feed-forward/feedback control combined with the profile shift technique significantly improved the response time.

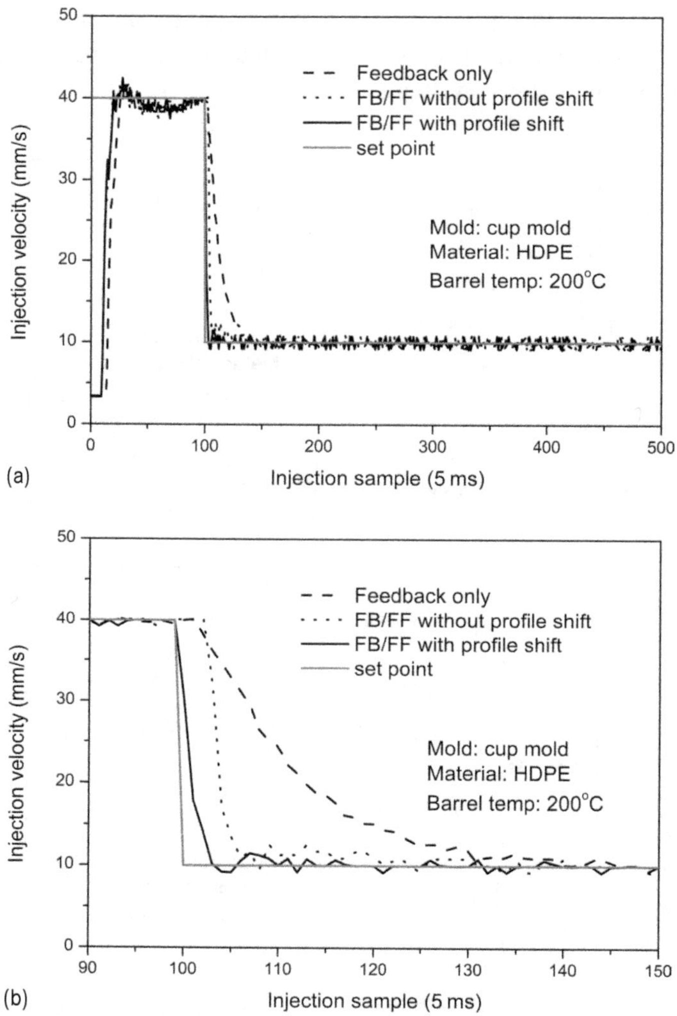


**Figure 2.14** A comparison of the velocity responses using adaptive feedback, feed-forward/feedback, and feed-forward/feedback with profile shift: (a) injection velocity response throughout a complete injection phase, and (b) the zoomed-in velocity response around the set-point step change

#### 2.3.4.3 Cycle-to-Cycle Adaptation

Injection molding is a cyclic process. The injection velocity adaptive controller needs to initialize the model parameters for the model estimator at the beginning of each injection cycle. A proper selection of the initial parameters for the model estimation is crucial to the initial control performance and the model convergence time. The initial values are commonly set at fixed values, which may result in a long convergence time and strong oscillation during the initial control periods. For injection molding, this problem can be severe because the molding machine needs about 50 cycles to be warmed up and stabilized; the process dynamics during the warm-up stage may vary significantly. On the other hand, the injection molding dynamics during the early injection may not change significantly for two consecutive cycles.

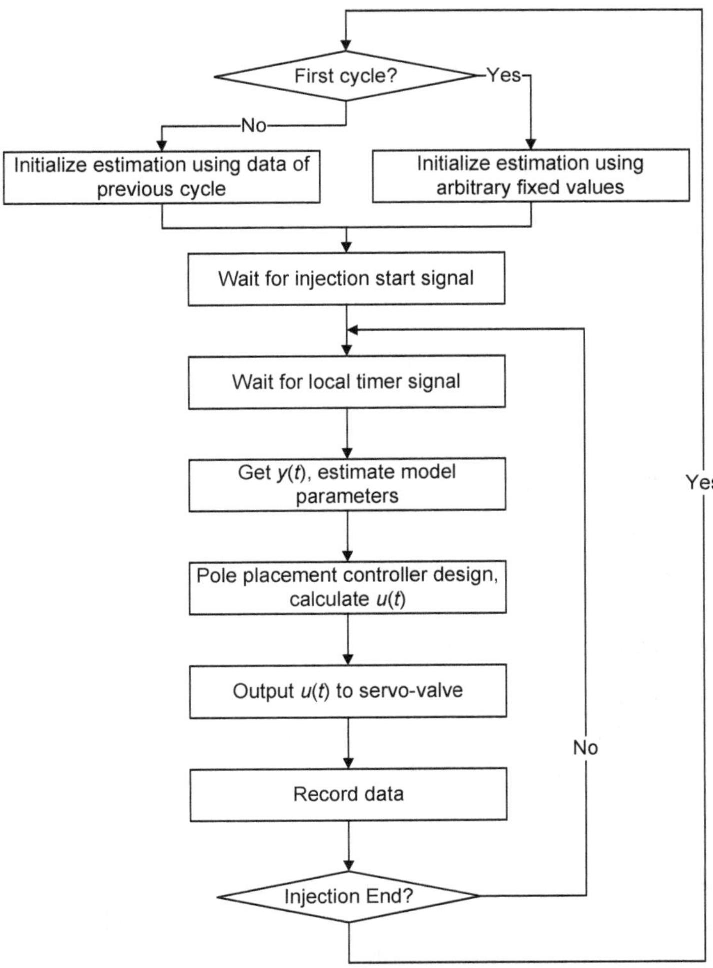


**Figure 2.15** Simplified flow chart diagram of the adaptive velocity controller

It is, therefore, possible to initialize the current model estimation based on the last cycle's information. The adaptation is thus not only within a cycle, but also from cycle to cycle. This is also the early attempt to improve the control performance by exploiting the dynamics along the cycle direction. A simplified flow chart of the injection adaptive controller is illustrated in Fig. 2.15.

Figure 2.16 illustrates the effectiveness of the cycle-to-cycle adaptation. The solid line represents the velocity response during the early injection period of the first cycle with arbitrarily set initial model parameters. It is clear that the response exhibits large oscillation and slow convergence. The second cycle velocity response is plotted as the dashed line. The initial values are set based on the model parameters from the first cycle. The convergence time is significantly reduced, and the overshoot is eliminated. This result also encourages the analysis of cycle-wise dynamics to improve the controller, as well as the two-dimensional design of injection molding control systems.

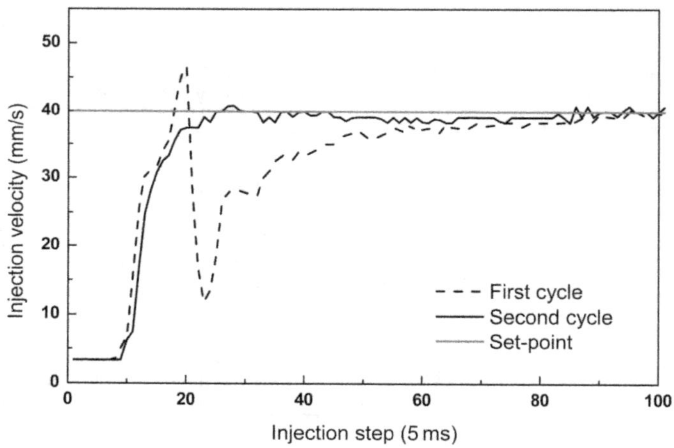


**Figure 2.16** Adaptive control with cycle-to-cycle adaptation for early stage of injection

### 2.3.4.4 Adaptive Control Results with Different Conditions

Injection velocity dynamics change significantly with different operating conditions. The robustness and set-point tracking performance of the velocity self-tuning controller is tested under different conditions in this section.

Set-point tracking ability is tested with different profiles. The nonlinear and time-varying characteristics suggest that the injection velocity dynamics change with the operating point. The control system is tested with a set-point profile with two step changes, one step change from 40 to 25 mm/s at 1 s and the other from 25 to 10 mm/s at 1.5 s of injection time. A ramp set-point profile is also tested, with the velocity change from 40 mm/s at 1 s to 10 mm/s at 1.5 s. The velocity responses for

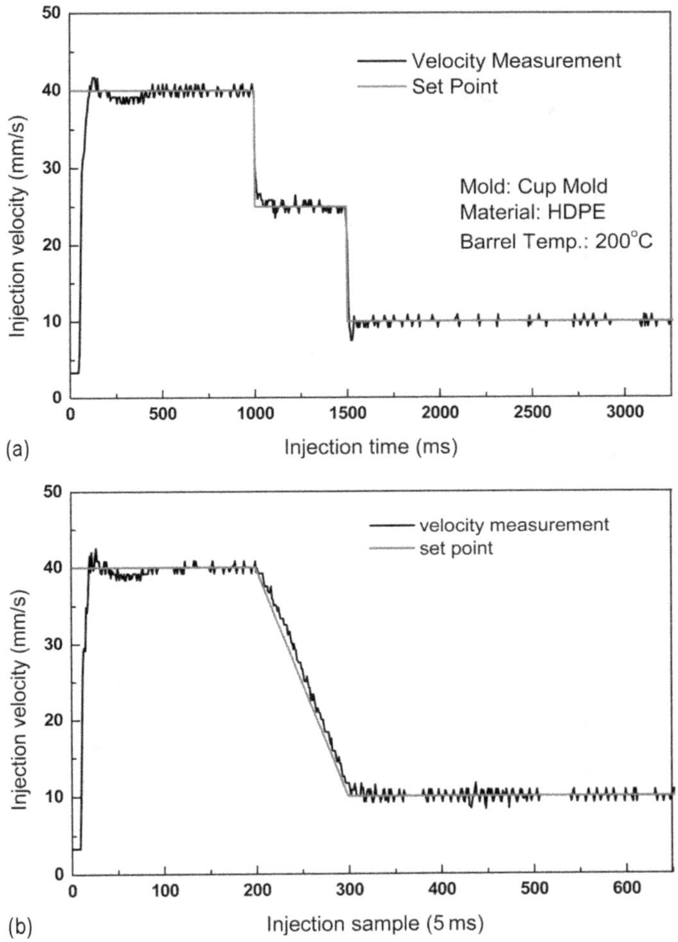


**Figure 2.17** Adaptive controller response to different injection velocity set-point profiles: (a) a two-step change profile and (b) a ramp profile

these two cases are plotted in Figs. 2.17(a) and (b), respectively. In both cases, the velocity closely follows the preset profile and proves that the control system works well under different set-point profiles.

The barrel temperature is another factor affecting the injection velocity dynamics. A change in the temperature leads to changes in the melt viscosity and the velocity dynamics. The control system is tested with different barrel front temperatures. The material used here is HDPE with a melt flow index of 20. Figures 2.18(a), 2.16, and 2.18(b) illustrate the velocity adaptive control responses for the barrel temperatures of 180, 200, and 220°C, respectively. A step change in the set-point profile is used, as shown by the dashed line. It can be observed that the control performance is consistently good with different barrel front temperatures.

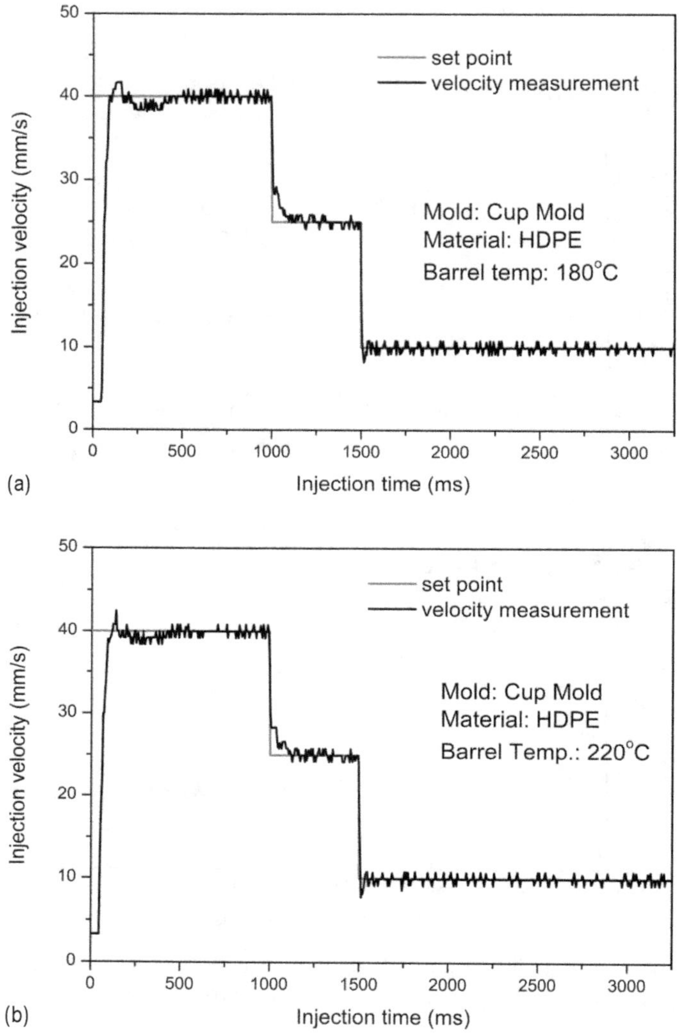


**Figure 2.18** Adaptive controller response to different barrel front temperatures: (a) barrel front temperature 180°C and (b) barrel front temperature 220°C

The injection velocity dynamics are highly dependent on the material used, so the adaptive control is tested with different materials, as shown in Fig. 2.19. The solid line represents the velocity control response with PS, while the dashed line denotes the response with ABS. The materials used have significantly different properties. The velocity follows the set-point step change closely for all the materials tested. This confirms that the adaptive controller works well for different materials.

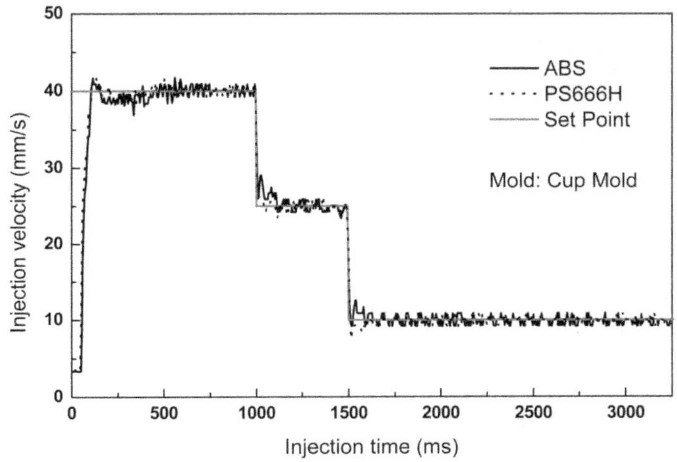


**Figure 2.19** Adaptive controller response to different materials

# 2.4 Model Predictive Control

## 2.4.1 Basic Principle of MPC and GPC

Model predictive control (MPC) [28] is a class of advanced process-control algorithms. It was originally developed for process industries such as chemical and petrochemical plants in the late 1970s. The research on MPC, both academic and industrial applications, grew rapidly during the last several decades. The wide application of MPC is mainly due to the following reasons:

(a) MPC can be used to deal with complicated process dynamics, including nonlinearity and time-varying characteristics, long time delay, and open-loop instability.

(b) MPC can deal with constraints in the process control naturally and systematically.

(c) MPC can be extended to multivariable control easily.

(d) Feed-forward control is inherently built in to the MPC design, so the process disturbances can be compensated for.

(e) MPC is suitable for batch processes since the reference trajectories of process settings are known before the cycle starts.

(f) MPC is a totally open control methodology following certain basic principles for further development and extension.

Due to the above advantages, various MPC algorithms have been proposed, for example, model algorithmic control (MAC), dynamic matrix control (DMC), generalized model predictive control (GPC), and predictive functional control (PFC). These designs all share the same basic features of MPC:

(a) Prediction of future outputs based on an internal dynamic model of the process.

(b) Calculation of an optimal control sequence by minimizing a predefined objective function.

(c) A receding horizon strategy that moves the control forward toward future sampling times.

The basic principle of model predictive control is schematically shown in Fig. 2.20, where the prediction, optimization, and receding of the MPC are clearly illustrated.

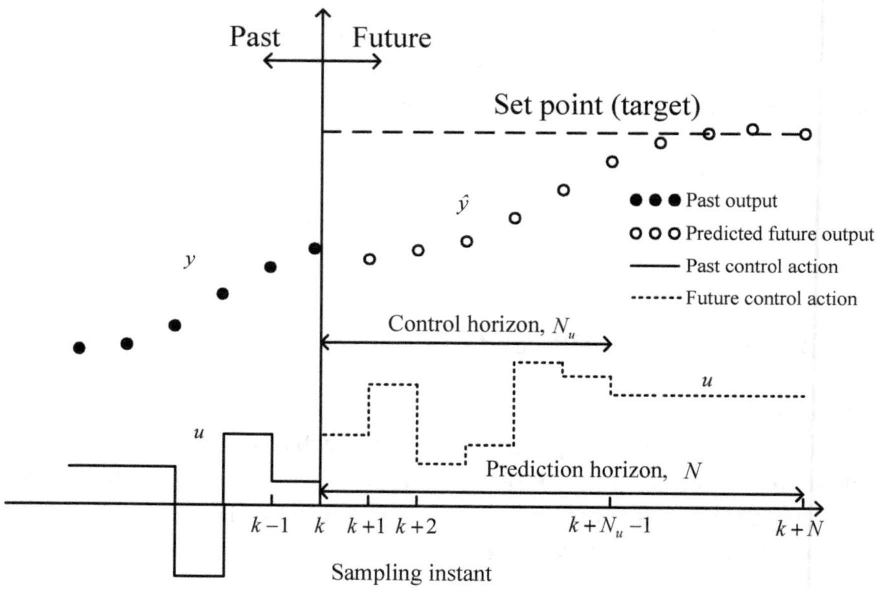


**Figure 2.20** Schematics of the model predictive control algorithm

In this book, the GPC design is first used to demonstrate the good performance of MPC. The GPC design was first proposed by Clark, Mohtadi, and Tuffs [29]. This control has been shown to be effective with model uncertainty in many process industry applications. To overcome the nonlinear time-varying characteristics of the injection molding process, an adaptive GPC scheme with a structure similar to Fig. 2.4 is selected, only with a GPC controller substituting for the pole-placement controller in the dashed block. Based on the estimated process model, the GPC determines an optimized sequence of future control moves by minimizing a cost function defined over a prediction horizon, as shown in Eq. 2.33:

$$J(N_1,N_2,N_u)=E\left\{\sum_{j=N_1}^{N_2} w_y(j)\left[\hat{y}(t+j\,|\,t)-w(t+j)\right]^2+\sum_{j=1}^{N_u} w_u(j)\left[\Delta u(t+j-1)\right]^2\right\} \tag{2.33}$$

where $\hat{y}(t+j\,|\,t)$ is the process output prediction over the next $j$ steps at time $t$, and $N_1$, $N_2$, and $N_u$ are the minimum, the maximum, and the control horizons, respectively. Here, $w_y$ and $w_u$ are the weighting factors; in this application, $w_y$ was set to be 1 and $w_u$ was set to be $trace(\mathbf{G})$. With these assumptions, the controller output can be derived as

$$\mathbf{u}=-\left(\mathbf{G}^T\mathbf{G}+\lambda\mathbf{I}\right)^{-1}\cdot\mathbf{G}^T\left(\mathbf{f}-\mathbf{w}\right) \tag{2.34}$$

where $\mathbf{u}=\left[\Delta u(t)\quad \Delta u(t+1)\cdots\Delta u(t+N_u-1)\right]^T$, $\mathbf{G}$ is the process step response matrix, $\mathbf{w}=\left[w(t+n_d)\quad w(t+n_d+1)\cdots w(t+n_d+N_u-1)\right]^T$ is the future set-point trajectory, and $\mathbf{f}$ is a group of terms that depend only on the past control input/output:

$$\mathbf{f}=\mathbf{F}\left(z^{-1}\right)y(t)+\mathbf{G}'\left(z^{-1}\right)\Delta u(t-1) \tag{2.35}$$

where

$$\mathbf{F}\left(z^{-1}\right)=\begin{bmatrix} F_{d+1}\left(z^{-1}\right) \\ F_{d+2}\left(z^{-1}\right) \\ \vdots \\ F_{d+N}\left(z^{-1}\right) \end{bmatrix} \tag{2.36}$$

and

$$\mathbf{G}'\left(z^{-1}\right)=\begin{bmatrix} \left[G_{d+1}\left(z^{-1}\right)-g_0\right]z \\ \left[G_{d+1}\left(z^{-1}\right)-g_0-g_1z^{-1}\right]z^2 \\ \vdots \\ \left[G_{d+N}\left(z^{-1}\right)-g_0-g_1z^{-1}-\ldots-g_{N-1}z^{-(N-1)}\right]z^N \end{bmatrix} \tag{2.37}$$

The step response matrix $\mathbf{G}$ was updated online in real time by estimating the process dynamics so that the nonlinearity and time-variation of the process can be traced. The GPC design is applied to the injection velocity control, and the result is compared with that of the pole-placement control to illustrate the advantages of the MPC algorithm.

### 2.4.2 Model Order Determination

The first step in controller implementation is the determination of model orders. The standard open-loop step response method was used to analyze the injection velocity dynamics. All of the following open-loop tests were repeated several times with good repeatability. Simple step changes in process input (servo-valve opening) were introduced to excite the process, as shown in Fig. 2.21(a). The servo-valve opening, which was a manipulated variable controlled by a digital controller, was changed from 20% to 40% at 0.3 s and then step-changed back from 40% to 20% at the injection time of 0.8 s. The corresponding open-loop velocity response can

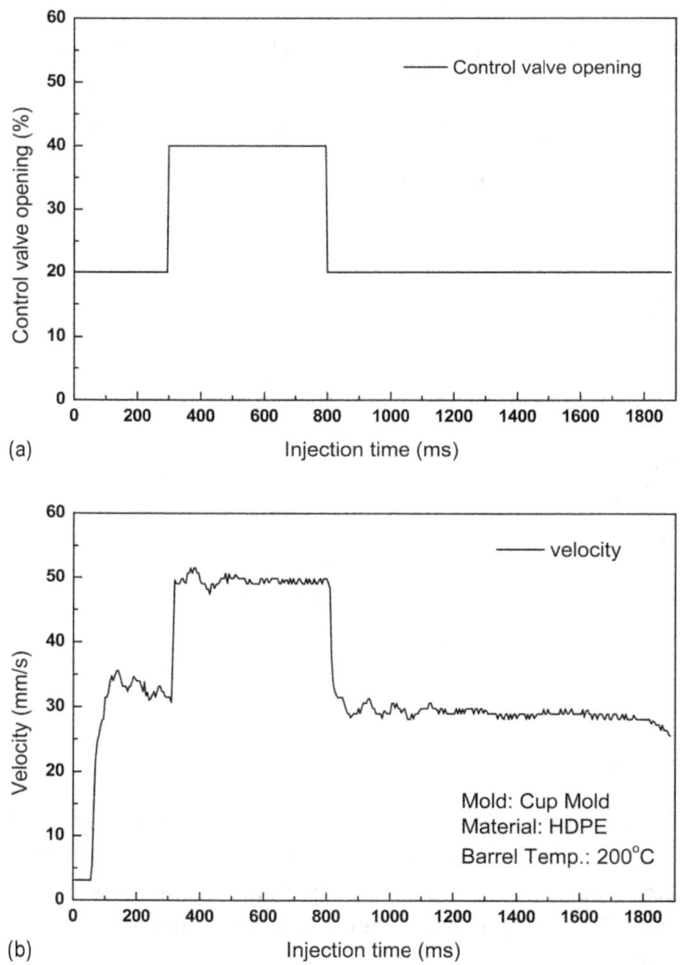


**Figure 2.21** Injection velocity open-loop test:
(a) servo-valve opening changes, and (b) injection velocity response

be recorded and analyzed, as plotted in Fig. 2.21(b). The response was fitted with an ARX model with different delay orders to determine the delay. Akaike's final predictive errors (FPE) [30] criterion was used to determine the degree of model matching. It was found that the lowest FPE value happens when $n_d = 3$. Therefore the delay of injection velocity was determined to be three steps, or 15 ms (because the sampling period equals 5 ms).

$N_a$, in practice, should be set equal to the number of dominant poles, and the effect of the insignificant poles can be neglected to reduce computation complexity. High orders of $n_a$ and $n_b$ may give a better match between the response data and the model, but they could also result in overfitting of the measurement error and increased design complexity. By using the FPE values and taking the above practical considerations into account, the orders of polynomials $A$ and $B$ were determined to be one for injection velocity in this example.

### 2.4.3 Comparison with Pole-Placement Control

The adaptive GPC controller is designed and implemented to control the injection velocity for demonstration and comparison purposes. The model orders of the GPC design were the same as that of the ARX model. To solve the GPC problem, the horizons and $N_u$, $N_1$, and $N_2$ must be determined. A clear design formulation for the determination of these parameters is lacking, so only some rules of thumb can be followed. The GPC also requires the computation of a $N_u \times N_u$ matrix inversion. A small $N_u$ reduces the computation load and at the same time restricts the optimality of the GPC. The selection of a control horizon is therefore a compromise between performance optimality and computational load. With this consideration, $N_u$ has been set as 4 in this work. The minimum and maximum costing horizons were determined to be $N_1 = n_d = 3$ and $N_2 = n_d + N_u = 7$, as the process output was only influenced by $u(t)$ after $n_d$ sampling periods.

The above designed adaptive GPC controller was experimentally tested, as shown in Fig. 2.22. The response followed the constant set point closely.

The next step is to show the performance and features of GPC through the comparison with the pole placement control.

It has been introduced in Section 2.3 that the original pole-placement design gives a slow response in the field application. The adaptive feed-forward and profile shift technologies have to be incorporated to enhance the response time of the pole-placement control. Further analysis shows that the pole-placement design is also sensitive to the model mismatch. An injection molding controller is required to work with different molding conditions because product change is fairly common for any molding machine. In industrial situations, it cannot be guaranteed that the

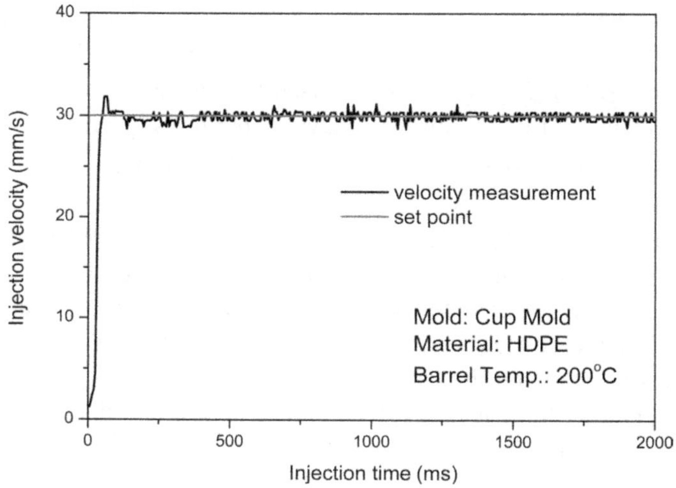


**Figure 2.22** Adaptive GPC responses to a constant set-point profile

model structure will closely match the process dynamics. Therefore, a controller with good robustness to model order mismatch is of great importance in injection velocity control. The adaptive pole-placement controller was found to be sensitive to delay order mismatch, as demonstrated in Fig. 2.23, where the solid line shows the pole-placement control without delay mismatch and the dotted line denotes the response with delay mismatch. This result clearly indicates that the pole placement design becomes oscillatory and unstable when a delay mismatch is introduced.

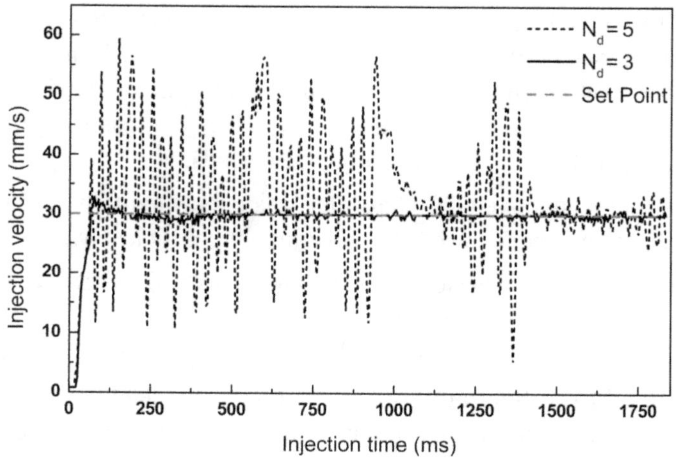


**Figure 2.23** Adaptive pole-placement responses with and without model mismatch

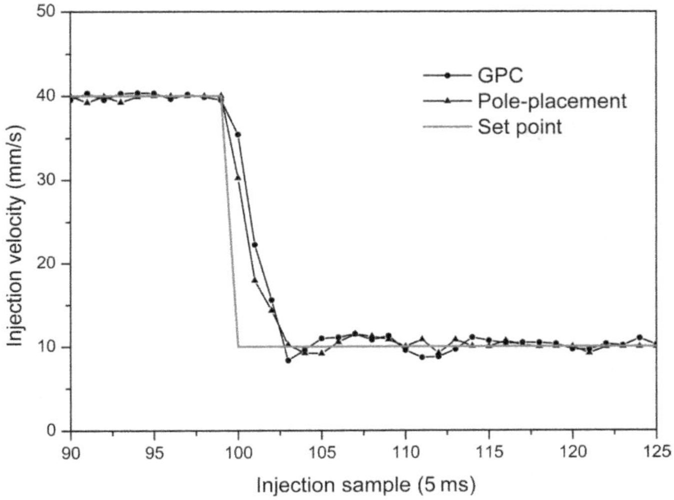


**Figure 2.24** Comparison of set-point, step-change responses for the adaptive GPC and the pole-placement controls

The set-point tracking ability of the adaptive GPC controller can be tested using a step-change profile in the set point. The controller has good tracking ability, as indicated by the velocity response, represented by the dark solid line in Fig. 2.24, which is a zoomed-in plot around the set-point step change. It is shown that the controller took only 20 ms to track a set-point step change of 30 mm/s. For the STR controller with a pole-placement design, an adaptive feed-forward control scheme and a profile shift technique had to be used to achieve a comparable performance, as illustrated with the dotted lines in Fig. 2.24. The natures of the set-point feed-forward and the compensation of process delay are inherently built into the GPC design, so they need not be separately considered, as in the case of pole-placement design.

It is demonstrated in Fig. 2.23 that the adaptive pole-placement control scheme was sensitive to the model mismatch. The same test was performed with the GPC design, as shown in Fig. 2.25. The dotted line and the solid line represent responses with and without delay mismatch, respectively. There was no significant difference in the velocity responses for the GPC design, suggesting that the GPC controller has good robustness to model mismatch. The GPC controller has also been found to be robust to other model order mismatches. Figure 2.26 compares the velocity responses of the pole-placement and the GPC designs with model mismatch.

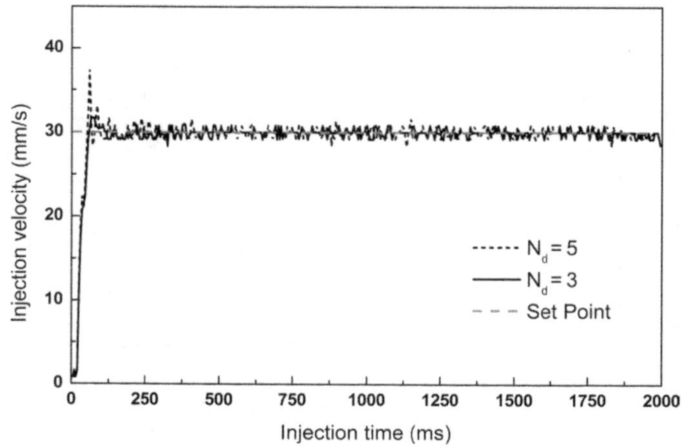


**Figure 2.25** Adaptive GPC responses with and without model mismatch

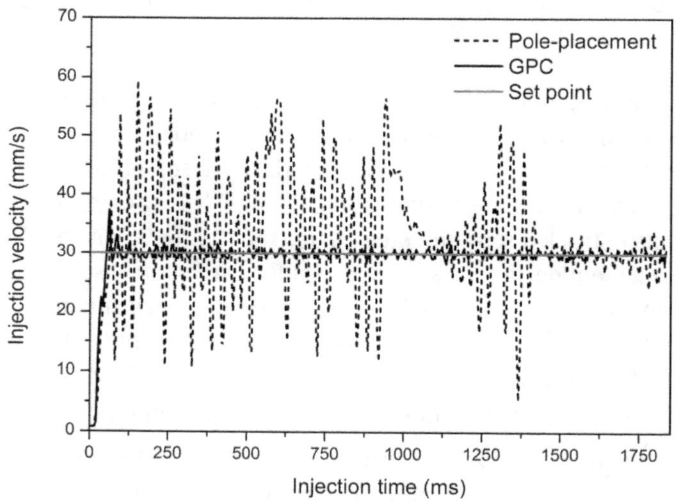


**Figure 2.26** Comparison of the pole-placement and the GPC designs with model mismatch

### 2.4.4 GPC Control with Different Conditions

The adaptive GPC controller was further tested with different operating conditions, starting with different barrel temperatures. Figure 2.27 shows that the velocity GPC control responses overlap each other for different barrel temperatures of 180, 200, and 220°C, indicating that consistently good control performance can be obtained with different barrel temperatures.

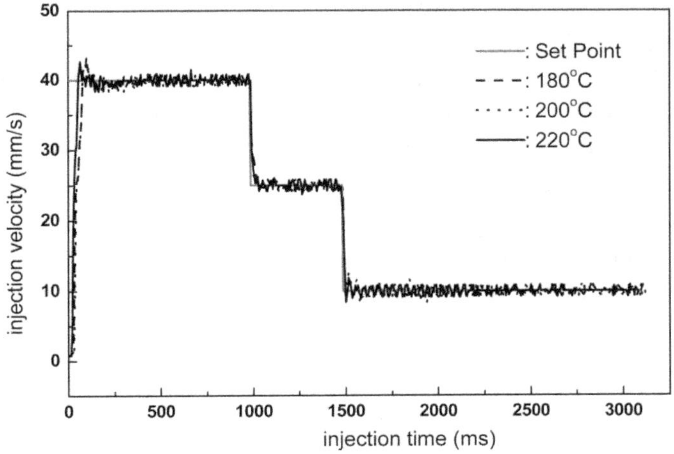


**Figure 2.27** Adaptive GPC responses with different barrel temperatures

Different materials were also tested, and the results are shown in Fig. 2.28. The solid line represents the velocity response with HDPE, while the dashed line denotes the response with ABS. These two materials have significantly different flow properties; for example, the ABS has a melt flow index of 6.5, and that of HDPE is 20. Despite the significant difference in the material properties, the injection velocity controlled by GPC followed the set point closely with both materials, indicating that the controller possesses good robustness to material variation.

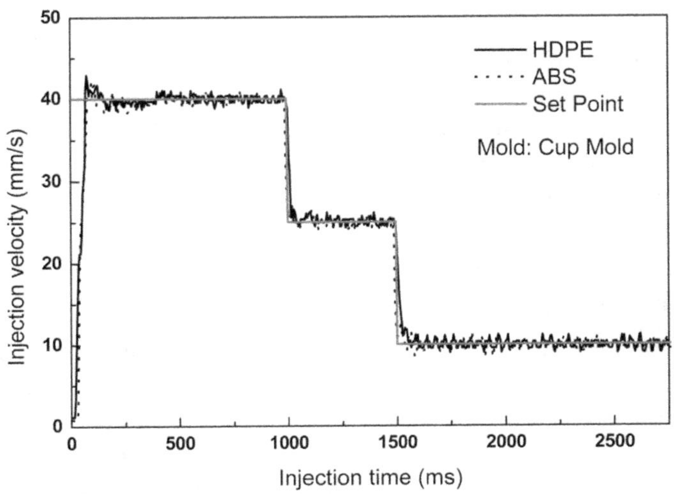


**Figure 2.28** Adaptive GPC responses with different materials

In this section, a self-tuning, regulator-type adaptive control scheme is introduced for the injection molding process control. A pole-placement and a generalized predictive controller design are both implemented and compared through experimental tests on injection velocity control. The results show that the pole-placement design, with several enhancements such as estimation antiwindup, adaptive feed-forward control, and cycle-to-cycle adaptation, can provide good control performance, but it is sensitive to model mismatch. The GPC, on the other hand, performs well over a wide range of molding conditions, and it has inherently good tracking performance and excellent tolerance to the model mismatch.

## 2.5 Fuzzy Systems in Injection Molding Control

The control applications introduced in the previous sections mainly focus on the set-point tracking of certain key process variables. However, it is important in the control engineering and injection molding process to develop an automatic decision-making system. In this section, a fuzzy-rule-based system is designed to give an accurate and automatic determination or decision of the velocity/pressure (V/P) transfer point. Before going further into the system design, it is necessary to give a brief introduction to the fuzzy system.

### 2.5.1 Fuzzy Inference System Background

A fuzzy system [31], constructed from human knowledge in the form of fuzzy IF-THEN rules, can easily perform a multi-input–single-output nonlinear mapping from a real-valued vector to a real-valued scalar. It is a knowledge-based system constructed from human knowledge in the form of fuzzy IF-THEN rules. A typical fuzzy system, as shown in Fig. 2.29, has four basic components: a fuzzifier, a fuzzy rule base, a fuzzy inference engine, and a defuzzifier.

The fuzzifier performs a mapping from a crisp value to a fuzzy set. The fuzzy set is characterized by a membership function that takes values in an interval of 0–1, representing the degree of belonging to a linguistic variable, such as large, medium, and small. Several types of fuzzifiers exist, such as singleton, Gaussian, and triangular fuzzifiers; details of the design and selection of a fuzzifier can be found in reference [31].

The fuzzy rule base consists of a set of fuzzy IF-THEN rules. This is the heart of a fuzzy system. The rules may be obtained from human expert knowledge. The key to fuzzy system design and implementation is accumulation and formulation of the knowledge.

The fuzzy inference engine performs a nonlinear mapping from fuzzy input sets to a fuzzy output set with the fuzzy IF-THEN rules in the rule base, using fuzzy logic principles. Among the choices of fuzzy inference engine, the product inference engine and the minimum inference engine are most commonly used for their computational simplicity.

The defuzzifier transfers the fuzzy output set to crisp output within a certain real-value range. Several types of defuzzifier also exist, for example, center of gravity, center average, and maximum defuzzifiers. The selection criteria can be found in reference [31].

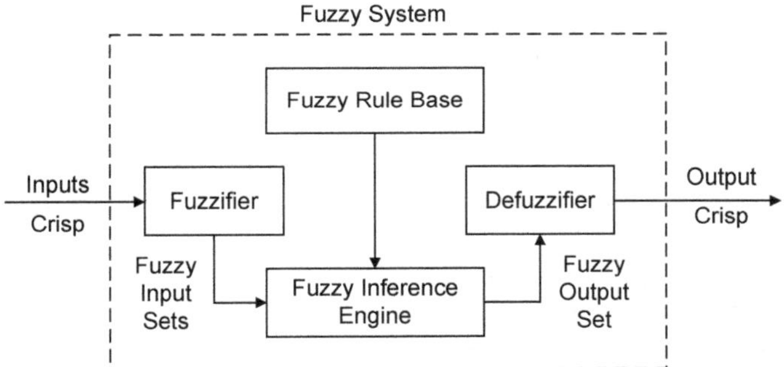


**Figure 2.29** Basic structure of a fuzzy system with fuzzifier and defuzzifier

## 2.5.2 Fuzzy V/P Switch-Over

This automatic switch-over decision-making algorithm is based on the observation of the nozzle pressure responses [32]. Figure 2.30(a) shows the nozzle pressure measurement around the V/P transfer points with different barrel temperatures, while Figure 2.30(b) illustrates the nozzle pressure variations with different injection velocities. It can be clearly seen that the changes in injection velocity and barrel temperature strongly affect the nozzle pressure profile, which suggests that all threshold-based methods need adjustments when the molding conditions change. It can also be observed that, for any molding, a significant nozzle pressure increase occurs when the cavity is completely filled. This observation may be easily described in linguistic terms, but it is difficult to describe it by mathematical expressions. A fuzzy inference system (FIS) is a proper tool to solve such a control decision problem. The FIS has a superior capability in formulating human knowledge and rules in a systematic manner and putting it into engineering systems [31]. It has been, therefore, proposed to design the V/P transfer system using the FIS.

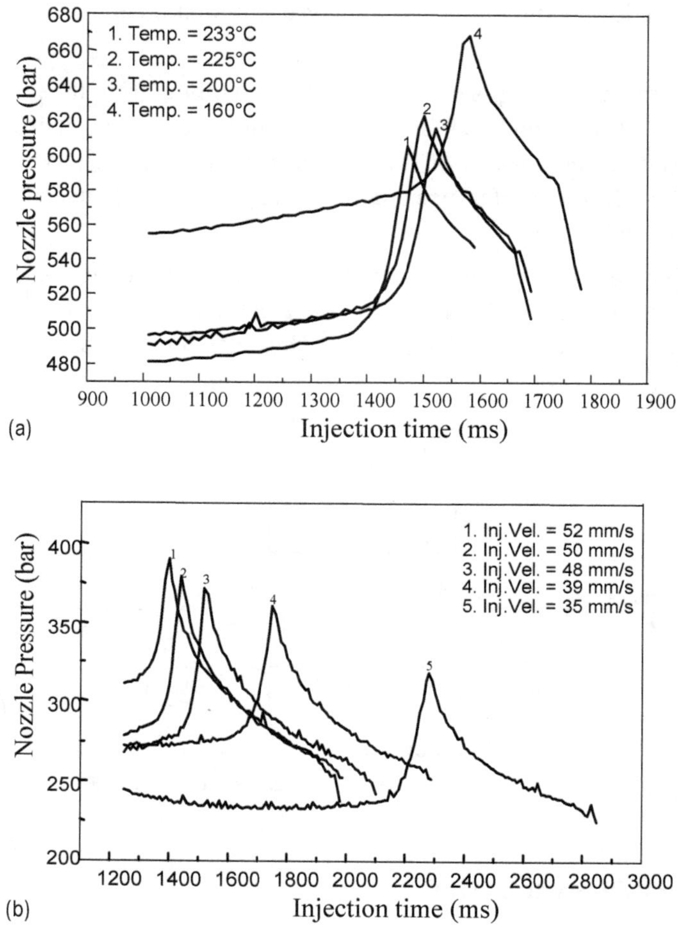


**Figure 2.30** Nozzle pressure measurement during filling and packing: (a) different melt temperatures, and (b) different injection velocities

As shown by Fig. 2.30, a sharp increase in the nozzle pressure can be observed when the last portion of a cavity is being filled. The sharp increase may be expressed by a relationship as

$$dP_N(k+1) > dP_N(k) > dP_N(k-1) \tag{2.38}$$

where

$dP_N(i) = P_N(i) - P_N(i-1)$ for $i = k+1, k, k-1$, and

$P_N(k)$ is the nozzle pressure measurement at the $k$th sampling instance.

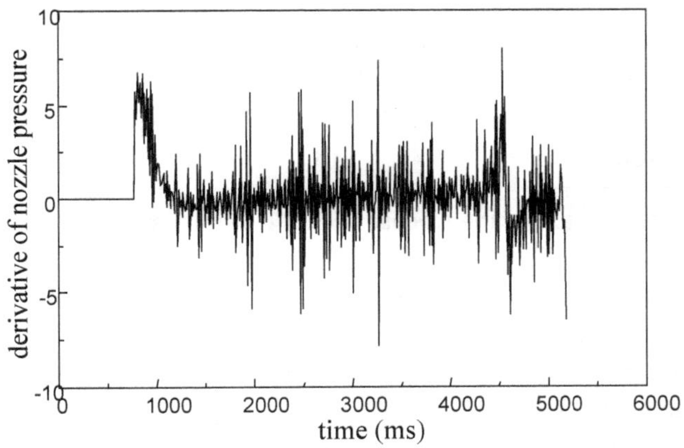


**Figure 2.31** Original nozzle-pressure derivative signal

To prevent false detection, an additional variable max_*avr* is defined to represent the trend of the nozzle pressure in the early stage of filling.

The $dP_N$ signal is, in essence, a differential of the nozzle pressure. The pressure differentiation leads to the amplification of the measurement noise. It is difficult to observe any clear trend from the direct $dP_N$ signal, as shown in Fig. 2.31. A low-pass filter is designed to remove the high frequency noise from the pressure signal before differentiation. The filter should not create significant phase shift, which obviously leads to a late detection of the V/P transfer. Several different filter schemes were compared by the fast Fourier transform (FFT) analysis. A second-order Chebyshev type I filter with a 0.5 dB ripple in the pass-band has been selected for its small phase lag and simplicity. Figure 2.32 shows the derivative of the nozzle pressure signal after filtering.

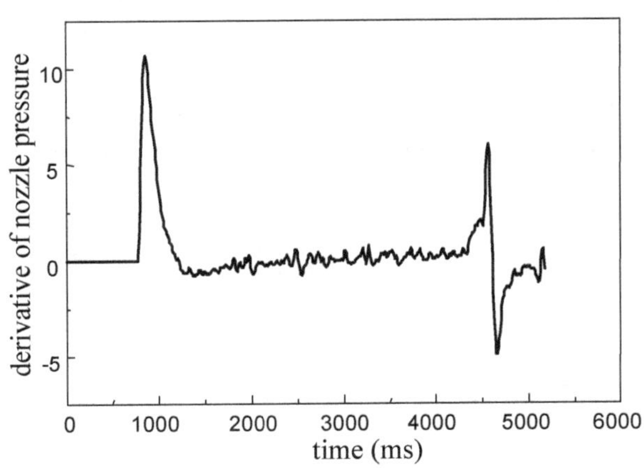


**Figure 2.32** Derivative of nozzle pressure signal after filtering

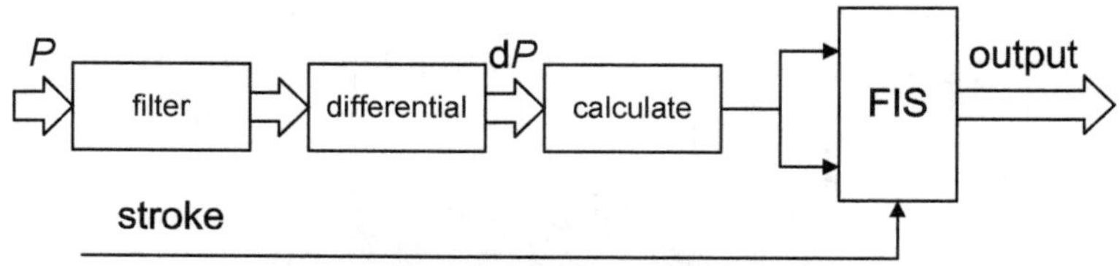


**Figure 2.33** Overall structure of the proposed FIS V/P transfer system

The block diagram of the proposed fuzzy system is shown in Fig. 2.33. The system uses the singleton fuzzifier, the center average defuzzifier, and the minimum inference engine. Two inputs have been chosen for the fuzzy system: $\mathrm{d}P_{\mathrm{N}}(k)/\mathrm{d}P_{\mathrm{N}}(k-1)$ and $\mathrm{d}P_{\mathrm{N}}(k)/\mathrm{max\_}avr$. The system has a single output indicating the optimal V/P transfer time. Through experimental data analysis, the ranges of the inputs are set to be 1.0–2.2 and 1.5–3.5, respectively. The range of the output is set to be 0–1, with the output close to 0.5 as the optimal V/P transfer point. An output value close to 0 indicates an early V/P transfer, while an output value close to 1 indicates a late V/P transfer. The linguistic term set for the two inputs is defined to be {small, medium, large} for output {early, exact, late}. The membership functions for both inputs are chosen to be a triangle shape for the linguistic terms "large" and "small" for simplicity, and a Gaussian function is selected for the "medium" for its narrow shape. The output membership function is chosen to be triangular in shape with the cross points at 0.5. The selections of the membership function shapes are based on the criteria discussed in Section 2.5.1. Figs. 2.34(a), (b), and (c) plot the membership functions for input 1, input 2, and output, respectively.

The experience and the experimental data analysis have dictated the rule base, as listed in Table 2.1. The values in parentheses are the weights of each rule. Care has been taken to ensure that any combination of inputs results in an appropriate output and that no rules with the same antecedent would have a different consequence. Therefore, the rule set of the fuzzy system has the properties of completeness and consistency.

The fuzzy system is constructed with a C language program developed in-house to detect in real time the proper V/P transfer for injection molding.

**Table 2.1** Rule Set for the Proposed FIS V/P Transfer System

| Input 1 \ 2 | Small | Medium | Large |
|---|---|---|---|
| Small | Early (1) | Early (1) | Early (1) |
| Medium | Early (0.5) | Exact (1) | Late (0.5) |
| Large | Early (0.5) | Exact (0.5) | Late (1) |

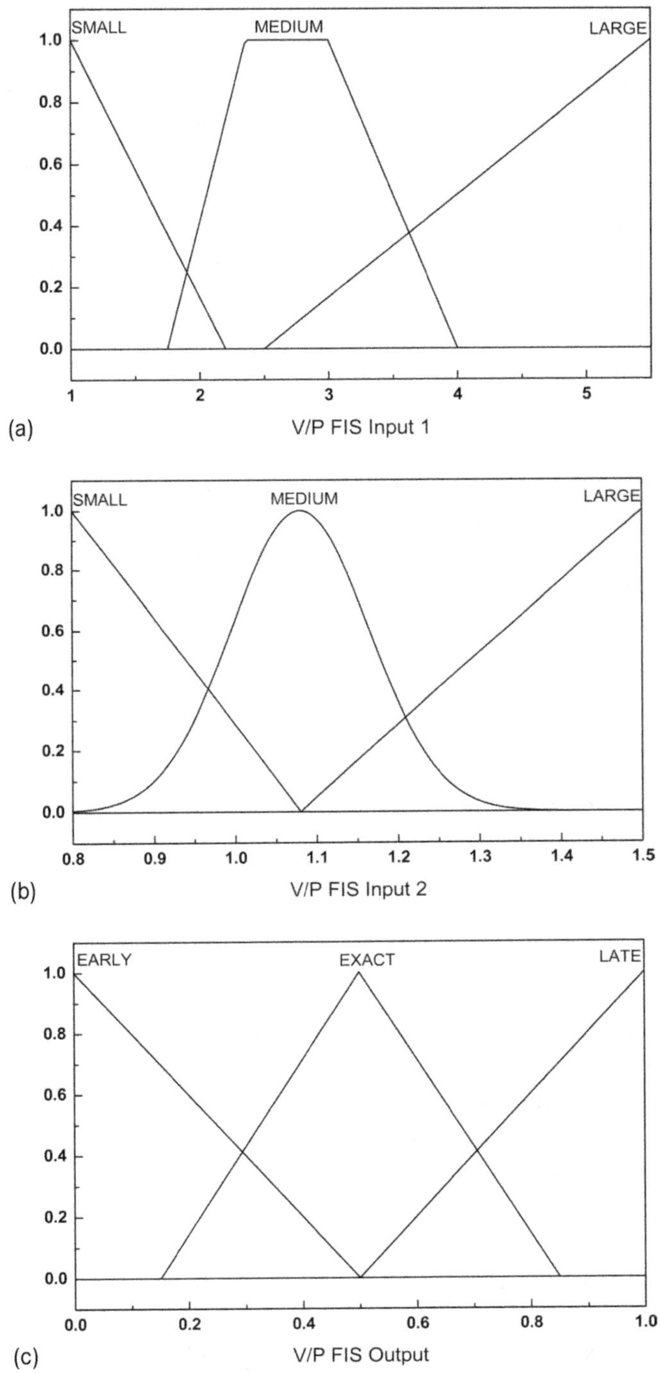


**Figure 2.34** Membership functions for the FIS system inputs and output: (a) input 1, (b) input 2, and (c) output

### 2.5.3 Fuzzy V/P System Experimental Test

For the first injection cycle, the operator is required to give a rough estimation of the injection stroke. The fuzzy system will start to work when the injection stroke reaches 50% of the estimated stroke. The system makes a V/P transfer decision when the output is close to 0.5. The injection stroke will be recorded as the current cycle injection stroke. To reduce the computation requirement, starting from the second cycle, the FIS will start working only after the injection stroke reaches 80% of the injection stroke. The injection stroke, therefore, may be considered as a third input to the fuzzy system; this input does not need to be fuzzified.

The fuzzy system developed is put to the test using the experimental data. The data for Fig. 2.30(a) and (b) were obtained from the injection molding experiments without using the proposed FIS V/P system. It is obvious that the V/P transfers have occurred too late. These experimental data were used as the inputs to the developed FIS V/P transfer system; the system is able to produce the correct V/P transfer for all cases. The V/P transfer times obtained by the FIS are given in Table 2.2. The cup mold was used for the above test. For a simple rectangular fan gate mold cavity, the V/P system was found to be able to automatically produce a proper V/P transfer time without any adjustment.

**Table 2.2** V/P Transfer Decisions Made by the FIS for Different Molding Conditions of Figs. 2.30(a) and (b)

| **Injection Velocity (mm/s)** | **52** | **50** | **48** | **39** | **35** |
|---|---|---|---|---|---|
| Recommended V/P by FIS (ms) | 1320 | 1400 | 1480 | 1660 | 2170 |
| Melt Temperature (°C) | 233 | 225 | 200 | 160 | |
| Recommended V/P by FIS (ms) | 1410 | 1450 | 1470 | 1530 | |

One hundred fan-gated parts are made with the original machine V/P transfer system. The part weights were found to vary within a standard deviation of 0.1039. Another 100 parts are produced with the FIS V/P transfer system. The part weights are found to vary within a standard deviation of 0.065, which is significantly smaller than that of the proposed FIS V/P transfer system.

### 2.5.4 Further Improvement

Further experiments were carried out to test the accuracy and robustness of the fuzzy V/P transfer system with different molding conditions. It was noticed that the system works well for most of the cases, including different molding machines without further modification, as shown in Fig. 2.35, in which the modified flat mold is used. However, the nozzle pressure may experience a change due to a sudden increase of the valve opening when the injection velocity set point has a step-up. The fuzzy system, in this case, may give a false detection of the V/P point. It is exemplified by Fig. 2.36, which illustrates the detection failure, with the same model as that of Fig. 2.35, due to a velocity set-point step-up from 20 to 40 mm/s introduced at the injection time of 750 ms. The nozzle pressure (the solid line) jumps up shortly after the step change of the injection velocity.

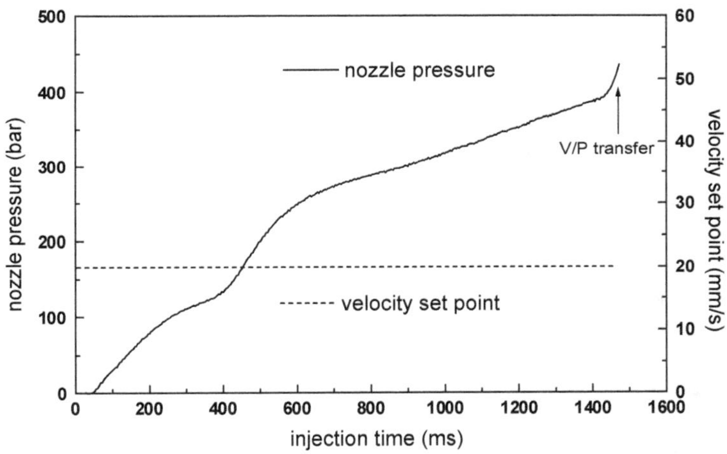


**Figure 2.35** Proper detection of the FIS V/P transfer system with a constant-velocity set point

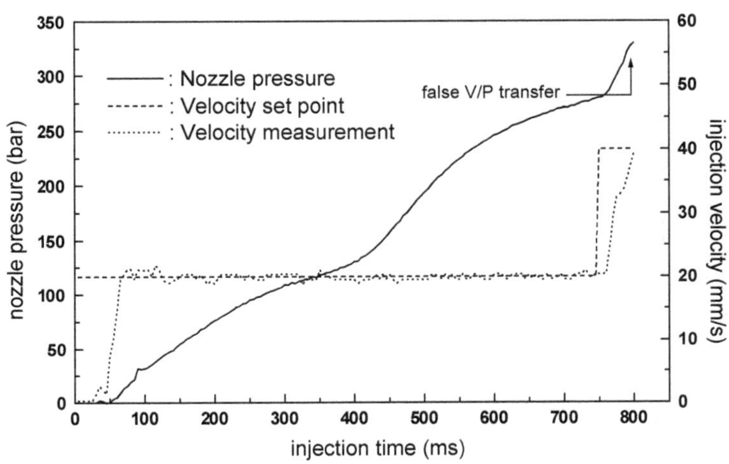


**Figure 2.36** Fuzzy V/P system fails to work with a velocity step change

A modification has been developed to extend the existing fuzzy V/P system to cover the profiled injection velocity cases. A true V/P transfer point should happen only when the mold cavity is filled. The injection velocity normally reduces at the true V/P transfer point, as shown in Fig. 2.37. The nozzle pressure with profiled injection velocity exhibits a different pattern: the pressure increase is accompanied by a velocity increase, as can be seen in Fig. 2.38. Because, in the fill phase of injection molding, the nozzle pressure and injection velocity are both manipulated by the same variable, the hydraulic servo-valve opening, the changes in the servo-valve opening lead to changes in both process variables. However, the valve-opening signal may not be available for certain cases. The velocity pattern therefore can be identified and used as an additional criterion.

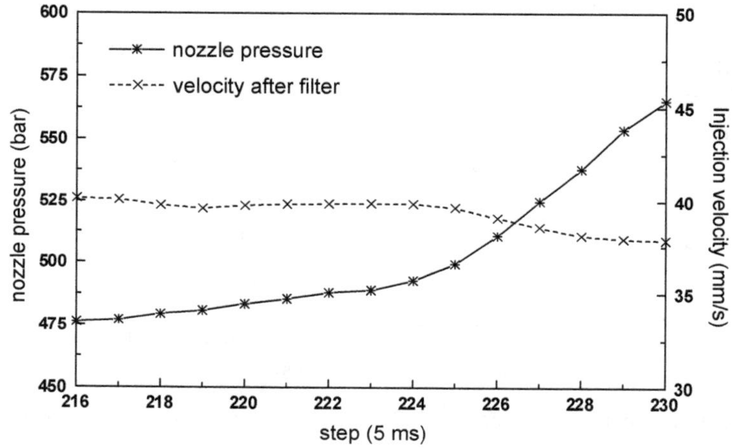


**Figure 2.37** Nozzle pressure increase caused by fully filling the mold cavity

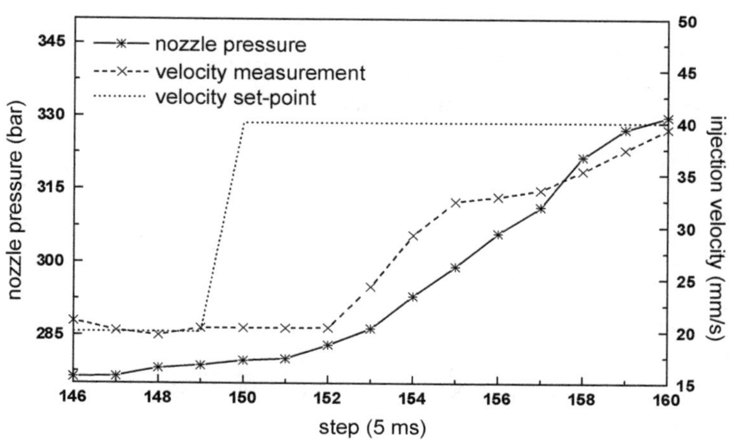


**Figure 2.38** Nozzle pressure increase caused by velocity set point change

Figure 2.39 shows the flow chart of the modified fuzzy V/P system. First, the injection velocity is passed through a Chebyshev filter to eliminate high-frequency noises of the measurement. The filtered injection velocity of each sample is then recorded. Once the fuzzy V/P system outputs a positive signal of detection, the program calculates the average value of the last 10 steps of injection velocity. This average value will be compared with the current step's velocity data: if the current velocity is larger than the average value, the output of the fuzzy V/P system is a false detection; otherwise the detection is true. The average value is used instead of the last step's data. On one hand, this can eliminate a sudden variation of the velocity; on the other hand, it can also avoid the effect of the system delay.

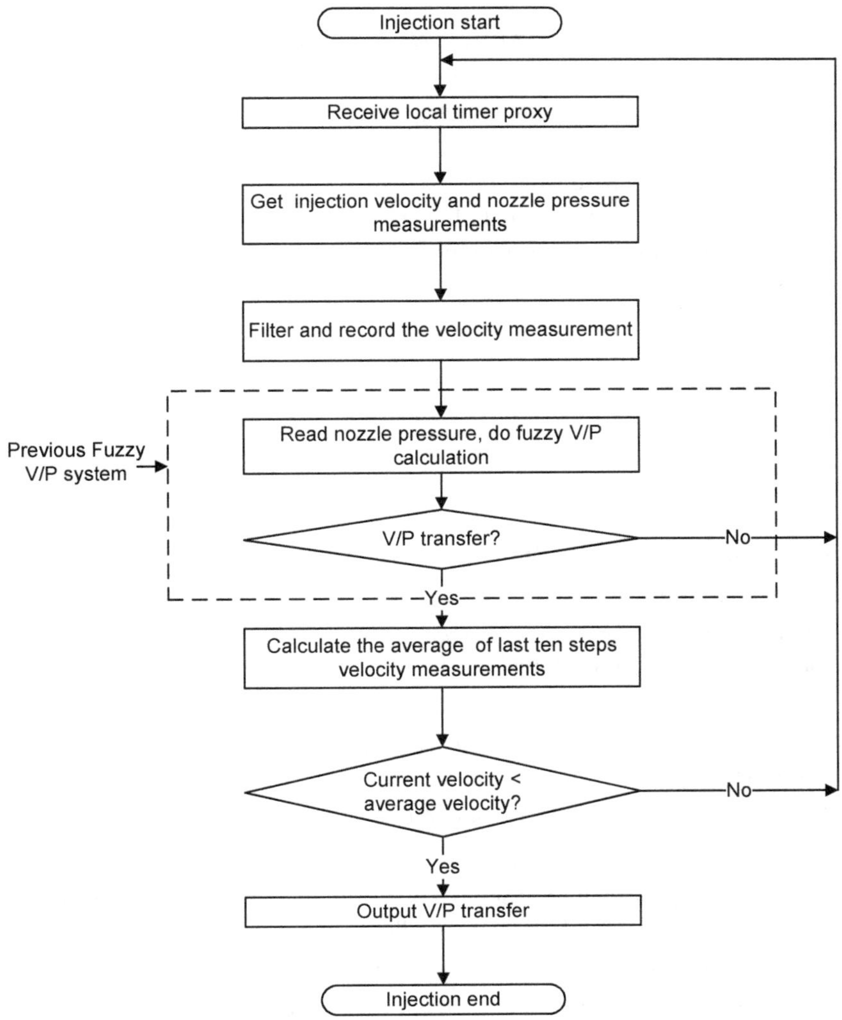


**Figure 2.39** Flow chart of the modified fuzzy V/P system

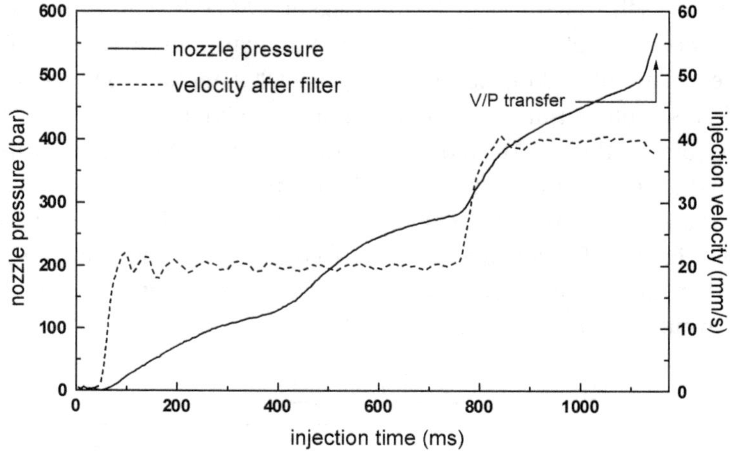


**Figure 2.40** Result of the modified fuzzy V/P system

Figure 2.40 shows the results of the modified fuzzy V/P system working with the same velocity set point profile as Fig. 2.36. The V/P transfer point is properly detected by the modified system. The fuzzy V/P system is successfully extended to the profiled velocity set point cases.

## References


[13] Y. Yang and F. Gao, *Control Eng. Pract.*, **8**, 1285-1296, 2000

[14] F. Gao, W.I. Patterson, and M.R. Kamal, *Polym. Eng. Sci.*, **36**, 1272-1285, 1996

[17] K.J. Aström and T. Hägglund, *PID Control-Theory, Design, and Tuning*, Instrument Society of America, North Carolina, 1995

[18] Y. Yang and F. Gao, *Int. Polym. Process.*, **XIV**, 196-205, 1999

[19] K.J. Aström and B. Wittenmark, *Adaptive Control* (2nd ed.), Addison-Wesley, Boston, MA, 1995

[20] D.E. Seborg, T.F. Edgar, and S.L. Shah, *AIChE Journal*, **32**, 881, 1986

[21] Y. Yang and F. Gao, *Polymer Eng. Sci.*, **39**, 2042-2063, 1999

[22] K. Warwick, *Implementation of Self-Tuning Controllers*, The Institution of Electrical Engineers, P. Peregrinus, London, UK, 1988

[23] S. Barnett, *Matrices: Methods and Applications*, Clarendon Press, Oxford, UK, 1990

[24] T.R. Fortescue, L.S. Kershenbaum, and E.E. Ydstie, *Automatica*, **17**, 831, 1981

[25] M.E. Salgado, G.C. Goodwin, and R.H. Middleton, *Int. J. Control*, **47**, 477, 1988

[26] J.E. Parkum, N.K. Poulsen, and J. Holst, *IFAC 11th Triennial World Congress*, Tallinn, Estonia, USSR, 1990

[27] M. Lundh and K.J. Åström, *Automatica*, **30**, 1649, 1994

[28] E.F. Camacho and C. Bordons, *Model Predictive Control in the Process Industry*, Springer-Verlag, London, UK, 1995

[29] D.W. Clark, C. Mohtadi, and P.S. Tuffs, *Automatica*, **23**, 137, 1987

[30] L. Ljung, *System Identification: Theory for the User*, Prentice-Hall, Englewood Cliffs, 1987

[31] L.X. Wang, *A Course in Fuzzy Systems and Control*, Prentice Hall, Professional Technical Reference (PTR), 1997

[32] F. Gao and C.H. Zhao, US Patent 6,309,571 B2, *Method and Apparatus for the Control of Injection Molding*, Oct. 30, 2001

# 3 Learning Type Control for the Injection Molding Process

## 3.1 Learning Type Control Background

In the injection molding process, raw plastic materials are fed to the machine through the hopper with a certain volume each cycle, are melted and transferred to the nozzle of the barrel, and then go through the injection, packing, and cooling stages to form the final product of limited amount. The whole cycle normally lasts for several seconds to minutes. This kind of production fits the definition of the batch process, as quoted below [33]:

*A process is considered to be batch in nature if, due to physical structuring of the process equipment or due to other factors, the process consists of a sequence of one or more steps (or phases) that must be performed in a defined order. The completion of this sequence of steps creates a finite quantity of finished product. If more of the product is to be created, the sequence must be repeated.*

The control of such a batch process, unlike that of the continuous processes, involves keeping a defined sequence of operations. At the same time, the control system must also control the key process variables to follow certain profiles repetitively and accurately to ensure a good part quality. The disturbances can come from the materials, the machine itself, the environments, and human interferences. Most disturbances cause the process to drift slowly over cycles instead of with abrupt changes. For example, the mold temperature may increase from cycle to cycle during the warm-up stage as a result of the continuous molding operation. Disturbances are typically difficult to accurately represent mathematically, but their impacts on the control system can be reflected in changes of the manipulated variables. Consider the unique features of the injection molding process: the application of traditional feedback control algorithms, like those introduced in Chapter 2, cannot achieve satisfactory performance directly, and they need to be tailor-made to fit the characteristics of the batch injection molding.

Another problem that has to be solved is a practical issue: the expenses of the control valves used for injection molding control. There are commonly two types of control valves used in the current molding industry, the servo-valve and the proportional valve, to provide directional and flow control of the hydraulic system. A servo-valve is a device that converts a low-power electrical input signal into a proportionally much higher hydraulic output with high accuracy. A servo-valve normally has complicated structure, including a torque motor, a flapper valve, a spool valve for the second hydraulic stage, and an internal feedback mechanism [34]. Using a servo-valve as the actuator for the closed-loop control, as indicated in the previous chapters, produces a good control performance with fast response. However, it increases not only the capital cost but also the operation cost. For example, it requires a much more sophisticated oil filtration system.

The proportional valve is another type of control valve developed to fill the wide gap between simple on/off valves and sophisticated servo-valves. Corresponding to the control signal, its proportional solenoid produces a proportional spool shift for flow controls [34]. The proportional valves have simpler structures and are less expensive than those of the servo-valves. It also has good tolerance of a small oil contamination. For these reasons, they are widely used in the molding industry. Compared to the servo-valve, the proportional valve has a relatively slower response, that is, a bigger time constant. The input-output linearity of proportional valves is also not as good as that of the servo-valve.

The controls of previous chapters were carried out with the fast-response servo-valves. For cases where proportional valves are used, the slow response could severely deteriorate the control performance. Considering the wide usage of proportional valves in the molding industry, it is desirable to develop a proper control algorithm that can achieve satisfactory control performances with both fast- and slow-response valves.

To solve the actuator response time problem and ensure the control performance, the control algorithm must be harmonious with the process characteristics. As a repetitive process, information on past molding controls may be explored for improvement of the current and future cycles. Triggered by this idea, there are several control strategies developed for processes with a repetitive nature like injection molding, including iterative learning control (ILC), repetitive control (RC), and run-to-run control (R2R), starting from the late 1970s. These strategies can be categorized into the learning-type control methods.

Iterative learning control is designed to mimic the human learning process. It was originally developed for the manipulation of industrial robots, in which it was required to repeat a given task with high precision [35–40]. By using the repetitive nature of the processes, ILC progressively and iteratively improves the control accuracy along two time dimensions for control input, that is, the trial (or batch) index $k$ from trial to trial, and the elapsed time index $t$ during a trial from step to step. This two-dimensional learning results in advantages over the conventional feedback control techniques where only one-dimensional input actions are made along the time axis. The key for a learning control design is to find an algorithm to ensure that the control input is generated for the next trial in such a way that the performance improves for each successive trial. The concept of iterative learning for generating such an input was first introduced by Uchiyama [35] and was later mathematically formulated by Arimoto et al. [36]. Since then, considerable efforts have been made on the development and analysis of iterative learning control [41–51]. Recently, ILC has been applied to many repetitive processes, such as batch reactor, batch distillation, and injection molding [52–54]. Comprehensive literature surveys on this subject can be found in reference [55].

Repetitive control was originally developed in a continuous process to track or reject periodic exogenous signals, in particular for a power supply facility in the 1980s. Later, Francis and Wonham [56] proposed an internal model principle and founded the basic theory for RC. The R2R control was developed by Sachs et al. [57] for batch processes like semiconductor manufacturing. The goal of R2R is quality control, and for most of the batch manufacturing process, the quality measurement rate is far slower than the process itself. The R2R control introduced the exponentially weighted moving average (EWMA) filter, similar to the EWMA control chart in traditional statistical process control, and developed a run-to-run control methodology. In fact, ILC, RC, and R2R control can be expressed by a uniform formulation [55]. In this book, ILC will be adopted to represent the learning-type control algorithm for injection molding control applications.

## 3.2 Basic Iterative Learning Control

### 3.2.1 PID-Type ILC

The basic structure of a learning controller is shown in Fig. 3.1. The conventional ILC scheme works as an open-loop feed-forward compensator. It generates the control signals of the future cycle $u_{k+1}$ by using the input, output, and the desired trajectory of the current cycle, $u_k$, $y_k$, and $r$, respectively. Though over the past few decades ILC has made a good advance, the main research efforts were focused on the simple P-type, D-type, and modified PD-type or PID-type learning algorithms. These learning algorithms have good physical insight, and they are relatively easy to implement. It has been found that an ILC not using within-cycle feedback tends to be sensitive to perturbation and tends to be slow in system convergence [58]. Furthermore, the tuning of such a PID-type ILC controller is a vague task, and the transient response cannot be guaranteed. A feedback combination is thus considered in ILC studies [58, 59].

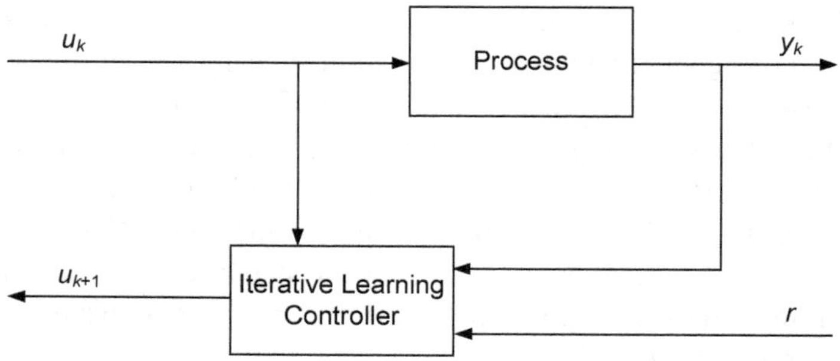


**Figure 3.1** Basic structure of the iterative learning controller

P-type learning, one of the simplest methods, updates the control signal of the complete injection stage $[0, N]$ for the next cycle based on the error profile of the current cycle, as shown in Eq. 3.1:

$$u_{k+1}(t) = u_k(t) + K_{\text{ILC-P}}\, e_k(t+1) \tag{3.1}$$

where $e_k(t) = y\_ref(t) - y_k(t)$ is the tracking error of the $t$th sampling period in the current cycle $k$, and $K_{\text{ILC-P}}$ is the learning rate. The P-type learning control algorithm has been found to be effective for certain fast-response systems.

A more general form of PID-type ILC can be expressed by the following equation:

$$u_{k+1}(t) = u_k(t) + K_{\text{ILC-P}}\, e_k(t+1) + K_{\text{ILC-I}} \sum_{j=1}^{t+1} e_k(j) + K_{\text{ILC-D}} \left[e_k(i+1) - e_k(i)\right] \tag{3.2}$$

where $K_{\text{ILC-P}}$, $K_{\text{ILC-I}}$, and $K_{\text{ILC-D}}$ are the proportional, integration, and derivative learning gains, respectively. Without loss of generality, in the following test, the P-type learning algorithm will be adopted for demonstration purposes.

### 3.2.2 Time-Delay Consideration

It is clear from the formula of a PID-type ILC that the control of the current cycle is based on the process input and the error of the last cycle in a point-to-point manner. Up to now, most of the ILC results were for systems without a time delay. However, for many batch processes such as injection molding, the effects of a time delay cannot be ignored. There is a relatively large delay between the control valve and the process variable responses. For the injection stage, the adjustment of the control valve needs to pass through the whole hydraulic system and the melt between the injection screw and the melt flow front to get the injection velocity response. Considering that the melt is slightly compressible under high pressure, and coupled with its complicated visco-elastic properties, changes in the control valve cannot affect the injection velocity instantaneously. During packing, since the material inside the mold is gradually converted from melt to solid, the pressure response exhibits significant time-varying characteristics. The long process delay, as well as variations of delay during filling and packing, makes it difficult to apply the simple point-to-point ILC method. To solve this problem, control law Eq. 2.39 can be modified to taken into consideration the delay term:

$$u_{k+1}(t) = u_k(t) + K_{\text{ILC-P}}\, e_k(t + t_\text{d} + 1) \tag{3.3}$$

where $t_\text{d}$ is an estimated delay time. In this equation, the control error at time $t + t_\text{d}$ is used to update the control input at time $t$ for the next cycle. The control law of Eq. 3.3 can be applied to cases where the time delay is exactly known. For processes with an uncertain delay, there is no guarantee that this control law will be convergent.

For a system with a varying delay bounded by $h$, Park et al. [60] proposed to hold the control input at a constant over the duration $h$, resulting in a modified learning control law:

$$\begin{aligned} &u_{k+1}(t) = u_k(mh) + \Gamma e_k(mh + dh + \xi) \\ &\forall t \in [mh, mh + h), \quad m \in \{0, 1, \ldots, M - d\} \end{aligned} \tag{3.4}$$

where $e_k(mh + dh + \xi) = y_d(mh + dh + \xi) - y_k(mh + dh + \xi)$, $\xi$ is the initial remainder, and $dh + \xi$ the upper limit of delay. The system divides the process time span by the size of the time delay uncertainty $h$. It has been shown that the convergence can be maintained by this method [60]. This idea is adopted here for injection molding process control.

### 3.2.3 P-Type ILC for Injection Velocity

A simple P-type ILC has been developed and implemented to control the injection velocity for demonstration purposes. The sampling rate of the velocity controller is determined to be 10 ms, and the proportional learning gain $K_{\text{ILC-P}}$ is selected to be 0.004 using a trail-and-error method. For the first cycle of injection, the control valve is controlled open-loop based on the previous test results. The velocity control results are plotted in Fig. 3.2.

It is obvious that the velocity is well controlled by the P-type ILC: after several cycles of learning, the velocity is converged to the set-point profile, and even after 100 cycles of test, the control is still stable and exhibits good tracking performance. To show the converge procedure of the learning control, Figure 3.3 shows the sum of squared error (SSE) of the first 50 injection cycles, which clearly illustrates the good performance of the ILC.

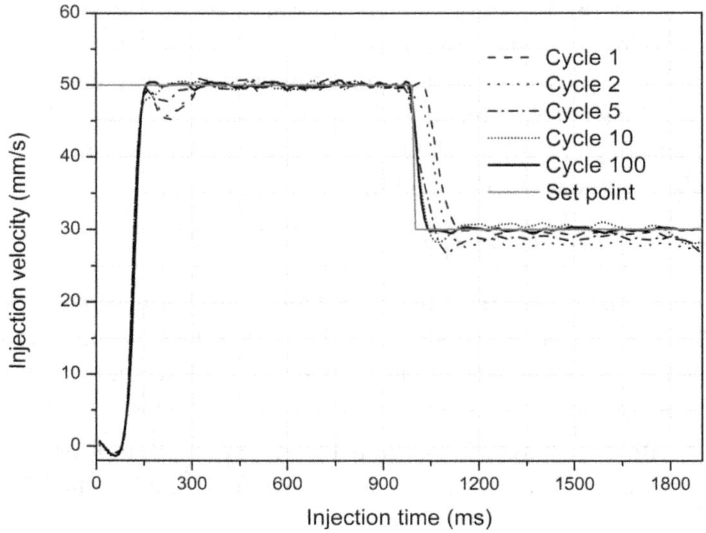


**Figure 3.2** Injection velocity P-type ILC result

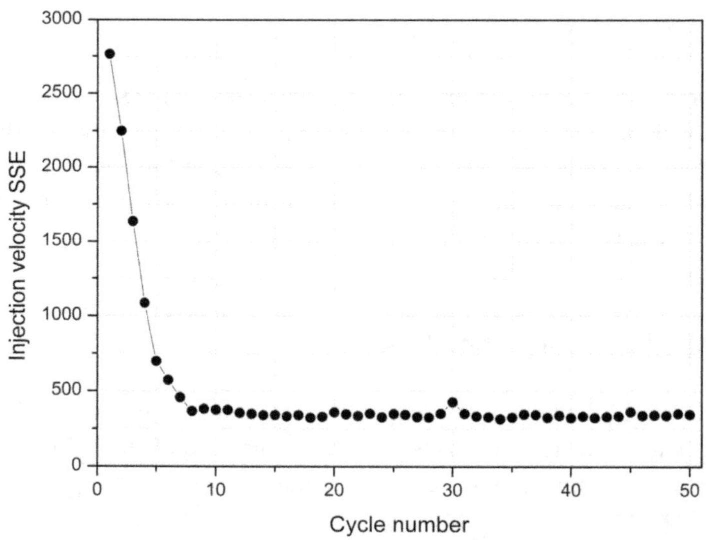


**Figure 3.3** SSE converge procedure of the P-type ILC for injection velocity

### 3.2.4 P-Type ILC for Packing Pressure

To further demonstrate the performance of iterative-learning-type control algorithms, the P-type ILC is extended to the packing pressure control. Due to the severe time-varying characteristics of the packing pressure dynamics and relatively larger time constant, the sampling rate of the controller is determined to be 50 ms, much longer than that of the injection velocity. The proportional learning rate is selected to be 0.001, also with a series of trail-and-error tests. The designed ILC is applied to experimentally control the packing pressure, and the results are shown in Fig. 3.4.

The first cycle's control valve opening is set to be open-loop again, and the pressure response is far from the set-point profile. After five cycles of learning, the sixth cycle's pressure is close to the set point, especially for the first half of the packing.

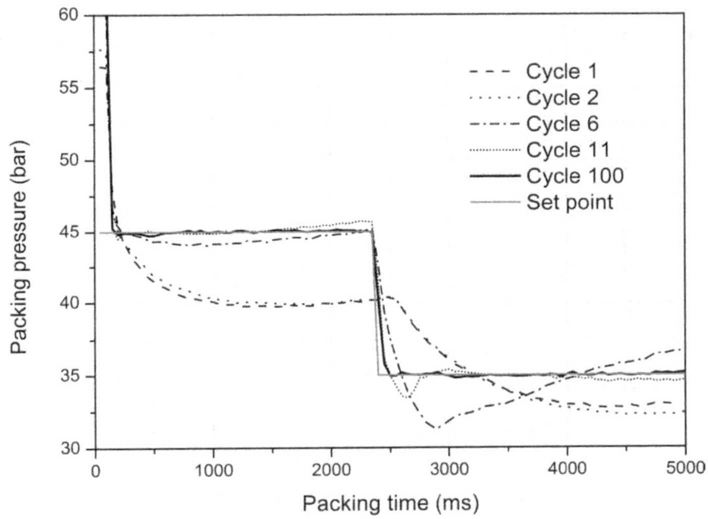


**Figure 3.4** Packing pressure P-type ILC result

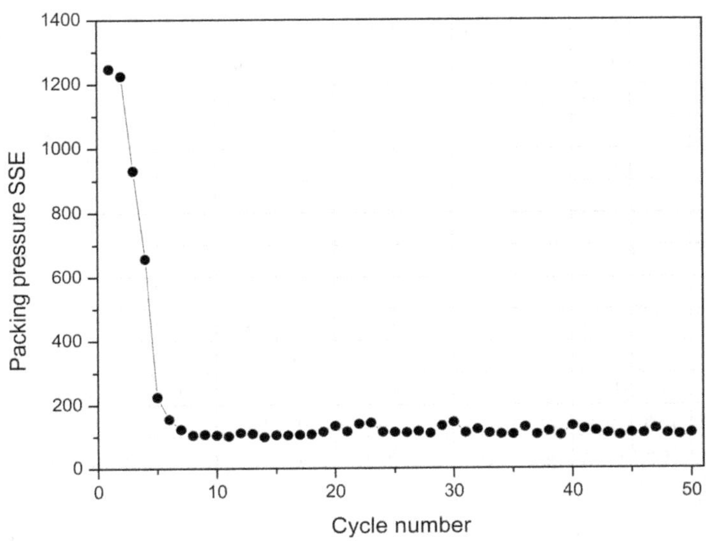


**Figure 3.5** SSE converge procedure of the P-type ILC for packing pressure

The SSE converge procedure is shown in Fig. 3.5. The successful application of packing pressure control again proved the good potential of iterative-learning-type control algorithms.

## 3.3 Optimal Iterative Learning Control

The basic PID-type ILC faces several problems in industrial applications, especially for the injection molding process. The first problem is the tuning of the learning rate. It must be based on the process gain, but a too-high learning rate could make the controller sensitive to disturbances, while a too-small learning rate reduces the convergence speed. It is difficult to balance and find a proper learning rate. Therefore, in the injection velocity and packing pressure applications, the learning rates were determined through a trail-and-error method. The second problem of PID-type ILC is that the control is purely dependent on the error information of previous cycles, and there is no within-cycle feedback in the basic ILC. Such a cycle-wise feed-forward control could cause poor control when the disturbance is severe, which is common for injection molding operations.

Recently, a new ILC algorithm has been proposed by Amann et al. [61], combining the Riccati feedback with the typical ILC feed-forward based on an optimization principle. The scheme has the advantage of step-size automatic determination and hence guarantees an exponential convergence. Simulations have shown that Amann's optimal learning algorithm is effective in comparison to the traditional ILC schemes; this raises the expectation of wider applications in industrial settings. In reality, uncertainties always exist in the process disturbances, and it is also very likely that the process initialization may not be exactly repeatable. These practical issues are important to industrial implementations and are therefore being addressed to Amann's original work in this book for injection molding application purposes.

In the following section, the optimal iterative learning control algorithm of Amann et al. [61] is applied to the injection molding process with uncertain disturbances where exact initial resetting is unavailable. A sufficient and necessary condition to ensure that the ILC has the bounded-input-bounded-output (BIBO) stability is established. An analysis on the selection of the weighting matrices for the cost function is consequently performed. Simulation and experimental applications to control injection molding filling velocity as an illustration, using the introduced optimal iterative learning control, are given to demonstrate the effectiveness of the proposed algorithm.

A background introduction to the optimal learning algorithm is given below.

### 3.3.1 Problem Formulation

The process of interest is assumed to be described by the following sampled-time linear system with disturbances:

$$\begin{aligned} x_k(t+1) &= Ax_k(t) + Bu_k(t) + \nu_k(t), & 0 \le t \le N, \quad k = 0,1,2,\cdots \\ y_k(t) &= Cx_k(t) + \omega_k(t), & x_k \in R^n \quad u_k \in R^m \quad y_k \in R^p \end{aligned} \tag{3.5}$$

where the subscript $k$ denotes the iteration number of the operation corresponding to the trial index; for example, $y_k(t)$ is the value of the system output at time $t$, $0 \le t \le N$, at the $k$th operation. Here $\nu_k(t)$ and $\omega_k(t)$ denote the bounded state and external disturbance. Note that the exact state initialization of Eq. 3.5 for each iteration is not required in this work. The robustness to initial state variation and external disturbances will be discussed in this chapter. The state-space matrices $A$, $B$, and $C$ are assumed to be time-invariant for simplicity. It is possible, without any technical difficulties, to extend all results of this chapter to time-varying systems. Based on the linear system theory, the following solution to Eq. 3.5 can be deduced:

$$y_k(t) = \sum_{l=0}^{t-1} CA^{t-1-l}Bu_k(l) + \eta_k(t) \tag{3.6}$$

$$\eta_k(t) = CA^t x_k(0) + \sum_{l=0}^{t-1} CA^{t-1-l}\nu_k(l) + \omega_k(t) \tag{3.7}$$

It can be observed in the above that the initial actions of each trial and disturbance appear in the plant, extending the work of Amann et al. [61] to a more general case. In each trial, finite time intervals are involved. Equation 3.6 can be made in a vector form by building supervectors $y_k$, $u_k$, and $\eta_k$ from $y_k(t)$, $u_k(t)$, and $\eta_k(t)$ as below:

$$y_k = Gu_k + \eta_k \tag{3.8}$$

where

$$y_k = \begin{bmatrix} y_k(1) \\ y_k(2) \\ \vdots \\ y_k(N) \end{bmatrix}, \quad u_k = \begin{bmatrix} u_k(0) \\ u_k(1) \\ \vdots \\ u_k(N-1) \end{bmatrix}, \quad \eta_k = \begin{bmatrix} \eta_k(1) \\ \eta_k(2) \\ \vdots \\ \eta_k(N) \end{bmatrix}, \quad G = \begin{bmatrix} CB & 0 & \cdots & 0 \\ CAB & CB & \cdots & 0 \\ \vdots & \vdots & \ddots & \vdots \\ CA^{N-1}B & CA^{N-2}B & \cdots & CB \end{bmatrix}.$$

The supervectors are presented with the omission of the argument time $t$. During implementation of the iterative learning control, $y_k$ and $u_k$ of previous trials need to be memorized for the computation of $u_{k+1}(t)$ of the current trial. Matrix $G$, a lower-triangular block matrix known as a Toeplitz matrix, can be determined from its first column. In this chapter, it is postulated that $\ker G^T = 0$, defined as the "regularity condition" by Amann et al. If process 3.5 has a relative degree of one, i.e., $CB \ne 0$, then $G$ is invertible in the SISO (Single-Input-Single-Output) case. Otherwise, if $CB = 0$, then a regularizing procedure can be performed, as detailed in the works of Amann et al. [61] and Silverman [62]. This regularity condition ensures that $G^T G$ (or $GG^T$) has at least one positive eigenvalue. Based on this postulation, a convergence proof different from Amann et al. [61] will be given in Section 3.3.3.

The objective considered here, for a tracking control problem, is that for any given reference trajectory denoted by $r(t)$, $1 \le t \le N$, an iterative learning controller is derived such that when applied to system 3.5 the closed-loop tracking error is reduced iteratively from trial to trial, even with the existence of initial errors and uncertain disturbances.

**Definition 3.1** [61]: An iterative learning control algorithm is causal if, and only if, the value of the control input to the system at time $t$ on the $(k+1)$th trial/experiment is computed only from the data that are available from the $(k+1)$th trial in the time interval $[0, t]$ and from previous trials.

### 3.3.2 Optimal Iterative Learning Controller

Consider the following nominal system composed of the coefficient matrices $A$, $B$, and $C$ of model 3.5:

$$\begin{aligned} &\hat{x}_k(t+1) = A\hat{x}_k(t) + Bu_k(t), && 0 \le t \le N, \quad k = 0, 1, 2, \cdots \\ &\hat{y}_k(t) = C\hat{x}_k(t), && \hat{x}_k \in \mathrm{R}^n \quad u_k \in \mathrm{R}^m \quad \hat{y}_k \in \mathrm{R}^p \end{aligned} \tag{3.9}$$

where the variables with superscript " $\hat{}$ " denote the nominal system outputs, and they are initialized by zeros. They represent the system outputs of Eq. 3.5 in the absence of any disturbances and initial errors. For the reference trajectory (or desired system output) $r(t)$, given over $1 \le t \le N$, in the $(k+1)$th trial, the nominal optimal iterative learning control law is obtained by minimizing the following quadratic performance index with respect to $u_{k+1}(t)$:

$$J_{k+1} = \sum_{t=1}^{N} \left[r(t) - \hat{y}_{k+1}(t)\right]^T Q(t)\left[r(t) - \hat{y}_{k+1}(t)\right] + \sum_{t=0}^{N-1} \left[\Delta u_{k+1}(t)\right]^T R(t)\left[\Delta u_{k+1}(t)\right] \tag{3.10}$$

where $\Delta u_{k+1}(t) = u_{k+1}(t) - u_k(t)$, and the weighting matrices $Q(t)$ and $R(t)$ are arbitrary, symmetric, positive definitive for all $t$. The index function Eq. 3.10 can be rewritten in matrix form as

$$J_{k+1} = \left[r - \hat{y}_{k+1}\right]^T Q\left[r - \hat{y}_{k+1}\right] + \Delta u_{k+1}^T R \Delta u_{k+1} \tag{3.11}$$

where $Q = diag\{Q(1), Q(2), \cdots, Q(N)\}$, $R = diag\{R(0), R(1), \cdots, R(N-1)\}$, and

$$\hat{y}_k = \begin{bmatrix} \hat{y}_k(1) \\ \hat{y}_k(2) \\ \vdots \\ \hat{y}_k(N) \end{bmatrix}, \qquad r = \begin{bmatrix} r(1) \\ r(2) \\ \vdots \\ r(N) \end{bmatrix} \tag{3.12}$$

By finding the partial derivative of Eq. 3.11 with respect to $u_{k+1}$, one obtains the nominal optimal control input

$$\hat{u}_{k+1} = \hat{u}_k + R^{-1}G^T Q\left[r - \hat{y}_{k+1}\right] \tag{3.13}$$

However, it can be observed that the algorithm Eq. 3.13 is not causal for the computation of $\hat{u}_{k+1}$ because, by this control law, $\hat{u}_{k+1}(t)$ would depend on values of $\hat{y}_{k+1}(t')$ for $t \le t' \le N$. Following Amann et al. [61], an equivalent form of Eq. 3.13 can be given:

$$S(t) = A^T S(t+1)\left\{I - B\left[B^T S(t+1)B + R(t+1)\right]^{-1} B^T S(t+1)\right\}A + C^T Q(t+1)C$$
$$t = 0, 1, \cdots, N-1; \quad S(N) = 0 \tag{3.14}$$

$$\varphi_{k+1}(t) = \left[I + S(t)BR^{-1}(t)B^T\right]^{-1}\left[A^T \times \varphi_{k+1}(t+1) + C^T Q(t+1)\hat{e}_k(t+1)\right]$$
$$t = 0, 1, \cdots, N-1; \quad \varphi_{k+1}(N) = 0 \tag{3.15}$$

where $\hat{e}_k(t+1) = r(t+1) - \hat{y}_k(t+1)$. The nominal input update law thus becomes

$$\hat{u}_{k+1}(t) = \hat{u}_k(t) - \left[B^T S(t)B + R(t)\right]^{-1} B^T \times S(t)A\left[\hat{x}_{k+1}(t) - \hat{x}_k(t)\right] + R^{-1}(t)B^T \varphi_{k+1}(t) \tag{3.16}$$

This indicates that a causal nominal control input could be obtained iteratively if the nominal states $\hat{x}_k$ and outputs $\hat{y}_k$ were introduced by the nominal system 3.9. This iterative learning control algorithm is also optimal when applied to Eq. 3.8 where $\eta_k = 0$, that is, the disturbance-free case. This chapter aims to develop an ILC algorithm in the presence of uncertain initials and disturbances. This can be achieved by calculating $u_k$ by replacing the nominal $\hat{x}_k$ and $\hat{y}_k$ in Eqs. 3.14, 3.15, and 3.16 with the measurements of $x_k$ and $y_k$ of the system Eq. 3.5. Therefore, a causal iterative learning control algorithm of interest can be summarized as

$$u_{k+1} = u_k + R^{-1}G^T Q e_{k+1} \tag{3.17}$$

$$\varphi_{k+1}(t) = \left[I + S(t)BR^{-1}(t)B^T\right]^{-1} \times \left[A^T \varphi_{k+1}(t+1) + C^T Q(t+1)e_k(t+1)\right]$$
$$t = 0, 1, \cdots, N-1; \quad \varphi_{k+1}(N) = 0 \tag{3.18}$$

$$u_{k+1}(t)=u_k(t)-\left[B^T S(t)B+R(t)\right]^{-1} B^T \times S(t)A\left[x_{k+1}(t)-x_k(t)\right]+R^{-1}(t)B^T\varphi_{k+1}(t) \tag{3.19}$$

where $S(t)$ is obtained by Eq. 3.14. It can be seen that the control algorithm consisting of Eqs. 3.13, 3.17, 3.18, and 3.19 is causal. In Eq. 3.17, $u_{k+1}(t)$ is obtained by improving the last trial input $u_k(t)$ by incorporating a feedback action of the current trial (second term on the right-hand side of Eq. 3.19) and a feed-forward action (third term of Eq. 3.19), which represents the information from previous trials.

In the work of Amman et al. [61], there is also lacking a guideline on the selection of the weighting matrices $Q$ and $R$ for the system convergence. This practical consideration is important in the design and application of the ILC to batch processes. The convergence and robustness analysis of the above method is conducted below with the uncertainties of initializations and disturbances.

### 3.3.3 Robust and Convergence Analysis

The robust bounded-input–bounded-output stability as defined below will be investigated for the proposed algorithm [63].

**Definition 3.2**: An iterative learning control system is said to be robustly BIBO (bounded-input-bounded-output) stable if, and only if, for a system with any uncertain initialization and bounded disturbance, the outputs are bounded for each trial for any bounded control reference.

The above definition considers the disturbances to the system and the initialization uncertainty along the trial axis. The robust BIBO stability together with the convergence and robustness of the iterative learning control algorithm is discussed.

**Theorem 3.1 (Robust BIBO Stability)**: The application of the iterative learning control algorithm of Eqs. 3.13, 3.17, 3.18, and 3.19 to plant 3.5 is robustly BIBO stable if, and only if, $I+GR^{-1}G^TQ$ and $I+R^{-1}G^TQG$ have all their eigenvalues outside the unit disc, or

$$\left\|I+R^{-1}G^TQG\right\|>1 \tag{3.20}$$

$$\left\|I+GR^{-1}G^TQ\right\|>1 \tag{3.21}$$

**Proof**: Premultiplying Eq. 3.16 by $G$ and in view of 3.7 and $e_k=r-y_k$, one obtains

$$e_{k+1}=e_k-GR^{-1}G^TQ\,e_{k+1}-\Delta\eta_{k+1} \tag{3.22}$$

where $\Delta\eta_{k+1}=\eta_{k+1}-\eta_k$. An iterative relationship for $e_k$ along the trial index $k$ is then followed as

$$e_{k+1} = \left(I + GR^{-1}G^TQ\right)^{-1} e_k - \left(I + GR^{-1}G^TQ\right)^{-1} \Delta\eta_{k+1} \tag{3.23}$$

Again substituting $e_{k+1} = r - y_{k+1}$ into Eq. 3.17 and using 3.8, it follows that

$$u_{k+1} = \left(I + R^{-1}G^TQG\right)^{-1} u_k + \left(I + R^{-1}G^TQG\right)^{-1} R^{-1}G^TQ\left(r - \eta_{k+1}\right) \tag{3.24}$$

The result follows by applying standard discrete time system theory.

**Theorem 6.2 (Convergence)**: Apply the iterative learning control algorithm of Eqs. 3.14, 3.17, 3.18, and 3.19 to plant 3.5, in which $R$ and $Q$ are chosen to satisfy Eq. 3.20 and 3.21. If all trials are repeated in the sense that all $x_k(0)$, and external disturbances $\nu_k(t)$ and $\omega_k(t)$ are the same along the trial index $k$, then the following convergence results hold:

$$\lim_{k \to \infty} u_{k+1} = \left(G^TQG\right)^{-1} G^TQ\left(r - \eta^*\right) \tag{3.25}$$

$$\lim_{k \to \infty} e_{k+1} = 0 \tag{3.26}$$

where $\eta^*$ is some constant vector.

**Proof**: If all trials are repeated, it follows from Eq. 3.7 that there exists a constant vector $\eta^*$ such that $\eta_k = \eta^*$ for all trial index $k$. Iteratively using Eqs. 3.24 and 3.23, one obtains

$$u_{k+1} = \left(I + R^{-1}G^TQG\right)^{-k} u_0 + \sum_{l=1}^{k} \left(I + R^{-1}G^TQG\right)^{-l} R^{-1}G^TQ\left(r - \eta_{k+2-l}\right) \tag{3.27}$$

$$e_{k+1} = \left(I + GR^{-1}G^TQ\right)^{-k} e_0 - \sum_{l=1}^{k} \left(I + GR^{-1}G^TQ\right)^{-l} \Delta\eta_{k+2-l} \tag{3.28}$$

respectively. Therefore, under the robust BIBO stable condition of **Theorem 3.1** one obtains

$$\begin{aligned}
\lim_{k \to \infty} u_{k+1} &= \left(I + R^{-1}G^TQG\right)^{-1} \left[I - \left(I + R^{-1}G^TQG\right)^{-1}\right]^{-1} R^{-1}G^TQ\left(r - \eta^*\right) \\
&= \left(R^{-1}G^TQG\right)^{-1} R^{-1}G^TQ\left(r - \eta^*\right) \\
&= \left(G^TQG\right)^{-1} G^TQ\left(r - \eta^*\right),
\end{aligned}$$

which completes Eq. 3.25.

Due to $\Delta\eta_{k+1} = 0$ and in view of Eq. 3.28, the limit of Eq. 3.26 follows readily.

### 3.3.4 Selection of the Weighting Matrices

The above stability derivation is based on the assumption that the initial error and disturbance are bounded. To achieve a reasonable transient performance, the weighting matrices $Q$ and $R$ must be selected carefully. Let $R=\lambda I,\ Q=\mu I$, where $\lambda$ and $\mu$ are positive design constants, and let $\rho=\mu/\lambda$. Notice that the performance of the optimal ILC is affected by the ratio of $Q$ and $R$ rather than their real values, as shown in Eqs. 3.20 and 3.21. A necessary condition has to be satisfied by $\lambda$ and $\mu$ to ensure robust bounded-input-bounded-output stability. This is straightforward from Eqs. 3.20 and 3.21 if $\lambda$ and $\mu$ are both positive and $G^TG$ or $GG^T$ has at least one positive eigenvalue. The following is to determine constants $\lambda$ and $\mu$ such that the resulting control system can not only reject uncertain disturbances but also track the desired reference with rapid convergence. It is derived from Eqs. 3.17, 3.18, and 3.19 that

$$u_{k+1}=\left(I+\rho G^TG\right)^{-k}u_0+\sum_{l=1}^{k}\rho\left(I+\rho G^TG\right)^{-l}G^T\left(r-\eta_{k+2-l}\right) \tag{3.29}$$

$$e_{k+1}=\left(I+\rho GG^T\right)^{-k}e_0-\sum_{l=1}^{k}\left(I+\rho GG^T\right)^{-l}\Delta\eta_{k+2-l} \tag{3.30}$$

Therefore, with fixed $GG^T$, a large value of $\rho$ (equivalently, large $\mu$) will be helpful in reducing error of the first trial $e_0$; that is, a quick convergence can be achieved trial by trial. However, from Eqs. 3.18 and 3.19 one obtains

$$\begin{gathered}S(t)=A^TS(t+1)\left\{I-B\left[B^TS(t+1)B+\lambda I\right]^{-1}B^TS(t+1)\right\}A+\mu C^TC\\ t=0,1,\cdots,N-1;\quad S(N)=0\end{gathered} \tag{3.31}$$

$$\begin{gathered}\varphi_{k+1}(t)=\lambda\left[\lambda I+S(t)BB^T\right]^{-1}\times\left[A^T\varphi_{k+1}(t+1)+\mu C^Te_k(t+1)\right]\\ t=0,1,\cdots,N-1;\quad \varphi_{k+1}(N)=0\end{gathered} \tag{3.32}$$

$$u_{k+1}(t)=u_k(t)-\left[B^TK(t)B+\lambda I\right]^{-1}B^T\times S(t)A\left[x_{k+1}(t)-x_k(t)\right]+\lambda^{-1}B^T\varphi_{k+1}(t) \tag{3.33}$$

It can be seen that a large $\rho$ (or large $\mu$) leads to a stronger feed-forward action on $u_{k+1}(t)$ through $\varphi_{k+1}(t)$, making the control system less sensitive to the variation of the output reference. A strong feed-forward action tends to accumulate stochastic errors that result from uncertainties and external disturbances, resulting in strong fluctuations in control input. On the other hand, it can be seen from Eqs. 3.7, 3.23, and 3.24 that when $A$ has eigenvalues outside the unit disc, the initialization uncertainties and external disturbances could cause a slow convergence or even oscillatory control. A varying weighting scheme is thus suggested here to take these practical considerations into account. Let $\rho_k = \mu_k / \lambda_k$ be a sequence approaching zero with the increase of cycle number *k*, i.e., $\rho_k \to 0$ (or $\mu_k \to 0$) when $k \to \infty$. Then Eqs. 3.14 and 3.32 become

$$S_k(t) = A^T S_k(t+1)\left\{I - B\left[B^T S_k(t+1)B + \lambda_k I\right]^{-1} B^T S_k(t+1)\right\}A + \mu_k C^T C$$
$$t = 0, 1, \cdots, N-1; \quad S_k(N) = 0 \tag{3.34}$$

$$\varphi_{k+1}(t) = \lambda_k\left[\lambda_k I + S_k(t)BB^T\right]^{-1} \times \left[A^T \varphi_{k+1}(t+1) + \mu_k C^T e_k(t+1)\right]$$
$$t = 0, 1, \cdots, N-1; \quad \varphi_{k+1}(N) = 0 \tag{3.35}$$

It is obvious that $S_k(t) \to 0$ and $\varphi_k(t) \to 0$ when $k \to \infty$, which indicates that a rapid convergence of $u_k(t)$ and $e_k(t)$ can be ensured through Theorem 3.1 and Eq. 3.19. The suggested scheme of selecting the weighting matrices *Q* and *R* is verified experimentally in the following sections.

### 3.3.5 Simulation Results

The simulation was conducted before the experiments to test the control algorithm with an ideal condition, without the disturbances and the initialization errors. There are two models used for the simulation. The first model is a first-order process with a small time constant plus time delay to simulate the injection velocity control with a fast-response servo-valve. The model can be represented in the *s* domain as follows:

$$\frac{Y(s)}{U(s)} = \frac{320}{10s+1}e^{-10s} \tag{3.36}$$

where the process gain, 320, time constant, 10, and the delay, 10, are determined based on the open-loop test results: step changes in the process input (servo-valve opening) were introduced to excite the process, and the corresponding injection velocity responses were recorded and then analyzed through system identification methods [17], as detailed in Chapter 2. Equation 3.36 can be easily transformed to the state-space model, as below:

$$x(t+1) = \begin{bmatrix} 0 & 1 & 0 \\ 0 & 0 & 1 \\ 0 & 0 & 0.61 \end{bmatrix} x(t) + \begin{bmatrix} 0 \\ 0 \\ 1 \end{bmatrix} u(t),$$
$$y(t) = \begin{bmatrix} 125.9 & 0 & 0 \end{bmatrix} x(t) \tag{3.37}$$

Only the measurement of the state variable $x_1$, the injection velocity, is used in the following simulation and experiments.

Another model is used to simulate the velocity control with a slow-response proportional valve. In the second model, a big time constant is added to slow down the response. The *s*-domain model representation is shown below:

$$\frac{Y(s)}{U(s)}=\frac{320}{(200s+1)(10s+1)} \tag{3.38}$$

Compared to model 3.36, a dominant pole with large time constant, 200, is added. The step responses of these two models are compared in Fig. 3.6. The *s*-domain model Eq. 3.38 can be transformed to the state space as follows:

$$x(t+1)=\begin{bmatrix}0 & 1 & 0 & 0\\ 0 & 0 & 1 & 0\\ 0 & 0 & 0 & 1\\ 0 & 0 & -0.59 & 1.58\end{bmatrix}x(t)+\begin{bmatrix}0\\0\\0\\1\end{bmatrix}u(t), \tag{3.39}$$
$$y(t)=\begin{bmatrix}1.42 & 1.69 & 0 & 0\end{bmatrix}x(t)$$

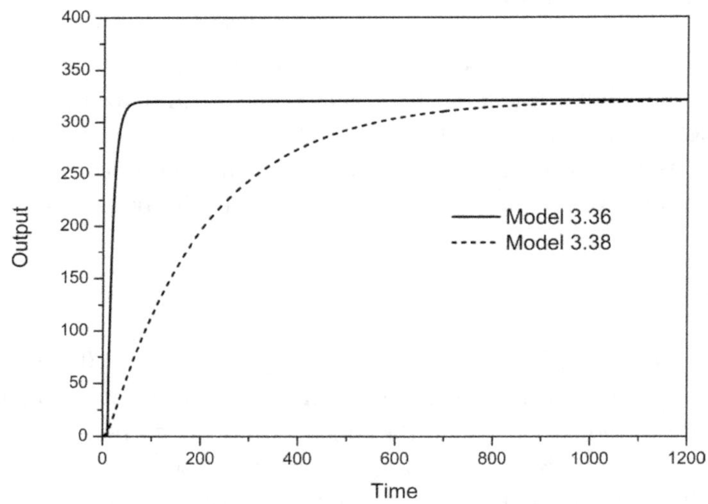


**Figure 3.6** Comparison of the step responses of models Eqs. 3.36 and 3.38

In all of the simulations, a step-change set-point profile was used: the output set point changes from 40 to 20 at the 150th step. The optimal learning control algorithm was applied to the fast-response system Eq. 3.36 first, with simple weighting factors $Q=R=1$. The control input of the first cycle was set to be 4%. The following trial control inputs and outputs are plotted in Fig. 3.7.

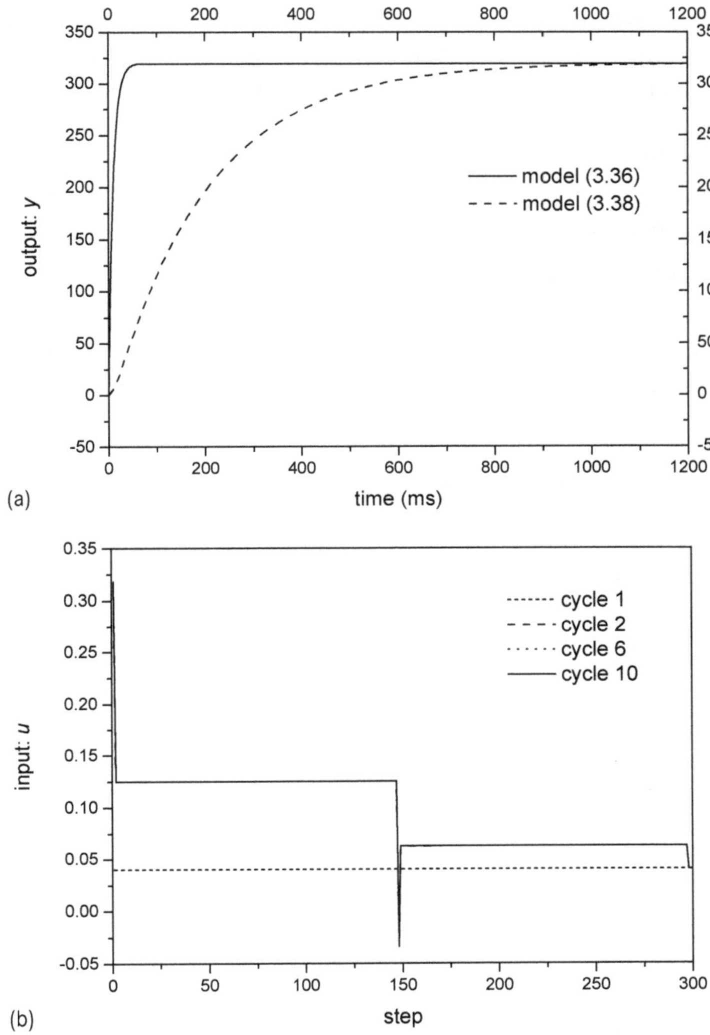


**Figure 3.7** Simulation results using model Eq. 3.38 and optimal learning control: (a) output *y*, and (b) corresponding input *u*

It is clearly shown in Fig. 3.7 that when applied to the fast-response system, the optimal learning control only used one cycle to converge to the set point. After the second iteration, all of the control outputs overlap the set point, indicating the quick convergence and good performance of the optimal control algorithm.

The optimal learning controller was then applied to slow-response system Eq. 3.38. Like other simulations, the control input of the first cycle is set to be 4%. The control outputs are shown in Fig. 3.8. Though the output of the first cycle is far from the set point, the system output response converges rapidly in the second cycle. The system outputs of the sixth and tenth cycles show that the set-point profile is perfectly followed, despite the slow response of the system. Figure 3.8 evidently demonstrated that the optimal ILC can control a process very well without disturbances and initialization errors. When a slow control valve is used, the optimal ILC should be able to control the injection velocity well as well.

Notice in all the simulations that the controllers change the control input several steps ahead of the set-point step change, leading to a perfect tracking without delay. This is an inherent advantage of the ILC.

With the simulation results as guidelines, the optimal ILC is applied experimentally to the real injection molding process, which is a nonlinear process with disturbances and initialization errors.

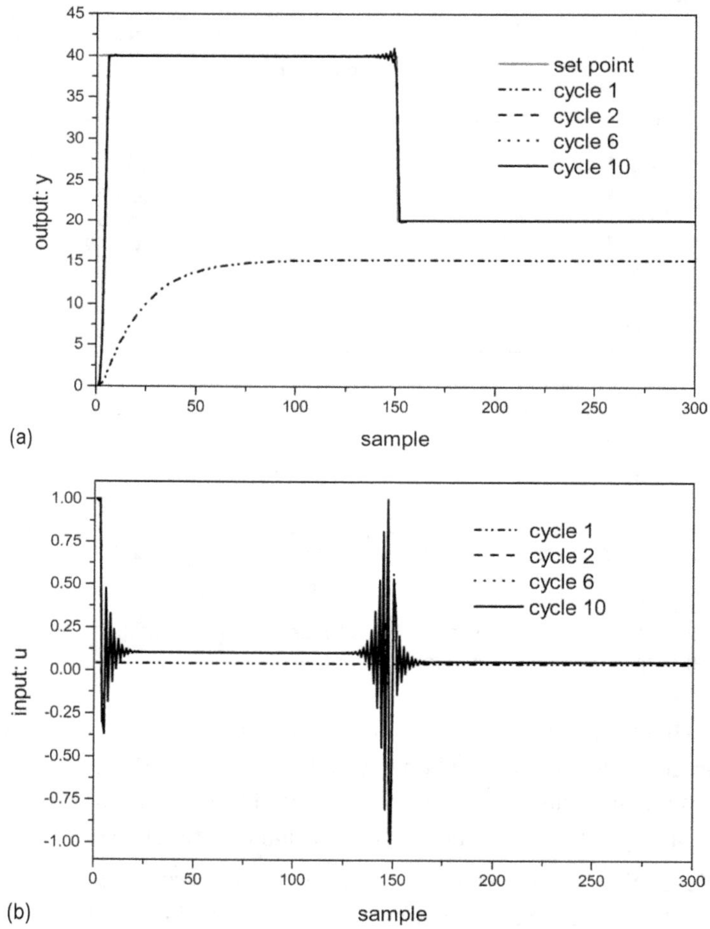


**Figure 3.8** Simulation results using model Eq. 3.38 and optimal learning control: (a) output $y$, and (b) corresponding input $u$

### 3.3.6 Experimental Results of Optimal ILC

Control using a proportional valve requires replacement of the servo-valves and modification of the hydraulic system. For a fair comparison, it is shown here using software to simulate the slow-response proportional valve, and the experiments can be done on one injection molding machine. As shown in Fig. 3.9, a first-order process with unit gain and a large time constant, $G_{add}$, was added after the controller. The dynamics of the control valve are thus changed from $G_v$ to $G_v' = G_{add} \cdot G_v$.

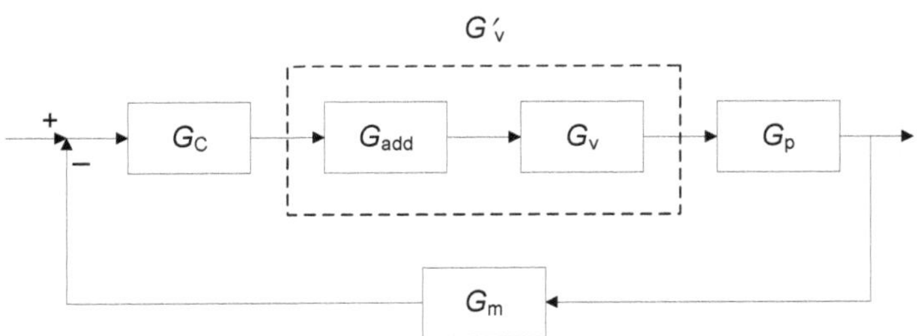


**Figure 3.9** Simulation of the slow-response valve

The optimal iterative learning control was first applied to the injection velocity control experimentally using the material HDPE. The weighting matrices $Q$ and $R$ were both selected to be 1, as in the simulation case. The injection velocity is controlled to follow a step-change profile. The initial input signal, that is, the first cycle's control input, is arbitrarily set to be 10%, as shown by the short-dashed line in Fig. 3.10(b). The control results are plotted in Fig. 3.10, where Fig. 3.10(a) shows the injection velocity responses (output) and Fig. 3.10(b) shows the corresponding servo-valve openings (input). It can be observed that the control responses became oscillatory with the increase of the cycle number $k$, contradicting the simulation result obtained earlier. The poor experimental control performance is attributed to the accumulation of initialization uncertainties and disturbances with the strong feed-forward action as chosen. In the optimal ILC design, a linear time-invariant model is used to approximate the dynamics of injection velocity, which is a nonlinear and time-varying process, so there inevitably exists a significant model mismatch. Due to the nature of the electrohydraulic system, the initial injection velocity response cannot be repeated exactly, resulting in the uncertainty of initialization error of the injection velocity control. Furthermore, there are disturbances during the molding process from different sources, such as variation of the material or operating conditions. With the existence of disturbances and model mismatch, a large $\rho$ leads to a strong feed-forward action and a weak feedback action. As a consequence, the error-rejection ability of the proposed learning controller is reduced.

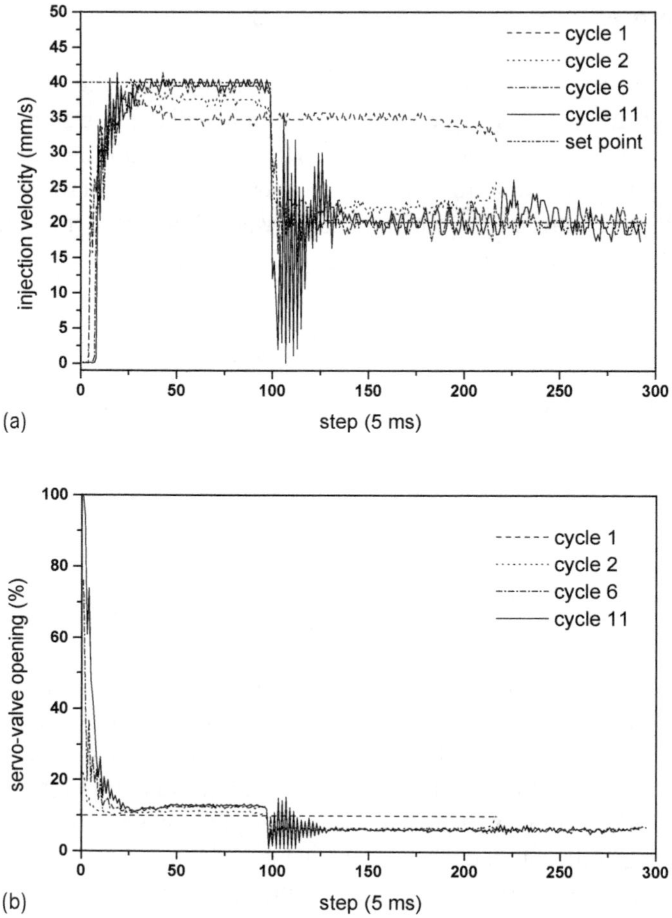


**Figure 3.10** Experimental results using optimal learning control with system Eq. 3.36 and constant weighting matrices (HDPE):
(a) injection velocity, and (b) corresponding valve opening

The method proposed above is implemented. Thus, the controller is modified with a varying $\rho$ to ensure the system convergence and to enhance the robustness of the optimal ILC. For the first cycle, the control input is set to be a constant value, 10%, the same as the last experiment. The gain matrix $S(t)$ and the feed-forward term in Eqs. 3.31–3.33 are then calculated with $\rho = 1.0$, to ensure a rapid control response convergence. For the following cycles, $\rho$ is set to decrease exponentially with the increase of the cycle number $k$, in a relation of $\rho = 0.6^{k-1}$. The material HDPE is used as in the last experiment, and the velocity is controlled to follow the same step-change profile as in the previous case. The resulting velocity responses are given in Fig. 3.11(a) with the corresponding valve opening shown in Fig. 3.11(b).

It can be observed that the velocity response of the second cycle converges rapidly, as shown by the dotted line in Fig. 3.11(a). The control of the sixth cycle has already converged, as plotted by the dash-dotted line. The solid line shows the result of the tenth cycle: the velocity tracks the set-point trajectory closely, despite the delay in the initial injection stage that was caused by the flow and charge of the hydraulic system. It is obvious that the control oscillation has been eliminated after the implementation of the proposed modification. The control response converges quickly, and the control system is stable with an increase in the cycle numbers.

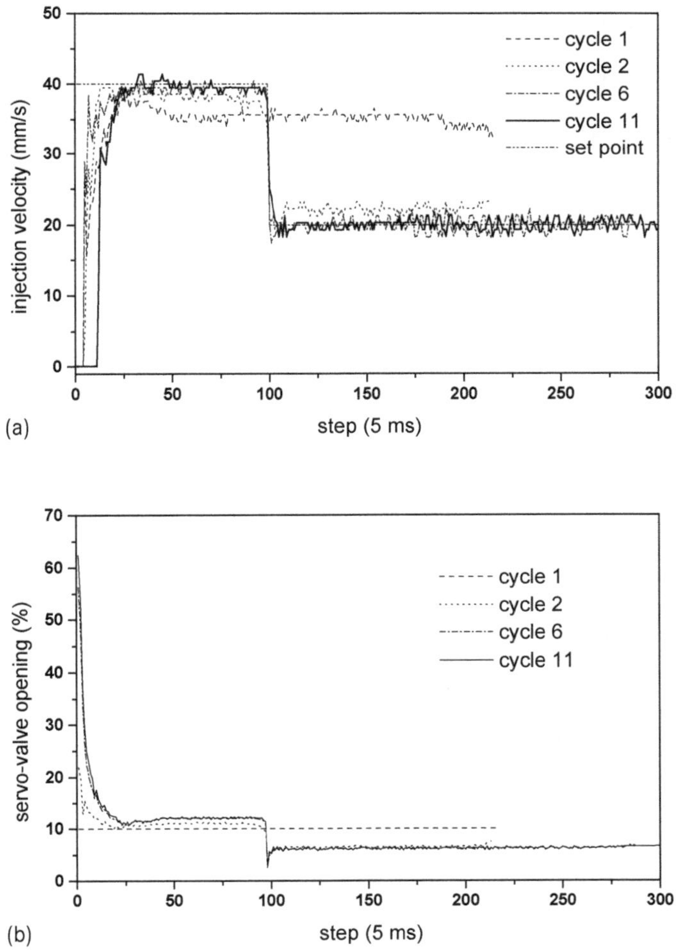


**Figure 3.11** Experimental results using optimal learning control with system Eq. 3.36 and changing weighting matrices (HDPE):
(a) injection velocity, and (b) corresponding valve opening

The optimal learning controller was also tested with the slow response system with HDPE as the processing material again. The first cycle's control signal was arbitrarily set to be 4%. The resulting injection velocity responses are shown in Fig. 3.12(a), and the corresponding control signals are shown in Fig. 3.12(b).

It is clear that the controls converge rapidly with the cycle number $k$. The results shown in Fig. 3.12 indicate that good control can be obtained with the optimal learning algorithm, even with the relatively slower valve.

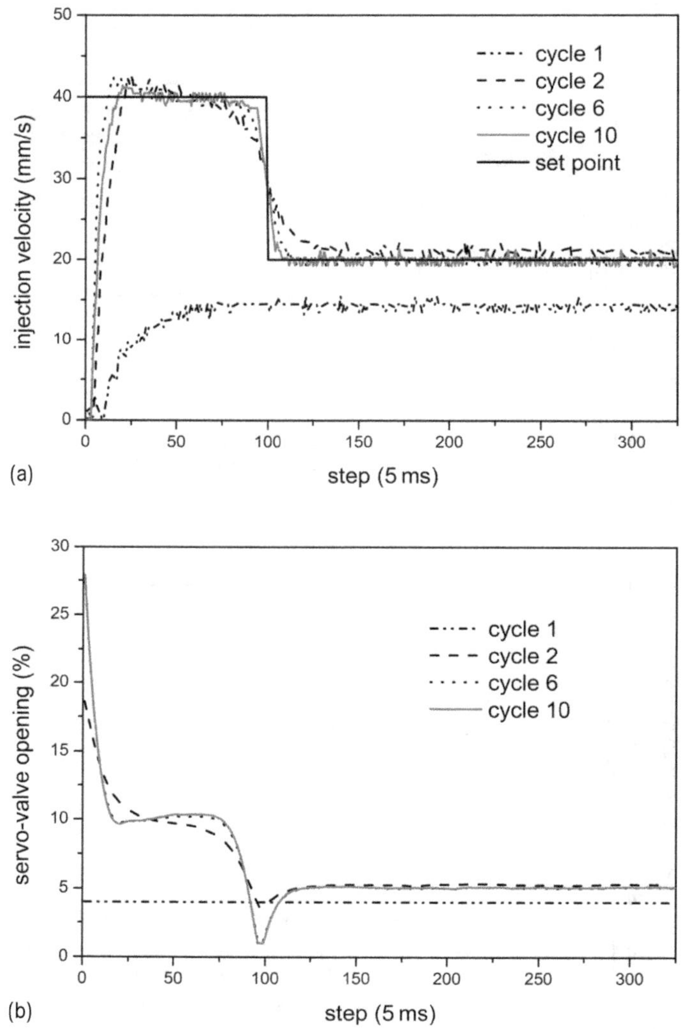


**Figure 3.12** Experimental results using optimal learning control with system Eq. 3.38 and constant weighting matrices (HDPE):
(a) injection velocity, and (b) corresponding valve opening

The nonlinear and time-varying characteristics suggest that injection velocity dynamics change with the operating point, and they highly depend on the material used in the molding process. The modified optimal ILC is further tested using a different material: PP, with a ramp set-point profile as shown by the black solid line in Fig. 3.13. The control input of the first cycle is randomly set to be 7%. The results are shown in Fig. 3.13. The velocity response converges rapidly; the responses of the 6th and 11th cycles overlap each other, indicating the good performance of the modified optimal ILC over different molding conditions.

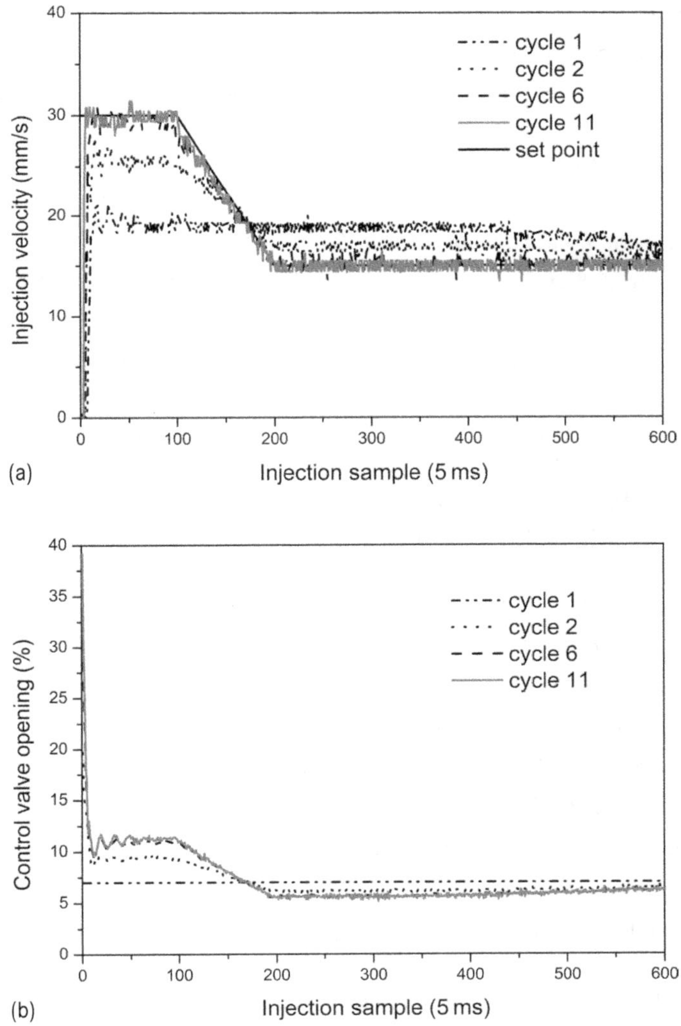


**Figure 3.13** Experimental results using optimal ILC with changing weighting matrices (PP): (a) injection velocity, and (b) corresponding valve opening

## References

[17] K.J. Aström and T. Hägglund, *PID Control-Theory, Design, and Tuning*, Instrument Society of America, North Carolina, 1995

[33] W.T. Shaw, *Computer Control of Batch Processes*, EMC Controls, Inc., Cockeysville, MD, 1982

[34] P. Rohner P, *Industrial Hydraulic Control* (4th ed.), John Wiley & Sons, Milton, Australia, 1995

[35] M. Uchiyama, *Trans. SICE (Society for Implementation and Control Engineers)*, **14**, 706, 1978

[36] S. Arimoto, S. Kawamura, and F. Miyazaki, *J. Robotic Syst.*, **1**, 123, 1984

[37] G. Casalino and G. Bartolini, *IASTED Symposium on Robotics and Automation*, **108**, 1984

[38] J.J. Craig, *Proceedings of American Control Conference*, San Diego, CA, June, 1984

[39] S. Arimoto, *Adaptive and Learning Systems*, Plenum Pub., New York, 379, 1986

[40] S. Arimoto, *Int. J. Adapt. Control Signal Process.*, **4**, 543, 1990

[41] D.-H. Hwang, Z. Bien, and S.-R. Oh, *IEE Proc.-D: Control Theory Appl.*, **138**(2), 39, 1991

[42] P.V. Kokotovic, *IEEE Control Syst. Mag.*, **12**(1), 7, 1992

[43] K.K.D. Young, *Variable Structure Control for Robotics and Aerospace Applications*, Elsevier, Amsterdam, 1993

[44] C. Ham, Z.H. Qu, and J.H. Kaloust, *Proceedings of American Control Conference*, 3024–3028, Seattle, WA, 1995

[45] Chien C.-J., *The Sampled-Data Iterative Learning Control for Nonlinear System*, Proceedings of the 35th Conference on Decision & Control, San Diego, CA, 4306–4311, 1997

[46] H.-S. Lee and Z. Bien, *Automatica*, **33**(8), 1591, 1997

[47] P. Lucibello, S. Panzier, and G. Ulivi, *Automatica*, **33**(4), 570, 1997

[48] J.-X. Xu, *Automatica*, **33**(11), 1905, 1997

[49] J.-X. Xu, *Automatica*, **33**(12), 2191, 1997

[50] B.M. Chen and S.R. Weller, *Automatica*, **34**(1), 111, 1998

[51] S. Ge, T. Lee, and G. Zhu, *IEEE Trans. Rob. Autom.*, **14**(1), 179, 1998

[52] K.S. Lee, S.H. Bang, S. Yi, J.S. Son, and S.C. Yoon, *J. Process Control*, **6**(4), 255, 1996

[53] H. Havlicsek and A. Alleyne, *IEEE/ASME Trans. Mechatronics*, **4**, 312, 1999

[54] E. Zafiriou, R.A. Adomaitis, and G. Gattu, *Proceedings of American Control Conference*, 1286–1288, Seattle, WA, 1995

[55] Y. Wang, F. Gao, and F.J. Doyle III, *J. Process Control*, **19**, 1589–1600, 2009

[56] B.A. Francis and W.M. Wonham, *Appl. Math. Optim.*, **2**, 170–194, 1975

[57] E. Sachs, R.-S. Guo, S. Ha, and A. Hu, *IEEE Trans. Semicond. Manuf.*, **4**, 134–144, 1991

[58] Z. Bien and J.-X. Xu, *Iterative Learning Control: Analysis, Design, Integration and Applications*, Kluwer Academic Publishers, Boston, MA, 1998

[59] K. Moore, M. Dahleh, and S.P. Bhattachayya, *J. Rob. Syst.*, **9**, 563, 1992

[60] K.-H. Park, Z. Bien, and D.-H. Hwang, *IEE Proc.-D, Control Theory Appl.*, **145**, 507, 1998

[61] N. Amann, D.H. Owens, and E. Rogers, *IEE Proc.-D, Control Theory Appl.*, **143**, 217, 1996

[62] L.M. Silverman, *IEEE Trans. Autom. Control*, **14**, 270, 1969

[63] F. Gao, Y. Yang, and C. Shao, *Chem. Eng. Sci.*, **56**, 7025-7034, 2001

# 4 Two-Dimensional Control Algorithms

## 4.1 Two-Dimensional Control Background

Injection molding is a typical batch process; it has its own characteristics in comparison to a continuous process. The obvious differences between a continuous process and a batch process like injection molding are (1) a batch process has a finite duration, (2) a batch process repeats itself until the specified amount of product has been made, and (3) a batch process is processed by an ordered set of activities. These characteristics make the control schemes proposed for a continuous process ill-suited for injection molding. Modifications of the original control algorithms have to be made to cope with these features. To summarize the difference between injection molding and a traditional continuous process, the distinctive nature of an injection molding process has three aspects:

1. Repetitive nature: the injection molding process repeats itself batch to batch to produce the same products;
2. Two-dimensional (2D) dynamic nature: there are within-batch and batch-to-batch dynamics in injection molding simultaneously; and
3. Multiphase nature: an injection molding process consists of more than one phase.

In this chapter, these aspects are studied extensively to introduce a new group of control algorithms developed specifically for batch processes like injection molding, that is, the 2D control algorithms.

Consider a digital injection molding controller in which for each cycle the control samples form a time series: if these different cycles' (with cycle duration $T$ and cycle number $K$) information is aligned as shown in Fig. 4.1, a two-dimensional system can be formed. Although cycle-wise dynamics do exist, for example, during the warm-up stages when the machine is slowly reaching steady operation, these different cycles' time series still lack a cycle direction causality due to the independent operation of each cycle.

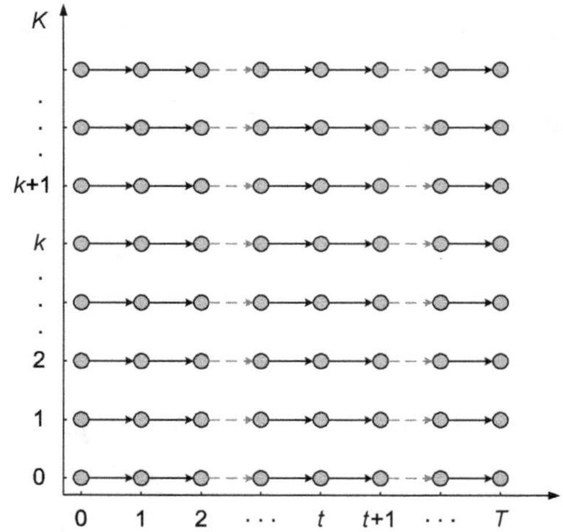


**Figure 4.1** Alignment of different cycles' information

Consider an injection molding process controlled by ILC. Any cycle's system output depends not only on the control action and output of the current cycle, but also the information from the last cycle. The cycle-wise dynamic is naturally introduced, and the 2D system becomes a causal system, as schematically shown in Fig. 4.2.

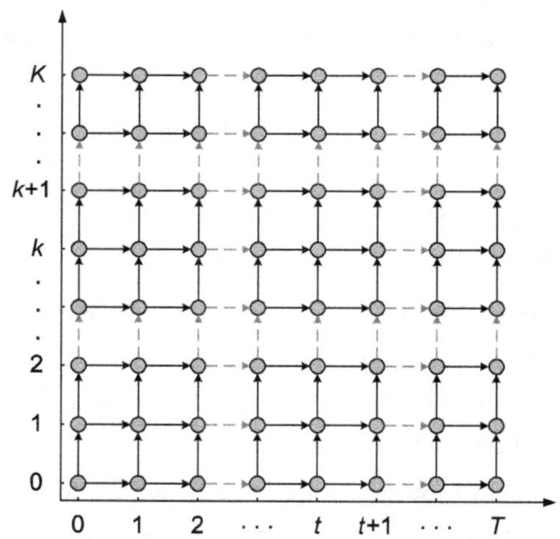


**Figure 4.2** Schematic of 2D causal system

The traditional 2D system and its analysis are designed for time-independent series, such as image processing. For this kind of 2D system, at current point $P(k,t)$, all of the information about points $P(i,j), i<k \cup j<t$, is known. The 2D system constructed with the ILC-controlled injection molding process is time dependent: only information about previous cycles or the current cycle's previous sampling periods

is known. Hence, the causality, controller design, and analysis of this time-dependent 2D system are totally different from those of the traditional one. To clearly show this difference, Fig. 4.3 plots the information-acquisition sequence of these systems, where Fig. 4.3(a) is the traditional 2D system and Fig. 4.3(b) is the 2D system of an injection molding process.

In this chapter, a 2D generalized, predictive, iterative learning control (2D-GPILC) based on a 1D model is introduced first, followed by a 2D dynamic matrix control (2D-DMC) also based on a 1D model. The system identification of the 2D model is applied, followed by a 2D generalized predictive control (2D-GPC) based on the 2D model.

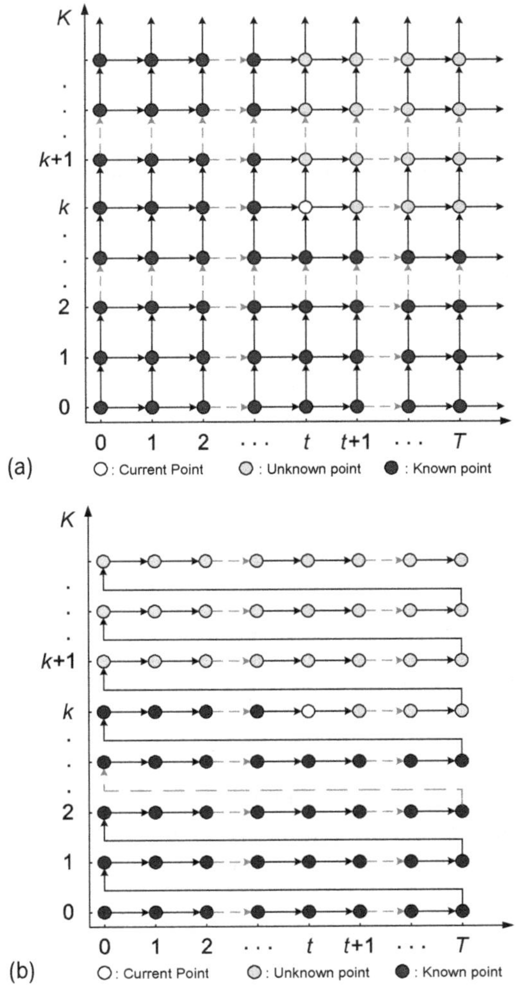


**Figure 4.3** Comparison of information-acquisition sequence:
(a) traditional 2D system, and
(b) 2D system of an injection molding process controlled by ILC

## 4.2 Two-Dimensional Generalized Predictive Iterative Learning Control

### 4.2.1 2D-GPILC Control Algorithm

The control algorithm is introduced briefly here. The detailed derivation and analysis were given in references [64–67].

A SISO batch process can be described by the following controlled autoregressive integrated moving-average (CARIMA) model:

$$\Sigma_P: \quad A\left(q_t^{-1}\right)y_k(t)=B\left(q_t^{-1}\right)\Delta_t\left(u_k(t)\right)+w_k(t) \tag{4.1}$$

$t=0,1,\ldots,T; \quad k=1,2,\ldots$

where $t$ and $k$ represent the discrete-time and cycle/batch index, respectively; $T$ is the time duration of each cycle; $u_k(t)$, $y_k(t)$, and $w_k(t)$ are the input, output, and disturbance of the process at time $t$ in the $k$th cycle, respectively; $q_t^{-1}$ indicates the time-wise unit backward-shift operator; $A\left(q_t^{-1}\right)$ and $B\left(q_t^{-1}\right)$ are both operator polynomials:

$$A\left(q_t^{-1}\right)=1+a_1q_t^{-1}+a_2q_t^{-2}+\cdots+a_nq_t^{-n} \tag{4.2}$$

$$B\left(q_t^{-1}\right)=b_1q_t^{-1}+b_2q_t^{-2}+\cdots+b_mq_t^{-m} \tag{4.3}$$

and $\Delta_t$ represents the time-wise backward difference operator, or
$\Delta_t\left(f(t,k)\right)=f(t,k)-f(t-1,k)$.

For the above repetitive process, introduce an ILC law with the form

$$\Sigma_{\text{ILC}}: \quad u_k(t)=u_{k-1}(t)+u_k(t-1)-u_{k-1}(t-1)+r_k(t) \tag{4.4}$$

$u_0(t)=0, \quad t=-1,0,1,\ldots,T$

where $r_k(t)$ is referred to as the updating law, to be determined online based on the MPC philosophy, and $u_0(t)$ is the initial profile of iteration. Letting $q_k^{-1}$ represent the cycle-wise unit backward-shift operator, $u_k(t)$ and $r_k(t)$ have the following relationship:

$$u_k(t)=\frac{1}{\left(1-q_k^{-1}\right)}\cdot\frac{1}{\left(1-q_t^{-1}\right)}\cdot r_k(t) \tag{4.5}$$

which is a 2D system with 2D integral transformation.

Control law Eq. 4.4 can be expressed equivalently as

$$\Delta_t\left(u_k(t)\right)=\Delta_t\left(u_{k-1}(t)\right)+r_k(t) \tag{4.6}$$

or

$$\Delta_k\left(u_k(t)\right)=\Delta_k\left(u_k(t-1)\right)+r_k(t) \tag{4.7}$$

where $\Delta_k$ represents the cycle-wise backward difference operator, or $\Delta_k\left(f(t,k)\right)=f(t,k)-f(t,k-1)$.

Substituting Eq. 4.7 into Eq. 4.1 leads to the following 2D system:

$$\Sigma_{2D-P}:\quad A\left(q_t^{-1}\right)y_k(t)=A\left(q_t^{-1}\right)y_{k-1}(t)+B\left(q_t^{-1}\right)r_k(t)+\Delta_k\left(w_k(t)\right) \tag{4.8}$$

Here $r_k(t)$, $y_k(t)$, and $\Delta_k\left(w_k(t)\right)$ represent the input, output, and disturbance of the system, respectively.

To ensure smooth operations along both the time and cycle directions, the following quadratic cost function is defined as the general control performance index over one cycle for the 2D system of Eq. 4.8:

$$\begin{aligned}
J(t,k,n_1,n_2)&=\begin{pmatrix}\hat{\mathbf{e}}_{k|k}\left(\left|_{t+n_1}^{t+1}\right|t\right)\\ \Delta_t\left(\mathbf{u}_k\left(\left|_{t+n_2-1}^{t}\right.\right)\right)\\ \Delta_k\left(\mathbf{u}_k\left(\left|_{t+n_2-1}^{t}\right.\right)\right)\end{pmatrix}^T\begin{pmatrix}\mathbf{Q}&0&0\\0&\mathbf{S}&0\\0&0&\mathbf{T}\end{pmatrix}\begin{pmatrix}\hat{\mathbf{e}}_{k|k}\left(\left|_{t+n_1}^{t+1}\right|t\right)\\ \Delta_t\left(\mathbf{u}_k\left(\left|_{t+n_2-1}^{t}\right.\right)\right)\\ \Delta_k\left(\mathbf{u}_k\left(\left|_{t+n_2-1}^{t}\right.\right)\right)\end{pmatrix}+\mathbf{r}_k^T\left(\left|_{t+n_2-1}^{t}\right.\right)\mathbf{R}\mathbf{r}_k\left(\left|_{t+n_2-1}^{t}\right.\right)\\
&=\hat{\mathbf{X}}_{k|k}^T(t)\bar{\mathbf{Q}}\hat{\mathbf{X}}_{k|k}(t)+\mathbf{r}_{k+l}^T\left(\left|_{t+n_2-1}^{t}\right.\right)\mathbf{R}\mathbf{r}_{k+l}\left(\left|_{t+n_2-1}^{t}\right.\right)
\end{aligned} \tag{4.9}$$

where the notations are

$$\begin{aligned}&\mathbf{f}_k\left(\left|_{t_2}^{t_1}\right.\right)=\left(f_k(t_1)\quad f_k(t_1+1)\quad\cdots\quad f_k(t_2)\right)\\ &\qquad f\in\{y,u,r,e\}\end{aligned} \tag{4.10}$$

and $\hat{\mathbf{f}}_{k|k}\left(\left|_{t+n_1}^{t+1}\right|t\right)$ indicates the prediction vector of the $k$th cycle over the time-wise prediction horizon based on the history information before time $t$ of the $k$th cycle, and

$$\hat{\mathbf{e}}_{k|k}\left(\left|_{t+n_1}^{t+1}\right|t\right)=\mathbf{y}_r\left(\left|_{t+n_1}^{t+1}\right.\right)-\hat{\mathbf{y}}_{k|k}\left(\left|_{t+n_1}^{t+1}\right|t\right) \tag{4.11}$$

$$\hat{\mathbf{X}}_{k|k}(t)=\begin{bmatrix}\hat{\mathbf{e}}_{k|k}\left(\left|_{t+n_1}^{t+1}\right|t\right)\\ \Delta_t\left(\mathbf{u}_k\left(\left|_{t+n_2-1}^{t}\right.\right)\right)\\ \Delta_k\left(\mathbf{u}_k\left(\left|_{t+n_2-1}^{t}\right.\right)\right)\end{bmatrix} \tag{4.12}$$

$$\mathbf{Q} = diag\{\eta(1), \ldots, \eta(n_1)\} \tag{4.13}$$

$$\mathbf{R} = diag\{\alpha(0), \ldots, \alpha(n_2 - 1)\} \tag{4.14}$$

$$\mathbf{S} = diag\{\beta(0), \ldots, \beta(n_2 - 1)\} \tag{4.15}$$

$$\mathbf{T} = diag\{\gamma(0), \ldots, \gamma(n_2 - 1)\} \tag{4.16}$$

$$\bar{\mathbf{Q}} = diag\{\mathbf{Q}, \mathbf{S}, \mathbf{T}\} \tag{4.17}$$

By the procedure proposed in references [64–66], the best prediction model for the 2D system of Eq. 4.8 can be formulated as follows:

$$\hat{\mathbf{y}}_{k|k}\left(\Big|_{t+n_1}^{t+1}\Big|t\right) = \mathbf{Gr}_k\left(\Big|_{t+n_2-1}^{t}\right) + \mathbf{y}_{k-1}\left(\Big|_{t+n_1}^{t+1}\right) + \mathbf{F}_k(t) \tag{4.18}$$

The detailed formulation of matrices **G** and $\mathbf{F}_k(t)$ can be found in references [64–66]. To minimize the cost function Eq. 4.9, the following relationships between variables $\Delta_t\left(\mathbf{u}_k\left(\Big|_{t+n_2-1}^{t}\right)\right)$, $\Delta_k\left(\mathbf{u}_k\left(\Big|_{t+n_2-1}^{t}\right)\right)$, and updating variable $\mathbf{r}_k\left(\Big|_{t+n_2-1}^{t}\right)$ are required:

$$\Delta_t\left(\mathbf{u}_k\left(\Big|_{t+n_2-1}^{t}\right)\right) = \mathbf{r}_k\left(\Big|_{t+n_2-1}^{t}\right) + \Delta_t\left(\mathbf{u}_{k-1}\left(\Big|_{t+n_2-1}^{t}\right)\right) \tag{4.19}$$

$$\mathbf{r}_k\left(\Big|_{t+n_2-1}^{t}\right) = \mathbf{H}\Delta_k\left(\mathbf{u}_k\left(\Big|_{t+n_2-1}^{t}\right)\right) - \Delta_k\left(u_k(t-1)\right)\Big|_{n_2} \tag{4.20}$$

where

$$\mathbf{H} = \begin{pmatrix} 1 & 0 & 0 & \cdots & 0 \\ -1 & 1 & 0 & \cdots & 0 \\ 0 & -1 & 1 & \cdots & 0 \\ \vdots & \vdots & \vdots & \ddots & \vdots \\ 0 & 0 & 0 & \cdots & 1 \end{pmatrix}_{n_2 \times n_2}$$

$$\Delta_k\left(u_k(t-1)\right)\Big|_{n_2} = \begin{pmatrix} \Delta_k\left(u_k(t-1)\right) \\ 0 \\ 0 \\ \vdots \\ 0 \end{pmatrix}_{n_2 \times 1} \tag{4.21}$$

Let $\mathbf{V} = \mathbf{H}^{-1}$, then

$$\Delta_k\left(\mathbf{u}_k\left(\Big|_{t+n_2-1}^{t}\right)\right) = \mathbf{Vr}_k\left(\Big|_{t+n_2-1}^{t}\right) + \mathbf{V}\Delta_k\left(u_k(t-1)\right)\Big|_{n_2} \tag{4.22}$$

Together with Eqs. 4.18, 4.19, and 4.22, the following generalized 2D prediction model is obtained:

$$\hat{\mathbf{X}}_{k|k}(t)=\bar{\mathbf{G}}\mathbf{r}_k\left(\left|_{t+n_2-1}^{t}\right.\right)+\bar{\mathbf{A}}\mathbf{X}_{k-1}(t)+\mathbf{W}_k(t) \tag{4.23}$$

where

$$\bar{\mathbf{G}}=\begin{pmatrix}-\mathbf{G}\\ \mathbf{I}\\ \mathbf{V}\end{pmatrix},\quad \bar{\mathbf{A}}=\begin{pmatrix}\mathbf{I}&0&0\\0&\mathbf{I}&0\\0&0&0\end{pmatrix} \tag{4.24}$$

$$\mathbf{X}_{k-1}(t)=\begin{pmatrix}\mathbf{e}_{k-1}\left(\left|_{t+n_1}^{t+1}\right.\right)\\ \Delta_t\left(\mathbf{u}_{k-1}\left(\left|_{t+n_2-1}^{t}\right.\right)\right)\\ \Delta_k\left(\mathbf{u}_{k-1}\left(\left|_{t+n_2-1}^{t}\right.\right)\right)\end{pmatrix},\quad \mathbf{W}_k(t)=\begin{pmatrix}-\mathbf{F}_k(t)\\ \mathbf{0}\\ \mathbf{V}\Delta_k\left(u_k(t-1)\right)\Big|_{n_2}\end{pmatrix} \tag{4.25}$$

$$\mathbf{e}_{k-1}\left(\left|_{t+n_1}^{t+1}\right.\right)=\mathbf{y}_r\left(\left|_{t+n_1}^{t+1}\right.\right)-\mathbf{y}_{k-1}\left(\left|_{t+n_1}^{t+1}\right.\right) \tag{4.26}$$

Based on the above prediction model, it results from the optimization algorithm that the quadratic cost function of Eq. 4.11 is minimized by the following optimal control law:

$$\mathbf{r}_k^*\left(\left|_{t+n_2-1}^{t}\right.\right)=-\left(\mathbf{R}+\bar{\mathbf{G}}^T\bar{\mathbf{Q}}\bar{\mathbf{G}}\right)^{-1}\bar{\mathbf{G}}^T\bar{\mathbf{Q}}\left(\bar{\mathbf{A}}\mathbf{X}_{k-1}(t)+\mathbf{W}_k(t)\right) \tag{4.27}$$

It follows from the definitions 4.9, 4.12, and 4.24 that

$$\begin{aligned}\mathbf{r}_k^*\left(\left|_{t+n_2-1}^{t}\right.\right)&=\left(\mathbf{R}+\mathbf{G}^T\mathbf{Q}\mathbf{G}+\mathbf{S}+\mathbf{V}^T\mathbf{T}\mathbf{V}\right)^{-1}\mathbf{G}^T\mathbf{Q}\left(\mathbf{e}_{k-1}\left(\left|_{t+n_1}^{t+1}\right.\right)-\mathbf{F}_k(t)\right)\\ &-\left(\mathbf{R}+\mathbf{G}^T\mathbf{Q}\mathbf{G}+\mathbf{S}+\mathbf{V}^T\mathbf{T}\mathbf{V}\right)^{-1}\mathbf{S}\Delta_t\left(\mathbf{u}_{k-1}\left(\left|_{t+n_2-1}^{t}\right.\right)\right)\\ &-\left(\mathbf{R}+\mathbf{G}^T\mathbf{Q}\mathbf{G}+\mathbf{S}+\mathbf{V}^T\mathbf{T}\mathbf{V}\right)^{-1}\mathbf{V}^T\mathbf{T}\mathbf{V}\Delta_k\left(u_k(t-1)\right)\Big|_{n_2}\end{aligned} \tag{4.28}$$

Now, let $\mathbf{K}_1$ and $\mathbf{K}_2$ be the first rows of matrices $\left(\mathbf{R}+\mathbf{G}^T\mathbf{Q}\mathbf{G}+\mathbf{S}+\mathbf{V}^T\mathbf{T}\mathbf{V}\right)^{-1}\mathbf{G}^T\mathbf{Q}$ and $-\left(\mathbf{R}+\mathbf{G}^T\mathbf{Q}\mathbf{G}+\mathbf{S}+\mathbf{V}^T\mathbf{T}\mathbf{V}\right)^{-1}\mathbf{S}$, respectively, and $K_3$ be the upper-left element of matrix $-\left(\mathbf{R}+\mathbf{G}^T\mathbf{Q}\mathbf{G}+\mathbf{S}+\mathbf{V}^T\mathbf{T}\mathbf{V}\right)^{-1}\mathbf{V}^T\mathbf{T}\mathbf{V}$; then the 2D-MPILC (Model-Predictive-Iterative-Learning-Control) scheme is given as follows:

$$\begin{aligned}\Sigma_{\text{2D-GPILC}}:\quad \Delta_t\left(u_k(t)\right)&=\Delta_t\left(u_{k-1}(t)\right)+\mathbf{K}_1\left(\mathbf{e}_{k-1}\left(\left|_{t+n_1}^{t+1}\right.\right)-\mathbf{F}_k(t)\right)\\ &+\mathbf{K}_2\Delta_t\left(\mathbf{u}_{k-1}\left(\left|_{t+n_2-1}^{t}\right.\right)\right)+K_3\Delta_k\left(u_k(t-1)\right)\end{aligned} \tag{4.29}$$

### 4.2.2 Injection Velocity Control with 2D-GPILC

The above 2D-GPILC is applied to the injection molding process experimentally to control the injection velocity. The experiments are conducted on two injection molding machines, one with 55 tons of clamping force and the other 80 tons, both hydraulic machines with proportional control valves. Two molds are used in the experiments, with the geometry shown in Fig. 4.4. The material used is HDPE.

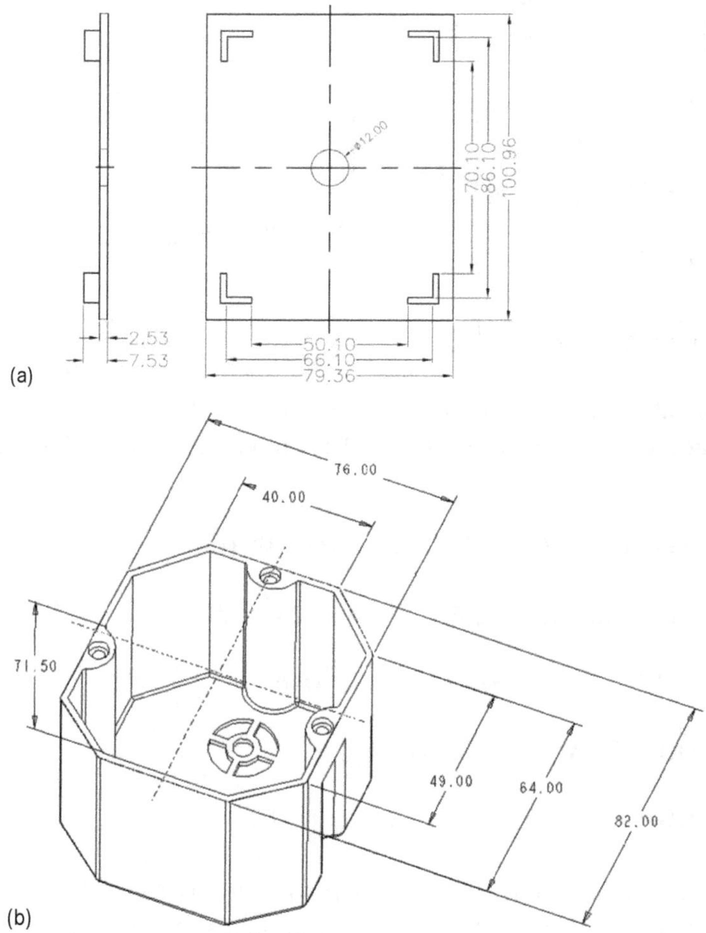


**Figure 4.4** Molds used in the experiments of 2D-GPILC

Before the controller design, an open-loop test was conducted on the 55-ton machine with the flat mold to get the injection velocity dynamics and guide the design of the controller. The result is plotted in Fig. 4.5. The dashed line in Fig. 4.5 shows the proportional control valve opening with two step changes, a step-up change from

40% to 45% and a step-down change from 45% to 30%. The corresponding injection velocity response is shown by the solid line. It is clearly shown by Fig. 4.5 that the injection velocity has a significant time delay and nonlinear dynamics. The open-loop test result was used for the 2D MPILC design, and the design parameters are listed in Table 4.1.

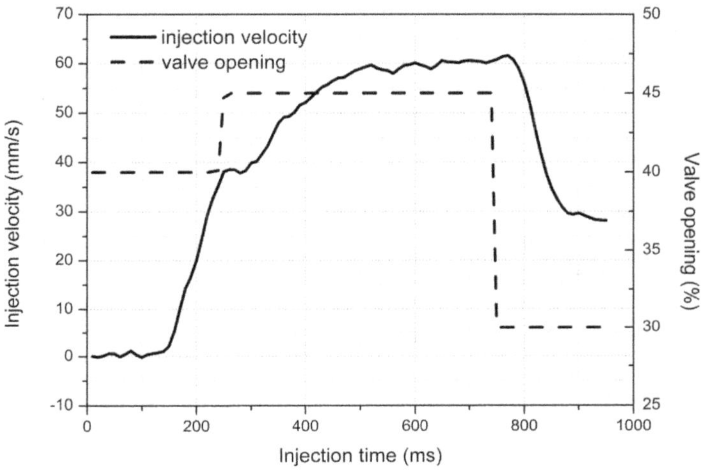


**Figure 4.5** Open-loop test of the injection velocity

**Table 4.1** List of 2D-MPILC Design Parameters

| Design Parameters | Value |
|---|---|
| **Delay Order** | 6 |
| **Model parameter A** | {-1.714975, 0.668412, 0.050785} |
| **Model parameter B** | {11.228274, -3.721156, -7.111082} |
| $\beta$ | 0.100000 |
| $\gamma(1), \gamma(2)$ | 2500, 5000 |
| $\eta$ | 6000 |
| **n1, n2** | 6, 3 |

The 2D-MPILC controller was tested first with a constant velocity setting, 35 mm/s, as plotted by the solid gray line in esponding control valve opening Fig. 4.6(a). The control result is shown in esponding control valve opening Fig. 4.6, where esponding control valve opening Fig. 4.6(a) shows the velocity responses and esponding control valve opening Fig. 4.6(b) plots the corresponding valve opening. It is obvious that after three cycles of learning, the controller can already provide a good control for constant velocity profiles, and after 10 cycles of learning, the velocity almost overlaps with the set point, indicating the good control performance during the constant set-point cases.

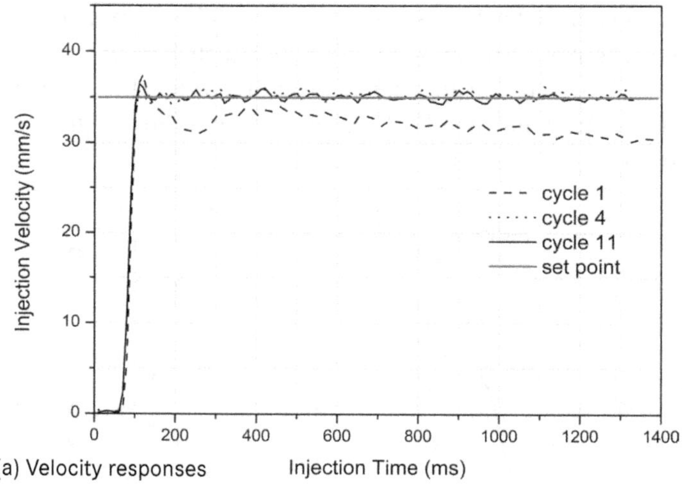


(a) Velocity responses

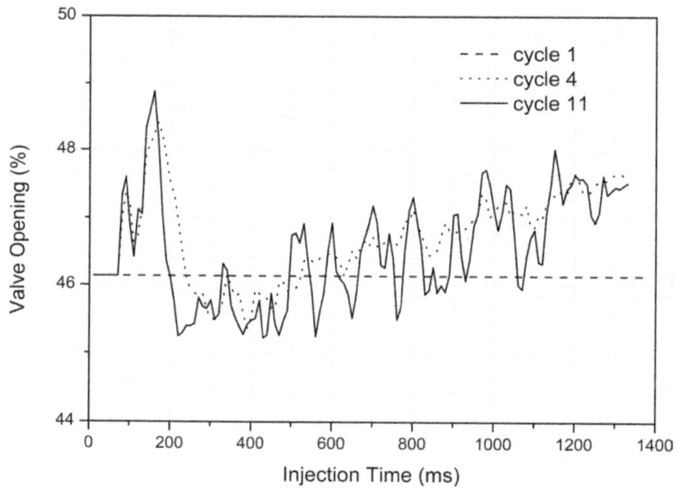


(b) Corresponding control valve opening

**Figure 4.6** Velocity control with a constant set point

The controller is further tested with a step-change profile. The experimental control results are shown in Fig. 4.7, where Fig. 4.7(a) shows the velocity control responses, Fig. 4.7(b) plots the corresponding valve opening, and for easy demonstration Fig. 4.7(c) shows the zoomed-in velocity plot around the step change. In Fig. 4.7(a), the solid gray line again denotes the set-point profile, and the dashed, dotted, and solid lines show the injection velocity responses of the first, fourth, and 11th cycle, respectively. It is shown in Fig. 4.7(a) that after three cycles of learning, the velocity can already reach the set point quickly in the initial stage and follows the set-point step change closely in the middle of the filling stage.

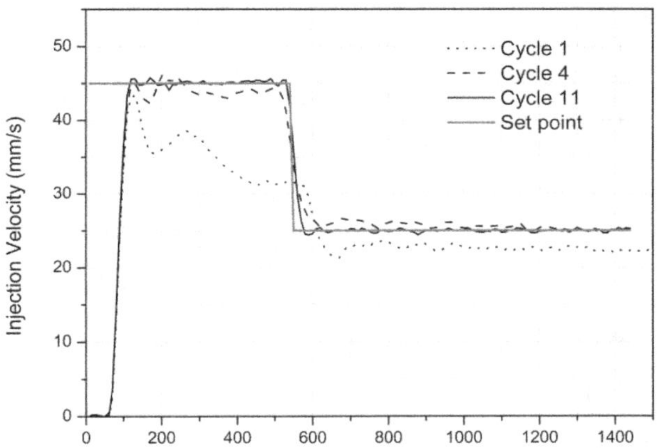


(a) Velocity responses

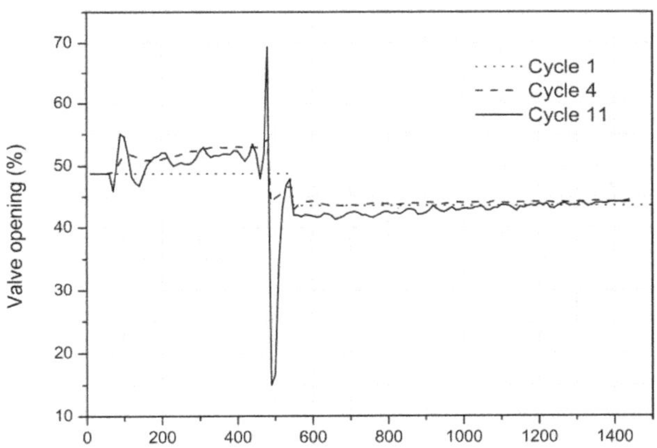


(b) Corresponding control valve opening

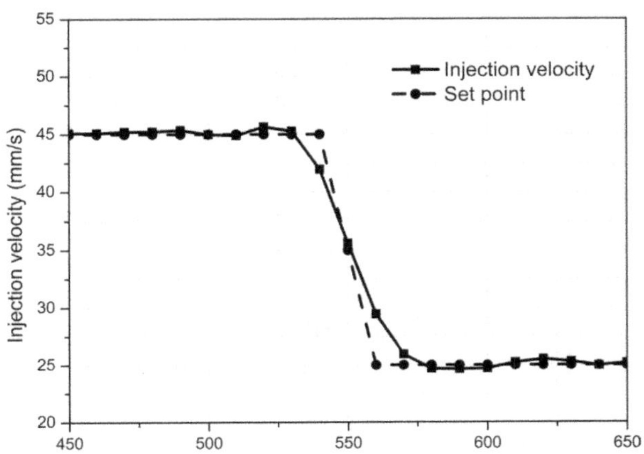


(c) Zoomed-in plot around step change

**Figure 4.7** Velocity control with step-change profile

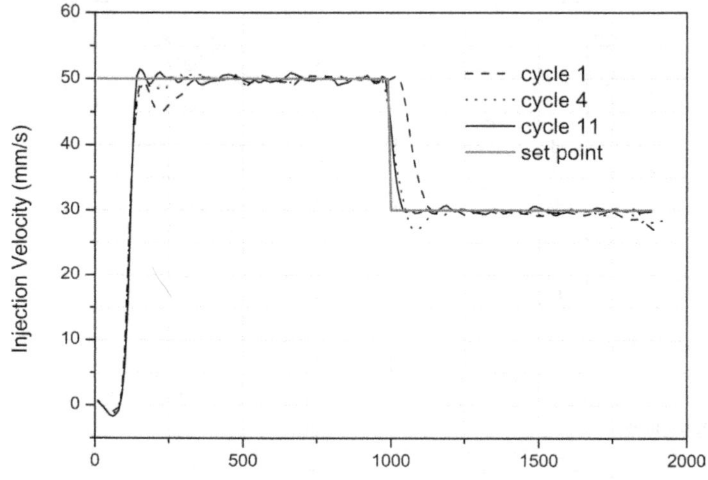


(a) Velocity responses

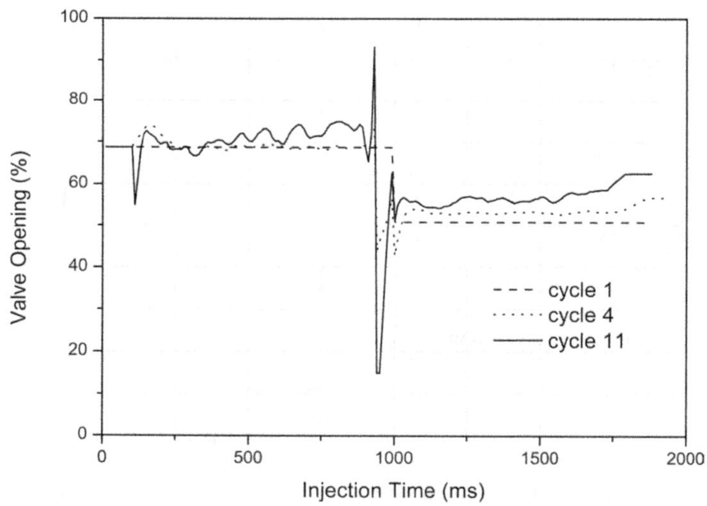


(b) Corresponding control valve opening

**Figure 4.8** Injection velocity control result with a different machine and mold

After 10 cycles learning, the velocity response almost overlaps with the set-point profile. A zoomed-in plot around the set-point step change, as shown in Fig. 4.7(c), clearly demonstrates that the 2D-MPILC only used five control steps, or 50 milliseconds, to follow the step change. This performance has reached the response time limit of the proportional control valve and proved the excellent control of the 2D-MPILC.

To test the robustness of the proposed control algorithm, the controller is applied to a totally different Korean injection molding machine. With a totally different mold, a cup mold as shown in Fig. 4.4(b), the controller is designed with a different set of parameters and experimentally tested again. A step-change profile is adopted, and the control results are shown in Fig. 4.8, where Fig. 4.8(a) shows the velocity control responses and Fig. 4.8(b) plots the corresponding valve opening. After three cycles of learning, the fourth cycle's velocity response, as shown by the dotted line in Fig. 4.8(a), is already very close to the set point. The 11th cycle's velocity, as shown by the solid line, almost overlaps with the gray set-point profile. This result indicates that the controller works well even with a totally different machine and mold.

## 4.3 Two-Dimensional Dynamic Matrix Control

In injection molding process control, disturbances can be divided into two types, repeatable or periodical and nonrepeatable, both of which are unavoidable in industrial environments. A 2D-DMC [68] is introduced in this section to deal with the nonrepeatable uncertainties and guarantees convergence and robustness under severe conditions.

### 4.3.1 Problem Formulation

The DMC algorithm [69] is an excellent choice for multivariable processes. Without loss of generality, the derivation depicted in this section is based on a single-input-single-output (SISO) discrete-time molding process. The results can be extended to the case of the multi-input–multi-output (MIMO) system. Assuming that the following SISO step response model operates repetitively over a finite period of time, see, for example, an injection cycle:

$$\Sigma_p : y(t,k) = \sum_{i=1}^{\infty} a_i \Delta_t u(t-i,k), \quad t = 0,1,\ldots, \quad T_k, k = 1,2,\ldots \tag{4.30}$$

where $t$ and $k$ represent the discrete time index and the batch index, respectively; $T_k$ is the time duration of the $k$th batch; $y(t,k) \in \mathrm{R}$ and $u(t,k) \in \mathrm{R}$ denote the controlled variable (output) and the manipulated variable (input); $a_i$ are the sampled output values for the unitary step input; and $\Delta_t$ indicates the time-wise backward difference operator:

$$\Delta_t f(t,k) = f(t,k) - f(t-1,k) \tag{4.31}$$

In an injection molding process, the input affects the output from time 0 to $t$ in the following way:

$$\Delta_t u(t,k) = \begin{cases} 0, t \in [-\infty, -1] \\ \Delta_t u(t,k), t \in [0, T_k] \end{cases} \tag{4.32}$$

Substituting Eq. 4.32 into Eq. 4.30, Eq. 4.30 can obtain

$$\Sigma_p : y(t,k) = \sum_{i=1}^{t} a_i \Delta_t u(t-i,k), \quad t = 0,1,\ldots, \quad T_k, k = 1,2,\ldots \tag{4.33}$$

Introduce the following notation:

$$g\left(\left|_{t_2}^{t_1}, k\right.\right) = \left[g(t_1,k), g(t_1+1,k), \cdots, g(t_2,k)\right]', \quad g \in \{y,u,r,w\} \tag{4.34}$$

where $g$ can be the output, input, updating law, or noise variable from time $t_1$ to $t_2$. Model 4.30 can be reorganized into a more compact form as follows:

$$\Sigma_p : y(t,k) = \sum_{i=1}^{t} a_i \Delta_t u(t-i,k) = A\left(\left|_{1}^{t}\right.\right) \times \Delta_t u\left(\left|_{t-1}^{0}, k\right.\right), \quad t = 0,1,\ldots, \quad T_k, k = 1,2,\ldots$$

where

$$A\left(\left|_{t_j}^{t_i}\right.\right) \triangleq \begin{bmatrix} a_{t_i} & \cdots & a_{t_j} \end{bmatrix} \tag{4.35}$$

**Remark 4.1:** In a continuous process, the process is required to be asymptotically stable for DMC to work; otherwise the output cannot be computed by the step-response model since the number of items included in the model will tend toward infinity with time. However, in a batch process such as injection molding, each batch operates over a finite period of time. Thus, the number of items included in the model is inherently finite. In this study, a sufficiently large number $N$ is selected to represent the truncated horizon.

### 4.3.2 Controller Design

To exploit the repetitive nature of the injection molding process, ILC is a good choice. To compensate for nonrepeatable uncertainties, DMC is adopted based on the step-response model. This combination will lead to the integration of a feed-forward controller (for ILC) and a feedback controller (for DMC), giving rise to 2D-DMC. In this section, the 2D equivalent model will first be obtained by utilizing the ILC control law so that it will have learning ability. The 2D prediction model for DMC will then be derived from the 2D equivalent model. Finally, the control law of 2D-DMC will be developed based on the 2D prediction model in the framework of MPC.

#### 4.3.2.1 2D Equivalent Model with Repetitive Nature

The ILC control law used in this study takes the following form:

$$\begin{aligned} r(t,k) &= u(t,k) - u(t-1,k) - [u(t,k-1) - u(t-1,k-1)] \\ &= \Delta_t u(t,k) - \Delta_t u(t,k-1) \end{aligned} \tag{4.36}$$

where $r(t,k)$ is the updating law to be determined by the DMC strategy. Different from the conventional ILC algorithm, Eq. 4.36 is a 2D control law with a time-wise integrator and a batch-wise integrator cascaded [67].

From Eq. 4.35, the input-output model can be derived as

$$y(t,k) = A\left(\Big|_{1}^{t}\right) \times \Delta_t u\left(\Big|_{t-1}^{0}, k\right) + w(t,k), \quad t = 0, 1, \ldots, \quad T_k, k = 1, 2, \ldots \tag{4.37}$$

where $w(t,k)$ is the uncertain and unmodeled dynamics. Substituting Eq. 4.37 into Eq. 4.36 leads to

$$\Sigma_{\text{2D-p}} : y(t,k) = y(t,k-1) + A\left(\Big|_{1}^{t}\right) \times r\left(\Big|_{t-1}^{0}, k\right) + \Delta_k\left(w(t,k)\right) \tag{4.38}$$

where $\Delta_k$ represents the batch-wise backward difference operator, $\Delta_k\left(f_k(t)\right) = f_k(t) - f_{k-1}(t)$. Equation 4.38 is the 2D equivalent model and has the input $r(t,k)$ instead of $u(t,k)$ in Eq. 4.37.

#### 4.3.2.2 2D Prediction Model

To obtain the updating law, a 2D prediction model needs to be derived to obtain the prediction of the process. At time *t* of batch *k*, according to the 2D equivalent model of Eq. 4.38, the prediction $\hat{y}(t+j|t,k)$ can be determined as follows:

$$\begin{aligned} \hat{y}(t+j|t,k) &= y(t+j,k-1) + A\left(\Big|_{1}^{t+j}\right) \times \hat{r}\left(\Big|_{t+j-1}^{0}, k\right) + \Delta_k\left(\hat{w}(t+j|t,k)\right) \\ &= y(t+j,k-1) + A\left(\Big|_{j+1}^{t+j}\right) \times r\left(\Big|_{t-1}^{0}, k\right) + A\left(\Big|_{1}^{j}\right) \times \hat{r}\left(\Big|_{t+j-1}^{t}, k\right) \\ &\quad + \Delta_k\left(\hat{w}(t+j|t,k)\right) \end{aligned} \tag{4.39}$$

Disturbances in the current injection cycle from time *t* to $t+j$ are considered to be constant and can be obtained from the 2D equivalent model Eq. 4.38 as

$$\begin{aligned} \Delta_k\left(\hat{w}(t+j,k)\right) &= \Delta_k\left(\hat{w}(t|t,k)\right) = y(t,k) - y(t,k-1) - A\left(\Big|_{1}^{t}\right) \times r\left(\Big|_{t-1}^{0}, k\right) \\ &= \Delta_k\left(y(t,k)\right) - A\left(\Big|_{1}^{t}\right) \times r\left(\Big|_{t-1}^{0}, k\right) \end{aligned} \tag{4.40}$$

Then the prediction at time $t+j$ can be estimated based on known past input $r\left(\Big|_{t-1}^{0}, k\right)$ and unknown current input and future input $\hat{r}\left(\Big|_{t+j-1}^{t}, k\right)$ as

$$\begin{aligned}\hat{y}(t+j|t,k) &= y(t+j,k-1) + A\left(\Big|_{j+1}^{t+j}\right)\times r\left(\Big|_{t-1}^{0},k\right) + A\left(\Big|_{1}^{j}\right)\times \hat{r}\left(\Big|_{t+j-1}^{t},k\right) \\ &\quad + \Delta_k\left(y(t,k)\right) - A\left(\Big|_{1}^{t}\right)\times r\left(\Big|_{t-1}^{0},k\right) \\ &= A\left(\Big|_{1}^{j}\right)\times \hat{r}\left(\Big|_{t+j-1}^{t},k\right) + f(t+j,k)\end{aligned} \tag{4.41}$$

where

$$f(t+j,k) = y(t+j,k-1) + \Delta_k\left(y(t,k)\right) + \left(A\left(\Big|_{j+1}^{t+j}\right) - A\left(\Big|_{1}^{t}\right)\right)\times r\left(\Big|_{t-1}^{0},k\right)$$

where $f(t+j,k)$ is the free response of the system.

Assume $p$ and $m$ are the prediction horizon and control horizon, respectively, and $p \geq m \geq 1$. The prediction model is

$$\begin{cases}\hat{y}\left(\Big|_{t+p}^{t+1}t,k\right) = G\times \hat{r}\left(\Big|_{t+m-1}^{t},k\right) + f\left(\Big|_{t+p}^{t+1},k\right) \\ f\left(\Big|_{t+p}^{t+1},k\right) = y\left(\Big|_{t+p}^{t+1},k-1\right) + I_{P\times 1}\times \Delta_k\left(y(t,k)\right) + H(t)\times r\left(\Big|_{t-1}^{0},k\right)\end{cases} \tag{4.42}$$

Define the system's dynamic matrices of unknown input $G$ and known input $H(t)$ as follows:

$$G = \begin{bmatrix} a_1 & 0 & \cdots & 0 \\ a_2 & a_1 & \cdots & 0 \\ \vdots & \vdots & \ddots & \vdots \\ a_m & a_{m-1} & \cdots & a_1 \\ \vdots & \vdots & \ddots & \vdots \\ a_p & a_{p-1} & \cdots & a_{p-m+1} \end{bmatrix} \tag{4.43}$$

$$H(t) = H_1(t) - H_2(t);$$

$$H_1(t) = \begin{bmatrix} a_{t+1} & a_t & \cdots & a_2 \\ a_{t+2} & a_{t+1} & \cdots & a_3 \\ \vdots & \vdots & \ddots & \vdots \\ a_{t+p} & a_{t+p-1} & \cdots & a_{p+1} \end{bmatrix},\quad H_2(t) = \begin{bmatrix} a_t & a_{t-1} & \cdots & a_1 \\ a_t & a_{t-1} & \cdots & a_1 \\ \vdots & \vdots & \ddots & \vdots \\ a_t & a_{t-1} & \cdots & a_1 \end{bmatrix}_{p\times t} \tag{4.44}$$

Note that $f\left(\Big|_{t+p}^{t+1},k\right)$ can be expanded as

$$f\left(\Big|_{t+p}^{t+1},k\right) = \underbrace{y\left(\Big|_{t+p}^{t+1},k-1\right)}_{\text{Effect of previous batch output}} + \underbrace{I_{P\times 1}\times \Delta_k\left(y(t,k)\right)}_{\text{Effect of changes in current batch output}} + \underbrace{H(t)\times r\left(\Big|_{t-1}^{0},k\right)}_{\text{Effect of pass control actions on current batch}} \tag{4.45}$$

while $\hat{y}\left(\Big|_{t+p}^{t+1}t,k\right)$ can be expanded as

$$\hat{y}\left(\left|_{t+p}^{t+1}\right| t,k\right) = \underbrace{G \times \hat{r}\left(\left|_{t+m-1}^{t}, k\right.\right)}_{\substack{\text{Effect of current and future}\\ \text{control actions on current batch}}} + \underbrace{f\left(\left|_{t+p}^{t+1}, k\right.\right)}_{\substack{\text{Effect of past control actions}\\ \text{on the previous and current batches}}} \tag{4.46}$$

#### 4.3.2.3 Cost Function and Control Law

To design the controller, a sequence of control effort ($r$) is computed by minimizing the tracking error over a prediction horizon. At the same time, the excessive input movement necessary for doing so should be penalized. In this study, the cost function is chosen as

$$J(t,k,m,p) = \left\| y_r\left(\left|_{t+p}^{t+1}\right.\right) - \hat{y}_k\left(\left|_{t+p}^{t+1}, t\right.\right)\right\|_Q^2 + \lambda \left\|\hat{r}\left(\left|_{t+m-1}^{t}, k\right.\right)\right\|_R^2 \tag{4.47}$$

where

$$y_r\left(\left|_{t+p}^{t+1}\right.\right) \triangleq \begin{bmatrix} y_r(t+1) \\ y_r(t+2) \\ \vdots \\ y_r(t+p) \end{bmatrix} \tag{4.48}$$

The explicit solution to the above problem provides the control law of the proposed 2D-DMC scheme:

$$\Sigma_{\text{2D-DMC}} : \hat{r}^*(t,k) = d^T\left(y_r\left(\left|_{t+p}^{t+1}\right.\right) - f\left(\left|_{t+p}^{t+1}, k\right.\right)\right) \tag{4.49}$$

where

$$\begin{aligned} d^T &= C^T\left(G^T Q G + \lambda R\right)^{-1} G^T Q = \left[d_1 \cdots d_p\right] \\ C^T &= \left[1 \quad 0 \quad \cdots \quad 0\right] \end{aligned} \tag{4.50}$$

In this updating law, $C^T$ with the dimension $m$ is used to obtain the first row of the matrix; $f\left(\left|_{t+p}^{t+1}, k\right.\right)$ and $G(t,k)$ can be obtained from Eqs. 4.44 and 4.43, respectively. Notice that the controller is actually an integration of prediction, control, and compensation in the framework of 2D systems, using the information in both the current batch and previous batches.

Recall that 2D-DMC is a 2D, step-response, model-based control strategy. It utilizes the idea of online optimization by tuning in real time. From the 2D system viewpoint, the control law Eq. 4.49 can be inherently decomposed as

$$\Delta_t\left(u(t,k)\right) = u_{\text{DMC}}(t,k) + u_{\text{ILC}}(t,k) \tag{4.51}$$

where $u_{\text{DMC}}(t,k)$ and $u_{\text{ILC}}(t,k)$ are respectively determined by

$$\Sigma_{\text{DMC}} : u_{\text{DMC}}(t,k) = d^T \left( y_r \left( \Big|_{t+p}^{t+1} \right) - I_{P\times1} \times y(t,k) - H(t) \times \Delta_t u \left( \Big|_{t-1}^{0}, k \right) \right) \tag{4.52}$$

and

$$\Sigma_{\text{ILC}} : u_{\text{ILC}}(t,k) = u_{\text{ILC}}(t,k-1) + d^T \left( e \left( \Big|_{t+p}^{t+1}, k-1 \right) \right) \tag{4.53}$$

Clearly $\Sigma_{\text{DMC}}$ has the same form as the usual DMC control strategy, which penalizes within-batch nonrepetitive disturbances. Note that $\Sigma_{\text{ILC}}$ is an ILC control strategy that adjusts the input using the error from the previous batch over the prediction horizon. This strategy can reject disturbances along the batch dimension over the horizon. Thus, both repeatable and nonrepeatable disturbances can be rejected in 2D-DMC. It is well known that MPC is a powerful tool for dealing with multiple variables and constraints [28]. The 2D-DMC is proposed in the framework of MPC, so it inherently can handle the interactions among variables and the constraints. The implementation procedure of 2D-DMC is given in the following design procedure.

**2D-DMC Design Procedure** [68]

(1) Obtain the step-response model coefficients $a_i$.

(2) Specify the design parameters by selecting the appropriate $Q$,$R$, $\lambda$, $m$, and $p$.

(3) Perform initialization: $k \leftarrow 1$ and $t \leftarrow 0$.

(4) Determine the dynamic matrices $G$ and $H(t)$ using Eq. 4.43 and Eq. 4.44.

(5) Calculate the control gains $d^T = [d_1 \cdots d_p]$ using Eq. 4.50 and $G$.

(6) Given the output of batch $k-1$, the current output $t$ of batch $k$ and the past input of batch $k$, calculate the free response $f\left(\Big|_{t+p}^{t+1}, k\right)$ based on $H(t)$ using Eq. 4.45.

(7) Given the set point, design the updating law at step $t$ of batch $k$, $\hat{r}^*(t,k)$, based on $d^T$ and $f\left(\Big|_{t+p}^{t+1}, k\right)$ using Eq. 4.49.

(8) If the end of batch $k$ is reached, jump to step 10; otherwise continue with step 9.

(9) Let $t \leftarrow t+1$ and go back to step 4.

(10) Let $k \leftarrow k+1$ and $t \leftarrow 1$ and go back to step 4.

In the 2D-DMC framework, the output prediction could exceed the duration of an operation batch as the process approaches the end of the cycle. For the open-loop stable process, the output prediction out of a batch can be taken as the same value as that at the end of a batch, and thus the prediction horizon will not shorten. For the open-loop unstable process, the prediction horizon will be reduced, and there is no control action for the output prediction out of a batch. In both cases, the duration of the current batch needs to be predetermined. Actually, the end of a batch can be determined by, for example, checking the time elapsed or certain conditions. The duration of a batch is available and constant if the time elapsed is taken as the end condition of the injection molding process. If a logic check of certain conditions is to be performed because the molding process operates repeatedly, the end point of the process can be predicted using information from past batches. For ease of implementation, the duration of the current batch can be taken to be the same as that of the previous batch. Note that, in the following experiment, the injection velocity process ends when the preset injection stroke is reached and good control performance is obtained.

#### 4.3.2.4 Analysis of Convergence and Robustness

In this section, both the convergence and robustness properties of 2D-DMC are investigated. When the uncertainties are repeatable, the sufficient conditions that guarantee the elimination of tracking error will be provided. When interval nonrepeatable uncertainties exist, the tracking error converges to a known bound, which indicates the robustness of the algorithm. The tracking error of the 2D-DMC evolving in two time dimensions complicates the analysis. In this work, the tracking errors in different batches will be combined to form an error vector. Thus, the 2D analysis problem is transformed into a one-dimensional (1D) one. The following definition and assumptions are made for the analysis.

**Definition 1.** The tracking error is defined as

$$e(t,k) \triangleq y_r(t) - y(t,k) \tag{4.54}$$

which is assumed to be zero, or

$$e(T+1,k) = e(T+2,k) = \cdots = e(T+p-1,k) = 0 \tag{4.55}$$

when the time index exceeds the duration of a batch.

**Assumption 1.** The durations of different batches are assumed to be constant, or

$$T_k \triangleq T \tag{4.56}$$

**Assumption 2.** The uncertainties are described in a general sense as follows:

$$w(t,k) = w_1(t) + w_2(t,k), \quad t = 0,1,\ldots, \quad T, k = 1,2,\ldots$$

and (4.57)

$$\upsilon_k^0 \triangleq \Delta_k\left(w_2(0,k)\right), \quad \upsilon_k \triangleq \Delta_k\left(w_2\left(\left|_{1}^{T}, k\right.\right)\right)$$

where $w_1(t)$ and $w_2(t,k)$ represent the repeatable and nonrepeatable uncertainties, respectively, and $\upsilon_k^0$ and $\upsilon_k$ denote the initial nonrepeatable uncertainties and the nonrepeatable uncertainties within a batch, respectively. They vary in the following interval:

$$\left\|\upsilon_{k+1}^{0}-\upsilon_{k}^{0}\right\| \leq b_{\upsilon 0} \quad \text{and} \quad \left\|\upsilon_{k+1}-\upsilon_{k}\right\| \leq b_{\upsilon} \tag{4.58}$$

#### 4.3.2.4.1 Model of the Closed-Loop Control System

Before moving to the convergence proof, the model of the closed-loop control system needs to be acquired first. Substituting Eq. 4.57 into the 2D equivalent Eq. 4.38, the following 2D equivalent model without repeatable uncertainties can be obtained:

$$\begin{aligned}\Sigma_{\text{2D-p}}: y(t,k) &= y(t,k-1)+A\left(\left|_{1}^{t}\right.\right)\times r\left(\left|_{t-1}^{0}\right.,k\right)+\Delta_{k}\left(w_{1}(t)\right)+\Delta_{k}\left(w_{2}(t,k)\right)\\ &= y(t,k-1)+A\left(\left|_{1}^{t}\right.\right)\times r\left(\left|_{t-1}^{0}\right.,k\right)+\Delta_{k}\left(w_{2}(t,k)\right)\\ t&=0,1,\ldots,\quad T,k=1,2,\ldots\end{aligned} \tag{4.59}$$

Substituting Eq. 4.59 into the control law Eq. 4.49, the relationship between the input and the error is obtained as

$$\begin{aligned}\Sigma_{\text{2D-DMC}}: r(t,k) &= d^{T}\left(y_{r}\left(\left|_{t+p}^{t+1}\right.\right)-y\left(\left|_{t+p}^{t+1}\right.,k-1\right)-I_{P\times 1}\right.\\ &\quad \left.\times\left(A\left(\left|_{1}^{t}\right.\right)\times r\left(\left|_{t-1}^{0}\right.,k\right)+\Delta_{k}\left(w_{2}(t,k)\right)\right)-H(t)\times r\left(\left|_{t-1}^{0}\right.,k\right)\right)\\ &= d^{T}\left(e\left(\left|_{t+p}^{t+1}\right.,k-1\right)-I_{P\times 1}\times\Delta_{k}\left(w_{2}(t,k)\right)-H_{1}(t)\times r\left(\left|_{t-1}^{0}\right.,k\right)\right)\end{aligned} \tag{4.60}$$

To obtain the model of the closed-loop control system, the explicit description of $r\left(\left|_{t-1}^{0}\right.,k\right)$ should be derived from the control law Eq. 4.60 and then substituted into Eq. 4.59. Taking $t$ from 0 to $t-1$ in 4.60 leads to

$$\begin{aligned}r(0,k) &= \sum_{i=1}^{p} d_{i}e(i,k-1)-\sum_{i=1}^{p} d_{i}\cdot\Delta_{k}\left(w_{2}(0,k)\right)\\ r(1,k) &= \sum_{i=1}^{p} d_{i}e(i+1,k-1)-\sum_{i=1}^{p} d_{i}a_{i+1}r(0,k)-\sum_{i=1}^{p} d_{i}\cdot\Delta_{k}\left(w_{2}(1,k)\right)\\ r(2,k) &= \sum_{i=1}^{p} d_{i}e(i+2,k-1)-\sum_{i=1}^{p} d_{i}a_{i+2}r(0,k)-\sum_{i=1}^{p} d_{i}a_{i+1}r(1,k)-\sum_{i=1}^{p} d_{i}\cdot\Delta_{k}\left(w_{2}(2,k\right.\\ &\vdots\\ r(t-1,k) &= \sum_{i=1}^{p} d_{i}e(i+t-1,k-1)-\sum_{j=0}^{t-2}\left(\sum_{i=1}^{p} d_{i}a_{i+t-1-j}\right)r(j,k)-\sum_{i=1}^{p} d_{i}\cdot\Delta_{k}\left(w_{2}(t-1,k\right.\end{aligned} \tag{4.61}$$

which can be rewritten in the vector form as follows:

$$\begin{bmatrix} r(0,k) \\ r(1,k) \\ \vdots \\ r(t-1,k) \end{bmatrix} = \begin{bmatrix} d_1 & d_2 & \cdots & d_p & 0 & \cdots & 0 \\ 0 & d_1 & d_2 & \cdots & d_p & \cdots & 0 \\ \vdots & \vdots & \ddots & \ddots & \ddots & \ddots & \vdots \\ 0 & 0 & 0 & d_1 & d_2 & \cdots & d_p \end{bmatrix}_{t*(t+p-1)} \times \begin{bmatrix} e(1,k-1) \\ e(2,k-1) \\ \vdots \\ e(t+p-1,k-1) \end{bmatrix} -$$

$$\begin{bmatrix} 0 & 0 & 0 & \cdots & 0 \\ \sum_{i=1}^{p} d_i a_{i+1} & 0 & 0 & \cdots & 0 \\ \sum_{i=1}^{p} d_i a_{i+2} & \sum_{i=1}^{p} d_i a_{i+1} & 0 & \cdots & 0 \\ \vdots & \vdots & \ddots & \ddots & \vdots \\ \sum_{i=1}^{p} d_i a_{i+t-1} & \sum_{i=1}^{p} d_i a_{i+t-2} & \sum_{i=1}^{p} d_i a_{i+1} & & 0 \end{bmatrix}_{t*t} \times \begin{bmatrix} r(0,k) \\ r(1,k) \\ \vdots \\ r(t-1,k) \end{bmatrix} - \begin{bmatrix} \Delta_k(w_2(0,k)) \\ \Delta_k(w_2(1,k)) \\ \vdots \\ \Delta_k(w_2(t-1,k)) \end{bmatrix} \cdot \sum_{i=1}^{p} d_i \tag{4.62}$$

Thus, the explicit description of $r\left(\big|_{t-1}^{0},k\right)$ is

$$r\left(\big|_{t-1}^{0},k\right) = DA(t)^{-1} \times D(t) \times e\left(\big|_{t+p-1}^{1},k-1\right) - DA(t)^{-1} \times \Delta_k\left(w_2\left(\big|_{t-1}^{0},k\right)\right) \cdot \sum_{i=1}^{p} d_i$$

where

$$DA(t) = \begin{bmatrix} 1 & 0 & 0 & \cdots & 0 \\ \sum_{i=1}^{p} d_i a_{i+1} & 1 & 0 & \cdots & 0 \\ \sum_{i=1}^{p} d_i a_{i+2} & \sum_{i=1}^{p} d_i a_{i+1} & 1 & \cdots & 0 \\ \vdots & \vdots & \ddots & \ddots & \vdots \\ \sum_{i=1}^{p} d_i a_{i+t-1} & \sum_{i=1}^{p} d_i a_{i+t-2} & \cdots & \sum_{i=1}^{p} d_i a_{i+1} & 1 \end{bmatrix}_{t*t}, \tag{4.63}$$

$$D(t) = \begin{bmatrix} d_1 & d_2 & \cdots & d_p & 0 & \cdots & 0 \\ 0 & d_1 & d_2 & \cdots & d_p & \cdots & 0 \\ \vdots & \vdots & \ddots & \ddots & \ddots & \ddots & \vdots \\ 0 & 0 & 0 & d_1 & d_2 & \cdots & d_p \end{bmatrix}_{t*(t+p-1)}$$

Here, $DA(t)$ is a lower triangular matrix with ones in the diagonal, so the inverse of $DA(t)$ is always valid. The model of the closed-loop system can be achieved by substituting Eq. 4.63 into Eq. 4.59, giving

$$\begin{aligned}\Sigma_{\text{2D-p-C}}: y(t,k) = y(t,k-1) + A\left(\Big|_1^t\right) \times DA(t)^{-1} \times D(t) \times e\left(\Big|_{t+p-1}^1, k-1\right) \\ - \sum_{i=1}^{p} d_i \cdot A\left(\Big|_1^t\right) \times DA(t)^{-1} \times \Delta_k\left(w_2\left(\Big|_{t-1}^0, k\right)\right) + \Delta_k\left(w_2(t,k)\right)\end{aligned} \tag{4.64}$$

Based on this model, the corresponding tracking error evolution law will be derived next.

#### 4.3.2.4.2 Tracking Error and Convergence Conditions

From Definition 1, it is easy to derive

$$\Delta_k\left(y(t,k)\right) = -\Delta_k\left(e(t,k)\right) \tag{4.65}$$

Combining Eq. 4.64 with Eq. 4.65 leads to

$$\begin{aligned}&\Sigma_{\text{2D-p-C}}: e(t,k) = LE_t \times e\left(\Big|_{t+p-1}^1, k-1\right) + LW_t \times \Delta_k\left(w_2\left(\Big|_{t-1}^0, k\right)\right) - \Delta_k\left(w_2(t,k)\right) \\ &\text{where} \\ &LE_t \triangleq \begin{bmatrix}0 & \cdots & 0 & 1_t & 0 & \cdots & 0\end{bmatrix}_{1^*t+p-1} - A\left(\Big|_1^t\right) \times DA(t)^{-1} \times D(t) \\ &LW_t \triangleq \sum_{i=1}^{p} d_i \times A\left(\Big|_1^t\right) \times DA(t)^{-1}\end{aligned} \tag{4.66}$$

Here $\begin{bmatrix}0 & \cdots & 0 & 1_t & 0 & \cdots & 0\end{bmatrix}_{1^*t+p-1}$ represents a $(t+p-1)$-dimensional row vector, where the $t$th element is one and all the other elements are zero. The tracking error in Eq. 4.66 is 2D with two indices $t$ and $k$. It will be transformed into 1D with one index $k$ in this study by combining the tracking errors within each individual batch. Taking $t$ from 1 to $T$

$$\begin{aligned}e(1,k) &= LE_1 \times e\left(\Big|_p^1, k-1\right) + LW_1 \times \Delta_k\left(w_2(0,k)\right) - \Delta_k\left(w_2(1,k)\right) \\ e(2,k) &= LE_2 \times e\left(\Big|_{p+1}^1, k-1\right) + LW_2 \times \Delta_k\left(w_2\left(\Big|_1^0, k\right)\right) - \Delta_k\left(w_2(2,k)\right) \\ &\vdots \\ e(T,k) &= LE_T \times e\left(\Big|_{p+T-1}^1, k-1\right) + LW_T \times \Delta_k\left(w_2\left(\Big|_{T-1}^0, k\right)\right) - \Delta_k\left(w_2(T,k)\right)\end{aligned} \tag{4.67}$$

which can be rewritten in the vector form

$$\begin{bmatrix} e(1,k) \\ e(2,k) \\ \vdots \\ e(T,k) \end{bmatrix} = \begin{bmatrix} LE_1(1) & LE_1(2) & \cdots & LE_1(p) & 0 & 0 & \cdots & 0 \\ LE_2(1) & LE_2(2) & \cdots & LE_2(p) & LE_2(p+1) & 0 & \cdots & 0 \\ \vdots & \vdots & \ddots & \vdots & \vdots & \vdots & \ddots & \vdots \\ LE_T(1) & LE_T(2) & \cdots & LE_T(p) & LE_T(p+1) & LE_T(p+2) & \cdots & LE_T(p+T-1) \end{bmatrix}$$

$$\times \begin{bmatrix} e(1,k-1) \\ e(2,k-1) \\ \vdots \\ e(p+T-1,k-1) \end{bmatrix} + \begin{bmatrix} LW_1(1) \\ LW_2(1) \\ \vdots \\ LW_T(1) \end{bmatrix} \times \Delta_k\left(w(0,k)\right)$$

$$+ \begin{bmatrix} -1 & 0 & \cdots & 0 & 0 \\ LW_2(2) & -1 & \cdots & 0 & 0 \\ LW_3(2) & LW_3(2) & \ddots & 0 & 0 \\ \vdots & \vdots & \ddots & -1 & 0 \\ LW_T(2) & LW_T(2) & \cdots & LW_T(T) & -1 \end{bmatrix} \times \begin{bmatrix} \Delta_k\left(w(1,k)\right) \\ \Delta_k\left(w(2,k)\right) \\ \vdots \\ \Delta_k\left(w(T,k)\right) \end{bmatrix} \tag{4.68}$$

Combining the above equation with Eq. 4.55 leads to the 1D (the batch dimension only) tracking error evolution law:

$$E_k = \phi \times E_{k-1} + \xi \times v_k^0 + \psi \times v_k$$

where

$$\phi = \begin{bmatrix} LE_1(1) & LE_1(2) & \cdots & LE_1(p) & 0 & 0 & \cdots & 0 \\ LE_2(1) & LE_2(2) & \cdots & LE_2(p) & LE_2(p+1) & 0 & \cdots & 0 \\ \vdots & \vdots & \ddots & \vdots & \vdots & \vdots & \ddots & \vdots \\ LE_{T-p}(1) & LE_{T-p}(2) & \cdots & LE_{T-p}(p) & LE_{T-p}(p+1) & LE_{T-p}(p+2) & \cdots & 0 \\ LE_{T-p+1}(1) & LE_{T-p+1}(2) & \cdots & LE_{T-p+1}(p) & LE_{T-p+1}(p+1) & LE_{T-p+1}(p+2) & \cdots & LE_{T-p+1}(T) \\ \vdots & \vdots & \ddots & \vdots & \vdots & \vdots & \ddots & \vdots \\ LE_T(1) & LE_T(2) & \cdots & LE_T(p) & LE_T(p+1) & LE_T(p+2) & \cdots & LE_T(T) \end{bmatrix}$$

$$\xi = \begin{bmatrix} LW_1(1) \\ LW_2(1) \\ \vdots \\ LW_T(1) \end{bmatrix}, \quad \psi = \begin{bmatrix} -1 & 0 & \cdots & 0 & 0 \\ LW_2(2) & -1 & \cdots & 0 & 0 \\ LW_3(2) & LW_3(2) & \ddots & 0 & 0 \\ \vdots & \vdots & \ddots & -1 & 0 \\ LW_T(2) & LW_T(2) & \cdots & LW_T(T) & -1 \end{bmatrix}, \quad E_k = e\left(\Big|_1^T, k\right) \tag{4.69}$$

where $\phi$, $\xi$, and $\psi$ are known constant matrices determined by the system step-response coefficients. For the process with completely repeatable uncertainties, where $b_{v0}$ and $b_v$ are both zero, the tracking error evolution law is simplified to

$$E_k = \phi \times E_{k-1} \tag{4.70}$$

Here $E_k$ is exponentially asymptotically convergent if the $\phi$ is stable, that is, the eigenvalue of $\phi$ is less than 1 [70]. The convergence condition can also be defined in the norm topology. Therefore, the convergence theorems can be stated in two different topologies as follows.

**Theorem 1.** For the system of Eq. 4.38 with Assumptions 1-2 and $b_{v0} = 0$, $b_v = 0$, using the control updating law Eq. 4.49, $E_k$ will exponentially asymptotically converge to zero as $k \to \infty$:

$$\lim_{k\to\infty} \left\| E_k \right\| = 0 \tag{4.71}$$

if $\left\| \lambda(\phi) \right\|_1 < 1$.

**Theorem 2.** For the system of Eq. 4.38 with the Assumptions 1-2 and $b_{v0} = 0$, $b_v = 0$, using the control updating law Eq. 4.49, $E_k$ will monotonically converge to zero as $k \to \infty$ in the $l_i, i \in \{1, 2, \infty\}$ norm topology:

$$\lim_{k\to\infty} \left\| E_k \right\| = 0 \tag{4.72}$$

if $\left\| \phi \right\|_i < 1$.

**Remark 2:** From Theorems 1 and 2, it can be seen that the 2D-DMC guarantees the ultimate tracking performance with repeatable uncertainties. Moreover, the performance does not depend on the preset trajectory in the derivation procedure of the theorems. From the application viewpoint, this algorithm can track the profile tightly even under model-plant mismatches, effectively combating the main deficiency of MPC.

Assumption 1 is used in Eq. 4.67 to obtain the 1D tracking error evolution law Eq. 4.69. If the duration $T_k$ of the batch varies, the closed-loop control system is guaranteed to converge from the beginning of the batch to $T_{\min}$ by Theorems 1 and 2, where $T_{\min}$ is the minimum duration of $k$ batch given by

$$T_{\min} \triangleq \min\{T_1, T_2, \cdots, T_k\} \tag{4.73}$$

For the output at $t > T_{\min}$, the closed-loop control system can still converge. That is, the control performance at time $t$ can be improved once when there is an output at time $t$ at any following batch. Actually, the duration of the batch will converge as well with the convergence of the closed-loop system. The difference is that the output generated during the time before $T_{\min}$ is up will converge faster than the output generated during the time after $T_{\min}$ is up. In the following experimental test, it should be noted that $T_k$ varies along the batch dimension, but the process can still be made to converge.

The 2D-DMC is proposed under the framework of DMC and ILC, the convergence of which can be guaranteed by Theorems 1 and 2. It should be noted that the convergence analysis conducted in this study is along the batch dimension. A smaller value of the convergence condition leads to a faster convergence rate. However, the tracking performance along the time dimension may be degraded during the convergence process. The guidelines for tuning the parameters of the proposed 2D-DMC are summarized as follows.

The weighting factor $\lambda$ allows the variable being manipulated to be weighted. Decreasing the value of $\lambda$ tends to increase the convergence rate of the closed-loop system along the batch dimension by reducing the value of the convergence condition. This will, however, lead to large fluctuations of the input. The weighting factor has the most important role in determining the control performance of 2D-DMC.

As the prediction horizon $P$ decreases, the controller tends to increase the convergence rate of the closed-loop system along the batch dimension by reducing the value of the convergence condition. This will, however, lead to poor control performance before the process is convergent. For example, the variable under control will oscillate in the dynamic process in the early molding cycles. The prediction horizon is often selected to cover the entire dynamics of the process.

Decreasing the value of the control horizon $M$ tends to decrease the convergence rate of the closed-loop system along the batch dimension by increasing the value of the convergence condition. Moreover, the computational load increases with the control horizon.

#### 4.3.2.4.3 Robustness Analysis

The convergence condition for the process with repeatable uncertainties has been studied above; however, nonrepeatable uncertainties inevitably exist in practice. In the following, the robustness of the control algorithm against nonrepeatable uncertainties will be analyzed.

First, introduce the following Lyapunov function for the tracking error in Eq. 4.69:

$$V_k = E_k^T F E_k \tag{4.74}$$

where $F$ is the solution of the following Lyapunov function:

$$\phi^T F \phi - F = -I \tag{4.75}$$

and $I$ is the unit matrix with a proper dimension. There is a feasible solution for $F$ if $\phi$ is stable, which means the convergence theorem derived in the previous subsection must be satisfied. Let $\lambda_{\min}(F)$ and $\lambda_{\max}(F)$ be the smallest and largest eigenvalues of $F$, respectively. Then

$$\lambda_{\min}(F)\|E_k\|^2 \le V_k \le \lambda_{\max}(F)\|E_k\|^2 \tag{4.76}$$

From the Lyapunov function 4.74 and Eq. 4.69,

$$\begin{aligned}\Delta_k V_{k+1} &= V_{k+1} - V_k \\ &= \left(\phi \times E_k + \xi \times v_{k+1}^0 + \psi \times v_{k+1}\right)^T F\left(\phi \times E_k + \xi \times v_{k+1}^0 + \psi \times v_{k+1}\right) - E_k^T F E_k \\ &= E_k^T \phi^T F \phi E_k - E_k^T F E_k + \left(v_{k+1}^0\right)^T \xi^T F \xi v_{k+1}^0 + v_{k+1}^T \psi^T F \psi v_{k+1} \\ &\quad + 2E_k^T \phi^T F \xi v_{k+1}^0 + 2E_k^T \phi^T F \psi v_{k+1} + 2\left(v_{k+1}^0\right)^T \xi^T F \psi v_{k+1}\end{aligned} \tag{4.77}$$

Based on **Assumption 2**, the following triangle inequalities can be obtained:

$$\begin{aligned}\left(v_{k+1}^0\right)^T \xi^T F \xi v_{k+1}^0 &\le \left\|\xi^T F \xi\right\| b_{v0}^2 \\ v_{k+1}^T \psi^T F \psi v_{k+1} &\le T^2 \left\|\psi^T F \psi\right\| b_v^2 \\ 2E_k^T \phi^T F \xi v_{k+1}^0 &\le \frac{1}{4} E_k^T E_k + 4T \left\|\xi^T F^T \phi\right\| b_{v0} \\ 2E_k^T \phi^T F \psi v_{k+1} &\le \frac{1}{4} E_k^T E_k + 4T^2 \left\|\psi^T F^T \phi\right\| b_v \\ 2\left(v_{k+1}^0\right)^T \xi^T F \psi v_{k+1} &\le 2T \left\|\xi^T F \psi\right\| b_{v0} b_v\end{aligned} \tag{4.78}$$

By using Eq. 4.78, one can transform Eq. 4.77 into

$$\begin{aligned}\Delta_k V_{k+1} &\le \left(-1 + \frac{1}{4} + \frac{1}{4}\right)\left\|E_k\right\| + f\left(b_{v0}, b_v\right) \\ f\left(b_{v0}, b_v\right) &= \left\|\xi^T F \xi\right\| b_{v0}^2 + T^2 \left\|\psi^T F \psi\right\| b_v^2 + 4T \left\|\xi^T F^T \phi\right\| b_{v0} \\ &\quad + 4T^2 \left\|\psi^T F^T \phi\right\| b_v + 2T \left\|\xi^T F \psi\right\| b_{v0} b_v\end{aligned} \tag{4.79}$$

On the basis of Eq. 4.76, it is easy to obtain

$$V_{k+1} - V_k \le -\frac{1}{2}\left\|E_k\right\| + f\left(b_{v0}, b_v\right) \le -\frac{V_k}{2\lambda_{\max}(F)} + f\left(b_{v0}, b_v\right) \tag{4.80}$$

It is also easy to find a constant $\varepsilon$ satisfying

$$\frac{1}{2\lambda_{\max}(F)} \ge \varepsilon > 0 \quad \text{and} \quad \varepsilon < 1 \tag{4.81}$$

Thus, Eq. 4.80 can be rewritten as

$$V_{k+1} - V_k \le -\varepsilon V_k + f\left(b_{v0}, b_v\right) \tag{4.82}$$

It is easy to obtain the following inequality from Eq. 4.82:

$$V_k \leq (1-\varepsilon)^k V_0 + \frac{1-(1-\varepsilon)^k}{1-(1-\varepsilon)} f(b_{v0}, b_v) \tag{4.83}$$

and since $0 < \varepsilon < 1$,

$$\lim_{k \to \infty} V_k \leq \lim_{k \to \infty} (1-\varepsilon)^k V_0 + \frac{1-(1-\varepsilon)^k}{1-(1-\varepsilon)} f(b_{v0}, b_v) = \frac{f(b_{v0}, b_v)}{\varepsilon} \tag{4.84}$$

Combining the Lyapunov functions 4.74 and Eq. 4.76 with Eq. 4.84, it is easy to achieve

$$\lim_{k \to \infty} \|E_k\| \leq \sqrt{\frac{f(b_{v0}, b_v)}{\varepsilon \lambda_{\min}(F)}} \tag{4.85}$$

Equation 4.85 gives a description of the tracking error in the system with nonrepeatable uncertainties. The system can still converge to a known bound. The following theorems summarize the above derivations.

**Theorem 3.** Based on **Assumptions 1–2** and the control updating law Eq. 4.49, the tracking error $E_k$ of the system 4.38 will exponentially asymptotically converge to a known bound as $k \to \infty$:

$$\lim_{k \to \infty} \|E_k\| \leq \sqrt{\frac{f(b_{v0}, b_v)}{\varepsilon \lambda_{\min}(F)}} \tag{4.86}$$

if $\|\lambda(\phi)\|_1 < 1$.

**Theorem 4.** Based on Assumptions 1–2 and the control updating law Eq. 4.49, the tracking error $E_k$ of the system 4.38 will monotonically converge to a known bound as $k \to \infty$ in the $l_i, i \in \{1, 2, \infty\}$ norm topology:

$$\lim_{k \to \infty} \|E_k\| \leq \sqrt{\frac{f(b_{v0}, b_v)}{\varepsilon \lambda_{\min}(F)}} \tag{4.87}$$

if $\|\phi\|_i < 1$.

**Remark 3:** From **Theorems 3** and **4**, it can be concluded that the 2D-DMC guarantees the convergence of error to a known bound. This bound depends on the value of the nonrepeatable uncertainty, while the value of $f(b_{v0}, b_v)$ increases with the uncertainty of the system. From Eq. 4.79, $b_v$ contributes more to $f(b_{v0}, b_v)$ than does $b_{v0}$.

### 4.3.3 Simulation Illustration

The injection velocity, the key controlled variable in the filling phase, is used in this section to demonstrate the performance of 2D-DMC [70]. The velocity model is assumed to be described by the following equation:

$$Y(s,k) = \frac{5.75}{3.262s+1}U(s,k) + W(s,k), \quad k = 1, 2, \ldots \tag{4.88}$$

Eq. 4.88 is the velocity model with the disturbance $W(s,k)$. Applying a step change to the control valve gives the response model

$$y(t,k) = \sum_{i=1}^{t} a_i \Delta_t u(t-i,k) + w(t,k), \quad t = 1, \ldots, \quad T_k, k = 1, 2, \ldots \tag{4.89}$$

where $T_k$ is 200 steps and is assumed to be invariant in this simulation; the coefficient sequences $a_i$ are shown in Table 4.2, "Coefficients of the Model Used in the Simulation," and the responses from the dashed line are in Fig. 4.9. The corresponding transfer function is

$$Y(s,k) = \frac{6.50e^{-s}}{4.482s+1}U(s,k), \quad k = 1, 2, \ldots \tag{4.90}$$

**Table 4.2** Coefficients of the Model Used in the Simulation

| $i$ | $a_i$ | $i$ | $a_i$ |
|---|---|---|---|
| 1 | 0.654 | 191 | 6.500 |
| 2 | 1.241 | 192 | 6.500 |
| 3 | 1.770 | 193 | 6.500 |
| 4 | 2.246 | 194 | 6.500 |
| 5 | 2.673 | 195 | 6.500 |
| 6 | 3.058 | 196 | 6.5001 |
| 7 | 3.404 | 197 | 6.500 |
| 8 | 3.716 | 198 | 6.500 |
| 9 | 3.996 | 199 | 6.500 |
| 10 | 4.247 | 200 | 6.500 |

The model-plant mismatch between Eq. 4.88 and the plant is illustrated in Fig. 4.9, which shows that the mismatch is obvious. The set point of $y(t,k)$ is chosen as the step change with a five-step transition time from time 101 to 104:

$$y_r(t,k)=\begin{cases}60, & t=1,2,\ldots,60\\ 60-(t-60)\times 20/3, & t=61,62\\ 40, & t=63,64,\ldots,140\\ 40+(t-140)\times 20/3, & t=141,142\\ 60, & t=143,144,\ldots,200\end{cases} \tag{4.91}$$

It must be noted that this response corresponds to a 30 ms transient time, smaller than the limit of a proportional control valve.

To demonstrate the effectiveness of the proposed control scheme, several cases are studied. In the following simulations and experiments, the sum of the squared error (SSE) between the set-point and the output measurement is adopted to evaluate the tracking performance:

$$SSE(k)=\sum_{t=1}^{T_k}\left(y_r(t,k)-y(t,k)\right)^2 \tag{4.92}$$

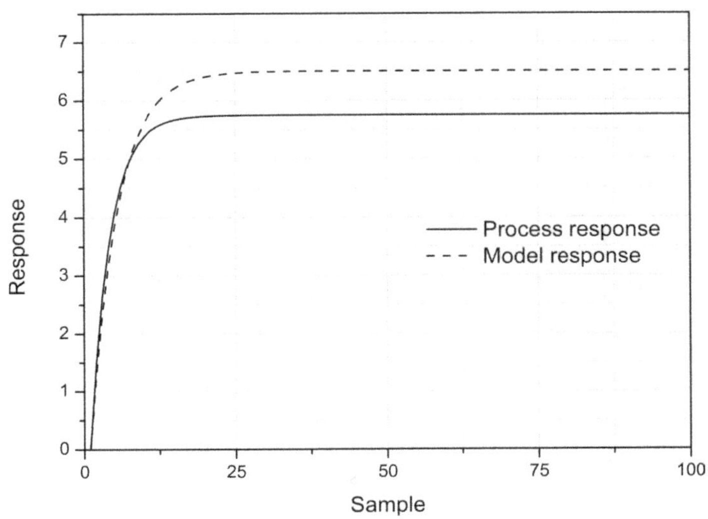


**Figure 4.9** Output responses of the model and plant in simulation

#### 4.3.3.1 Case 1: Convergence Test

The procedures for tuning the DMC parameters are given by Morari and Lee (1999) [71]. In this simulation, considering the trade-off between the convergence rate and the robustness of the controller, the following parameters are chosen: $Q = I$, $R = I$, $\lambda = 50$, $p = 10$, and $m = 10$. The disturbance $w(t,k)$ is set to zero for the convergence test.

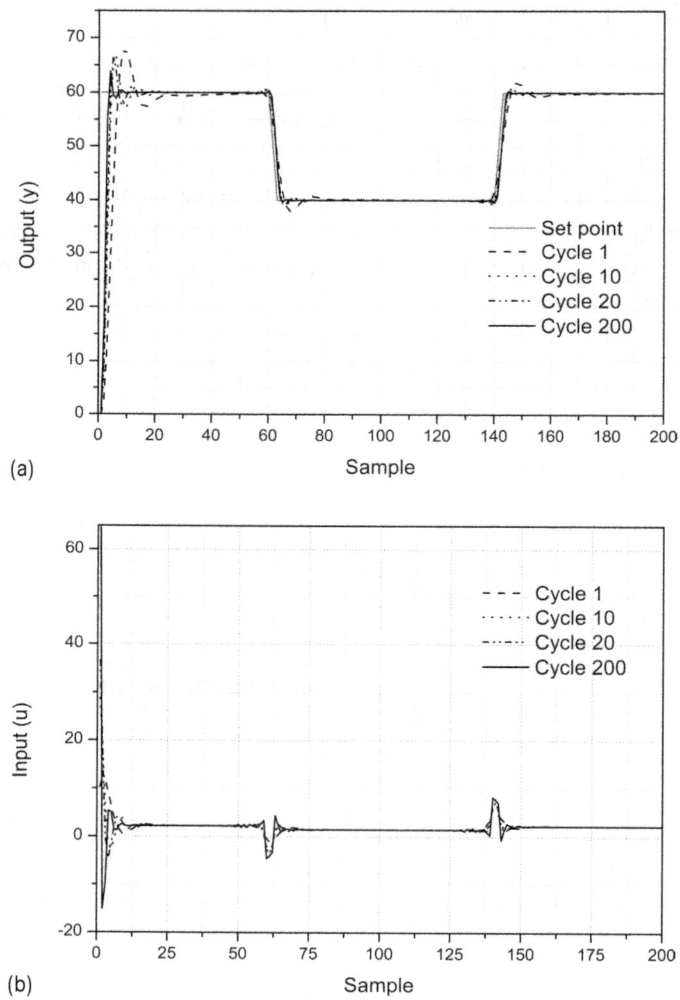


**Figure 4.10** 2D-DMC simulation result with model mismatches: (a) output response, and (b) control input

Figure 4.10 shows the output response and the input signal of the control system. It can be clearly seen that the set-point profile is perfectly tracked even with obvious model mismatch. Since this mismatch is exactly repeatable, it can be treated as a repetitive disturbance and thus eliminated by the ILC framework in the proposed control scheme. Figure 4.11 provides an enlarged view of the startup and step-response areas. It is shown that the dynamic tracking performance improves gradually from batch to batch. Figure 4.12 shows the fast convergence rate of the control system.

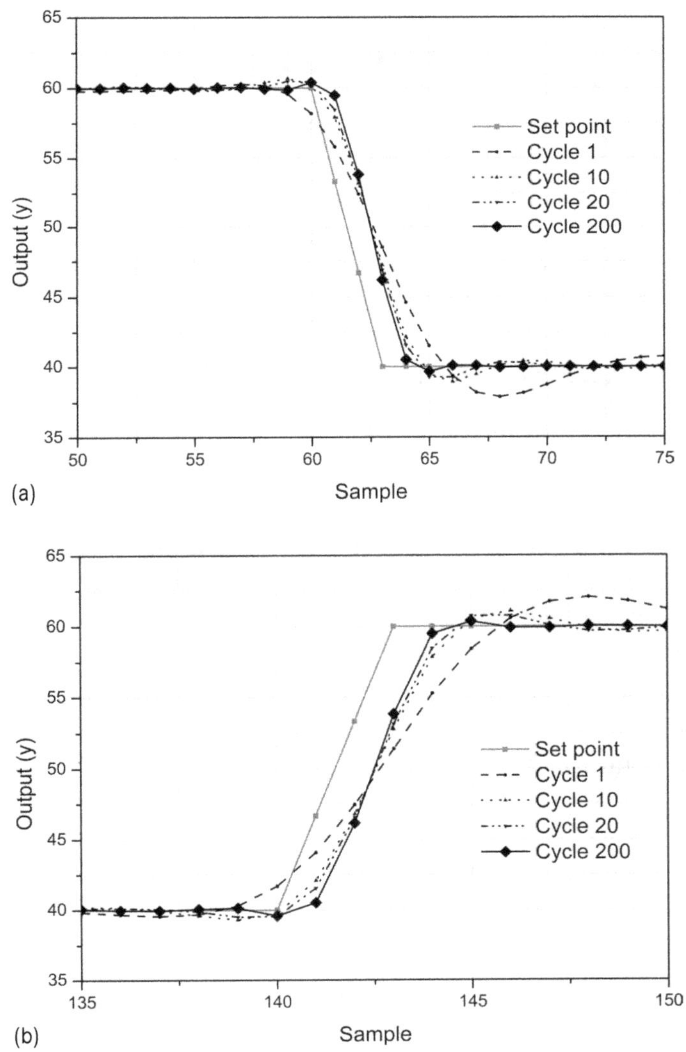


**Figure 4.11** Zoomed-in plot of simulation result:
(a) the step-down area, and (b) the step-up area

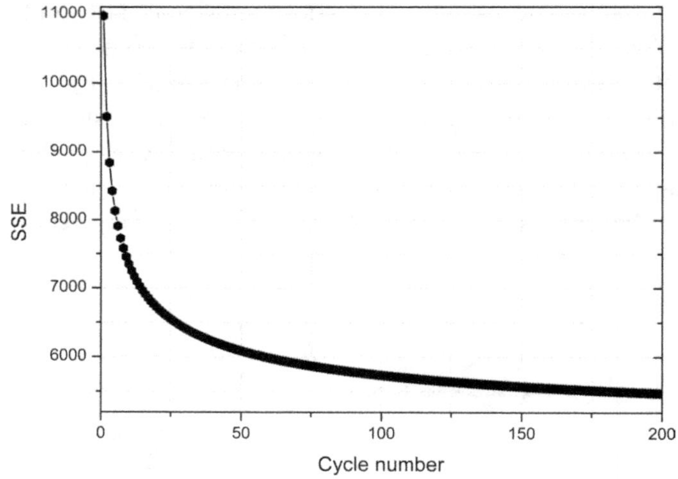


**Figure 4.12** Tracking performance without disturbance

Another case with sinusoidal repetitive disturbance is conducted to further test the robustness of the 2D-DMC control scheme.

#### 4.3.3.2 Case 2: Repetitive Disturbances

The disturbance in this case is chosen as

$$w(t,k)=5\times\sin(0.04\pi t) \tag{4.93}$$

where $w(t,k)$ changes in the time dimension only and can be simplified as $w(t)$. It represents the sinusoidal repeatable disturbances. Figure 4.13 shows that the output tracks the set point perfectly by adopting the input signal in a sinusoidal fashion to reject the sinusoidal disturbance.

Figure 4.14 shows the fast convergence rate, which converges monotonically. This indicates that the implementation of ILC in the 2D control scheme helps to handle the repetitive disturbances.

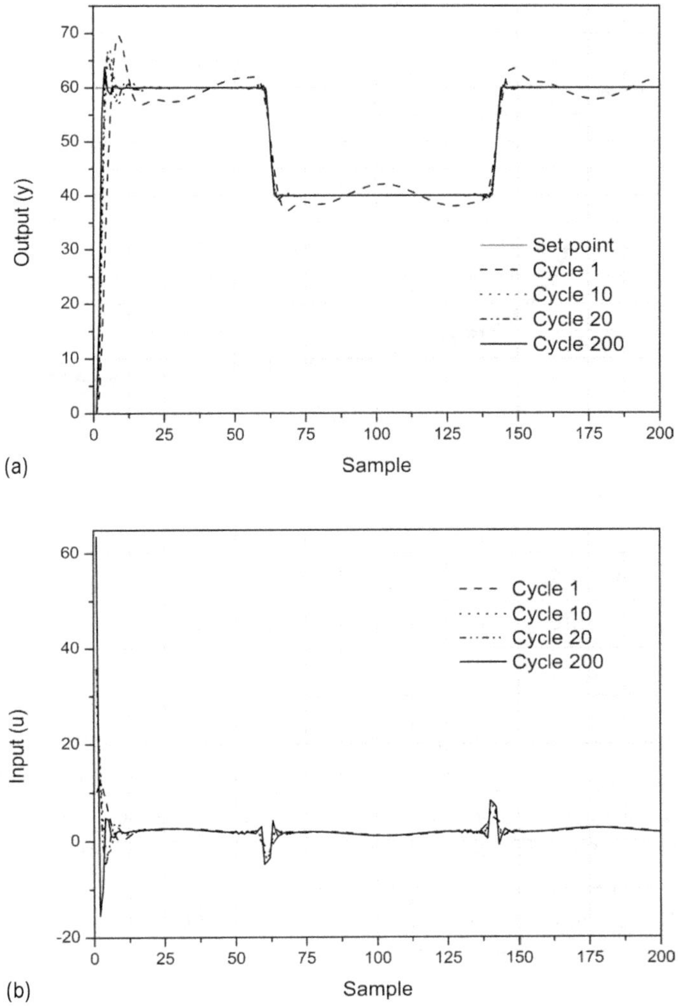


**Figure 4.13** Result of the simulation of the proposed 2D-DMC with repeatable disturbance: (a) controlled variable, and (b) control input

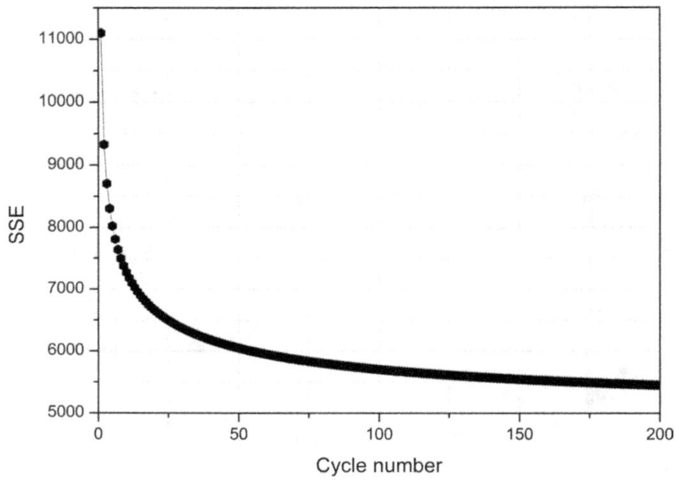


**Figure 4.14** Tracking performance with repeatable disturbance

#### 4.3.3.3 Case 3: Nonrepetitive Disturbances

The disturbance in this case is assumed as

$$w(t,k)=5\times\sin(0.04\pi t)+0.5\times\eta(t,k) \tag{4.94}$$

Compared with Case 2, an additional term $0.5\times\eta(t,k)$ is added to $w(t,k)$. Here $\eta(t,k)$ is a pseudorandom variable evenly distributed between the interval of $[-1,1]$ with zero mean and represents the nonrepetitive disturbances. The control results are illustrated in Figs. 4.15 and 4.16. There exist small perturbations in the output response and around the set-point change that are due to the feedback actions of the proposed control scheme, indicating the good control performance of the proposed control scheme in the presence of nonrepetitive disturbances.

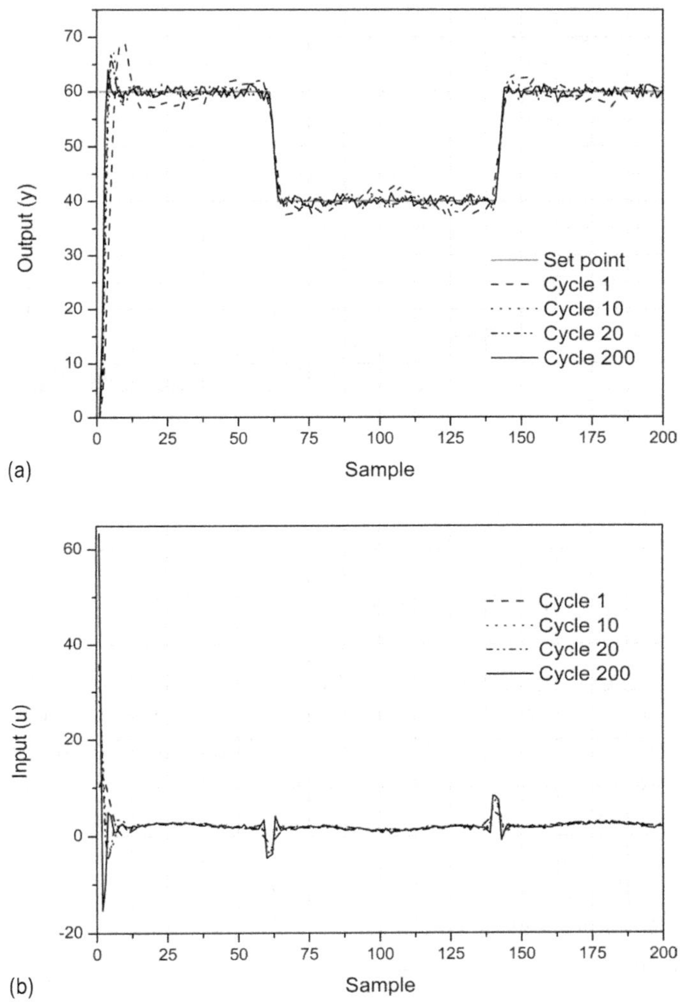


**Figure 4.15** Simulation results of the 2D-DMC with nonrepeatable disturbance: (a) controlled variable, and (b) control input

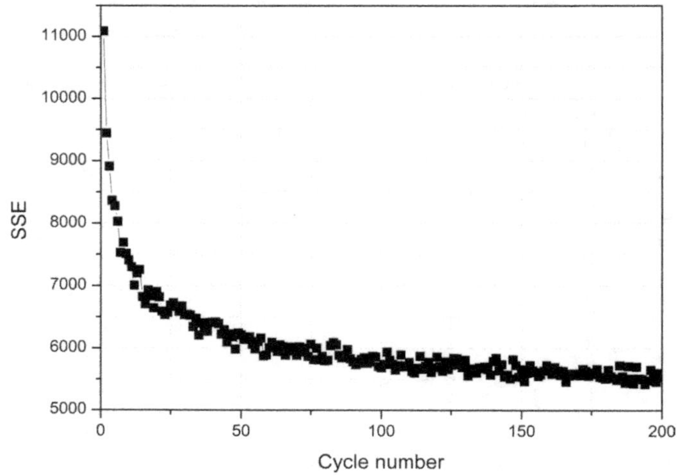


**Figure 4.16** Evolution of tracking performance over the batches with nonrepeatable disturbance

### 4.3.4 Experimental Test of 2D-DMC

The 2D-DMC design is applied experimentally to control the injection velocity on a 55-ton, industrial-grade, reciprocating-screw injection molding machine. The hydraulic system is controlled by Bosch-Rexroth proportional electrohydraulic relief and flow valves (Rexroth 0811402017 and Rexroth 0811403017, respectively). It should be noted that, in this application, there is no dedicated injection velocity sensor. A smart data-processing system is used for velocity measurement by differencing the displacement signal. A semicrystalline material, high-density polyethylene, is used for the experimental test.

The injection velocity (IV, $y(t,k)$) and the valve opening (VO, $u(t,k)$) of the control valves are the controlled variable and the manipulated variable, respectively. An open-loop test was conducted, with the valve opening step-down from 17.5% to 15.05% to estimate the velocity model for the controller design, as shown in Fig. 4.17:

A step-response model with a four-step delay can be obtained, and with this model, the model-plant mismatch during the corresponding operating range is illustrated in Fig. 4.18. Following a design procedure similar to that of the numerical simulation, the controller parameters in this application are chosen as $Q=I$, $R=I$, $\lambda=80000$, $p=15$, and $m=3$. The set point of $y(t,k)$ is chosen as a step change:

$$y_r(t,k)=\begin{cases}27.4, & t=1,2,\ldots,108\\ 13.44, & t=109,\ldots\end{cases} \tag{4.95}$$

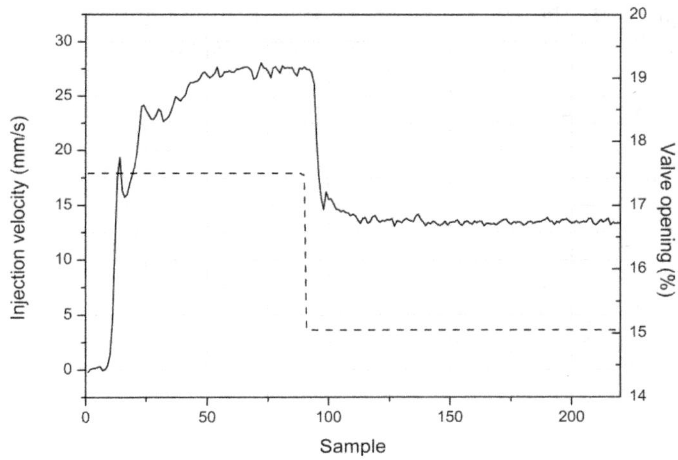


**Figure 4.17** Open-loop step response of the injection velocity

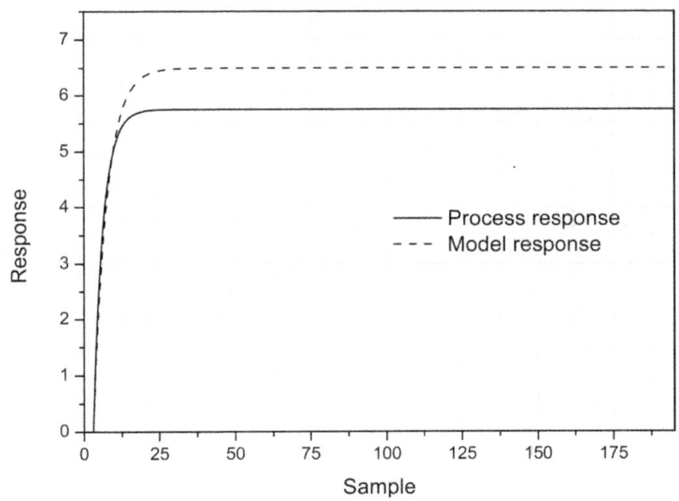


**Figure 4.18** Comparison of the process and model responses

Figure 4.19 presents the fast convergence rate of the proposed algorithm in this experiment. Convergence is achieved after several learning batches.

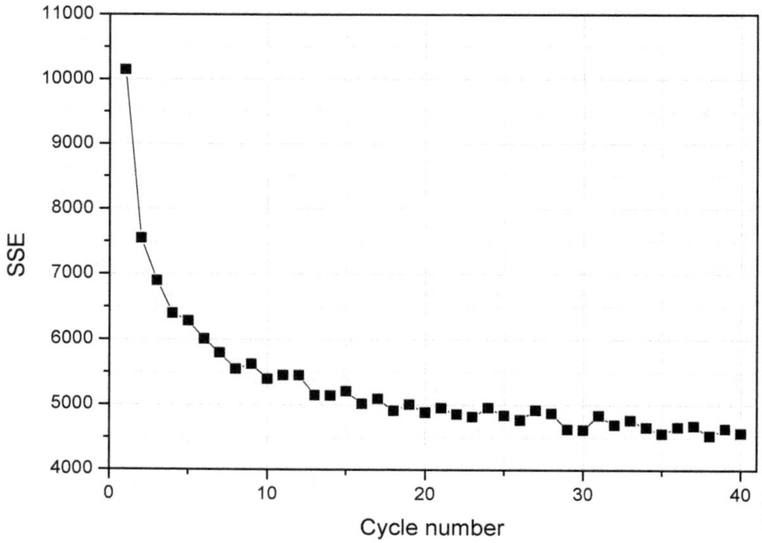


**Figure 4.19** Tracking performance in the application

The output tracking performance and the corresponding valve opening are depicted in Fig. 4.20. Figure 4.21 provides the output response after convergence. This result clearly shows the good set-point tracking performance of the 2D-DMC. It should be pointed out that the injection velocity control process terminates when the injection stroke reaches 40 mm in this study.

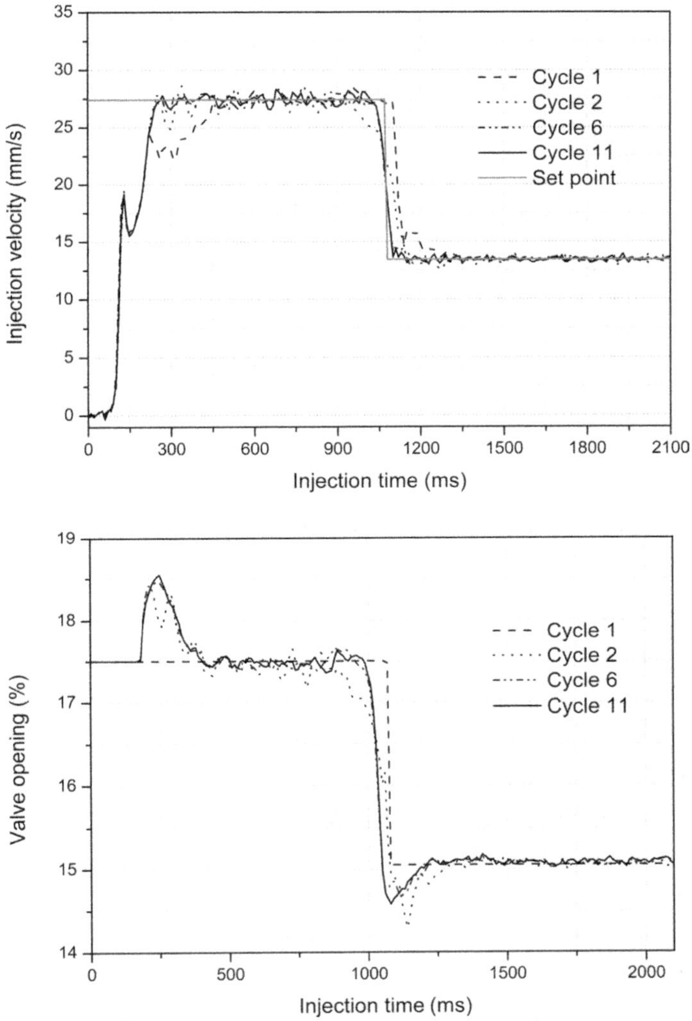


**Figure 4.20** Injection velocity experimental control results of the 2D-DMC

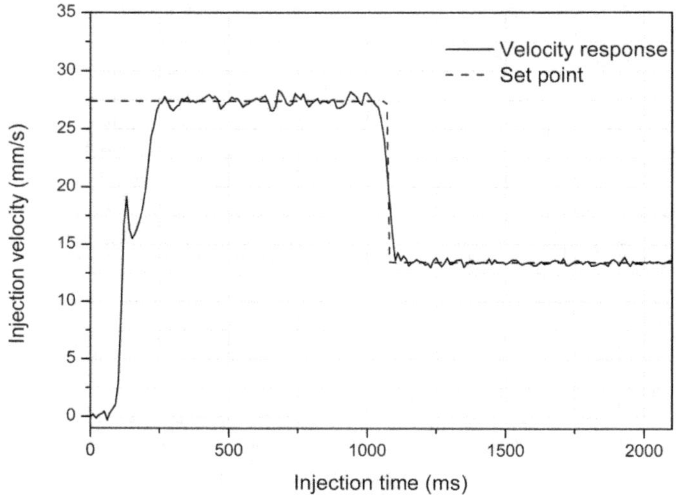


**Figure 4.21** Output responses of convergence in the application

## References

[28] E.F. Camacho and C. Bordons, *Model Predictive Control in the Process Industry*, Springer-Verlag, London, UK, 1995

[64] J. Shi, F. Gao, and T.J. Wu, *J. Process Control*, **15**, 907–924, 2005

[65] J. Shi, F. Gao, and T.J. Wu, *Ind. Eng. Chem. Res.*, **45**, 4603–4616, 2006

[66] J. Shi, F. Gao, and T.J. Wu, *Ind. Eng. Chem. Res.*, **45**, 4617–4628, 2006

[67] J. Shi, F. Gao, and T.J. Wu, *J. Process Control*, **17**, 715–727, 2007

[68] S. Mo, L. Wang, Y. Yao, and F. Gao, *J. Process Control*, **22**(5), 899–914, 2012

[69] C.R. Cutler and B.L. Ramaker, in *AICHE 86th National Meeting*, Houston, TX, 1979

[70] C.T. Chen, *Linear System Theory and Design* (3rd ed.), Oxford University Press, New York, 1999

[71] M. Morari and J.H. Lee, *Comput. Chem. Eng.*, **23**, 667–682, 1999

# 5 Statistical Process Monitoring (SPM) of Injection Molding: Basics

## 5.1 Process Monitoring

Process safety and product quality are two major focuses of attention in the injection molding industry as a typical and important manufacturing process. Process monitoring is a useful tool to guarantee process safety and to enhance product quality. In many references, process monitoring is also termed *fault detection and diagnosis* (FDD). In our views, the connotation of process monitoring is wider than fault detection and diagnosis. Besides real-time anomaly detection and troubleshooting, the main tasks of a general process-monitoring framework also include improvement of process understanding, assessment of end-product quality, and enhancement of production capability.

From the methodological point of view, methods for process monitoring can be divided into three categories: model-based methods, knowledge-based methods, and data-based methods.

Model-based process monitoring methods are popular in aerospace and mechanical systems, for which it is possible to obtain accurate first-principle models to describe a system's dynamic behaviors. However, for many large-scale modern manufacturing processes, characterization of first-principle models is usually difficult and costly, and sometime it is even impossible to build such models.

Knowledge-based process monitoring methods are based on available findings of process behaviors and experience of experts or field operators. Monitoring results are intuitive but somewhat subjective. Creation of process knowledge base is always a time-consuming and difficult task requiring a long-term accumulation of expert knowledge and experience.

Data-based process monitoring methods attract much interest in industrial processes, as they rely on only process measurements to detect process abnormalities, to diagnose possible causes, and to identify where improvements can be made with respect to process safety and product quality. The popularity of data-based process monitoring methods gets benefits from the wide utilization of distributed control systems (DCS) in process industries. Dynamic data collected by DCSs contain real-time information on the progression of disturbances, malfunctions, or faults in the plant, which help human operators assess and control plant performance, especially in the face of potential safety and quality problems.

Depending upon the type of measurements in DCSs, variables are usually divided into process variables and quality variables. Process variables are those online measurements that can track process dynamics. In injection molding processes, barrel and mold temperatures, injection and plastication pressures, screw rotation speed, valve openings, and so on, can be considered as process variables, since changes in their measurements or relationships are driven by physical or chemical laws. Quality variables are often estimated measurements related to product qualities, because for most quality variables, such as geometric dimensions, weight, and density of plastic products in injection molding processes, it is either too time-consuming or too costly to collect real-time quality measurements. Thus, online quality prediction is indispensable to process monitoring.

With the aid of DCSs, there are massive amounts of data accumulated in industrial processes. The first and most important step in data-based process monitoring is to model and interpret those data. Fortunately, significant progress has been made in the field of data mining and processing, which has provided many new modeling technologies for data-based process monitoring. Models are going from univariate to multivariate; focuses vary from linear and stationary to nonlinear and dynamic issues; applications cover from continuous to batch processes. Detailed discussions on various data-based process monitoring methods can be found in the books [72, 73] and the review papers [74–79].

## 5.2 Statistical Process Monitoring (SPM)

Statistical process monitoring (SPM) is dominant among those data-based process monitoring strategies, which applies statistical techniques to process analysis and process improvement. In fact, the use of statistical analysis originated in the area of statistical quality control (SQC) for detecting abnormal process conditions that may generate out-of-specification products.

The oldest SPM strategy dates back to 1931, when the Shewhart chart was developed to observe the magnitude and variance of a single product quality or key process variable at a given time. Exponentially weighted moving average (EWMA) and cumulative sum (CUSUM) control charts appeared subsequently in the 1950s for detecting small persistent process shifts. Since then, SQC charts had been widely recognized as a potentially powerful process monitoring tool [80].

The Shewhart chart and its modifications like EWMA and CUSUM are simple, easy to implement, and have been proven effective in monitoring an individual process variable or quality characteristic. However, those univariate methods are ineffective in dealing with processes with a lot of correlated process variables. After all, it is impractical to monitor hundreds of process variables individually. In addition, if the variables are correlated, as is often the case in real industrial plants, univariate control charts might miss alarming of a process abnormality.

The reason for this can be explained by an example in Fig. 5.1. In injection molding processes, melt temperature and viscosity are two correlated process variables, and a higher melt temperature will result in lower melt viscosity. Suppose $X_1$ and $X_2$ are melt temperature and viscosity, respectively, $\mu_1$ and $\mu_2$ are their mean values, and $UCL_1$, $LCL_1$, $UCL_2$, and $LCL_2$ are the upper and lower control limits in the corresponding individual Shewhart charts. In multivariate analysis, the normal operation region

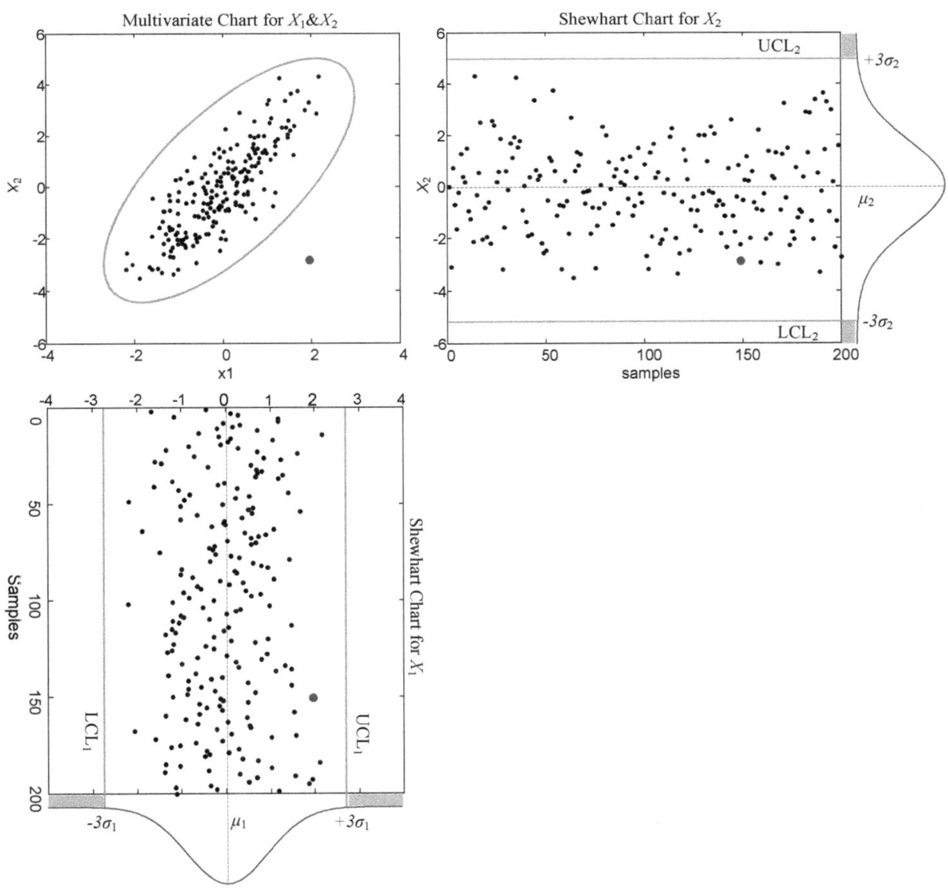


**Figure 5.1** Monitoring of two correlated variables using univariate and multivariate control charts

is defined by the joint distribution of $X_1$ and $X_2$, illustrated by the elliptical region in Fig. 5.1. Those small dots are normal samples, while the big dot is an abnormal sample caused by a sensor fault. From the figure, the big dot is “in-control” in the Shewhart charts for $X_1$ and $X_2$, but it can be recognized as an “out-of-control” sample in the multivariate control chart.

Process monitoring that relies on univariate control charts must carefully choose proper individual variables to monitor, and tighter control limits are required to reduce the number of missed alarms of process abnormalities. For instance, if tighter control limits $\underline{UCL}_1$, $\underline{LCL}_1$, $\underline{UCL}_2$, and $\underline{LCL}_2$ are set as in Fig. 5.2, the abnormal sample can be detected in the univariate control charts. But unfortunately, this will result in serious false alarms, because a lot of good samples are not acceptable any more.

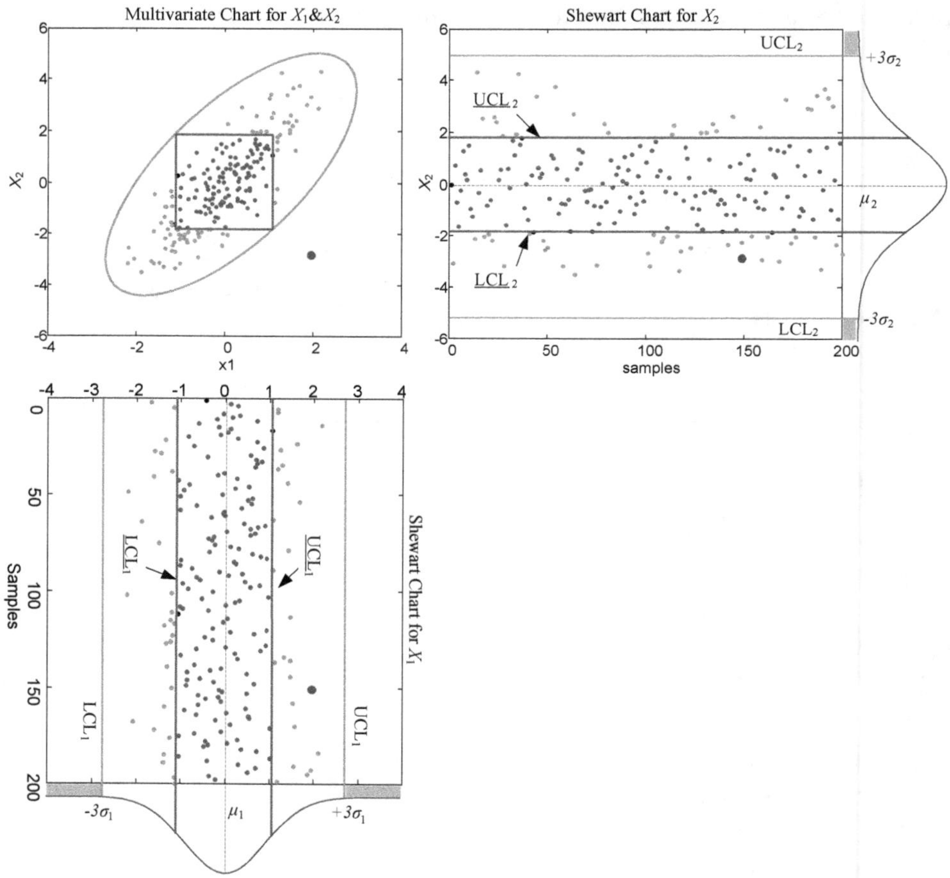


**Figure 5.2** Comparison of control windows in univariate and multivariate control charts

A high number of missed or false alarms are troublesome for most manufacturing processes that use inline univariate control charts for quality analysis and control. A missed alarm means the lack of awareness of a process abnormality, which may reduce process safety; while a false alarm may imply some undiscovered process knowledge that should be useful for process improvement.

In general, for correlated process variables, a multivariate control chart can define a more accurate "in-control" process window (the ellipse in Fig. 5.2) than the univariate control charts (the square in Fig. 5.2), so they can reduce significantly both missed and false alarms in process monitoring. Such an advantage comes from the fact that multivariate analysis methods can explore the correlation among multiple process variables, while univariate methods only monitor the magnitude and variation of single variables.

For example, Kazmer et al. [81] had compared the conventional control chart based monitoring methods and the multivariate analysis based process monitoring methods for an injection molding process. There were 12 different process faults imposed on the molding process, resulting in 24 defective products. The conventional-control-chart-based methods only detected 5 faults, while the multivariate methods detected 10 faults. In terms of prediction of defects, the conventional-control-chart-based methods detected 18 defects, while the multivariate methods detected 23 defects. Therefore, they made a conclusion that methods based on multivariate analysis outperformed the methods based on conventional control charts.

Nowadays, *statistical process monitoring* is often referred to as *multivariate statistical process monitoring*, which has been studied and applied extensively since the 1980s [82, 83]. Principal component analysis (PCA) and partial least squares (PLS) are the most widely used multivariate analysis techniques, which will be introduced in detail in Section 5.3. A lot of advanced methods on the basis of standard PCA and PLS were springing up subsequently, such as the nonlinear PCA/PLS, the dynamic PCA/PLS, the multiblock PCA/PLS, and various modified PCA/PLS methods for non-Gaussian processes, multimode processes, and so on. Kourti and MacGregor [84] published a tutorial paper on process analysis, monitoring, and diagnosis using multivariate projection methods in 1995. Kano et al. [85] had presented the comparison results of multivariate SPM methods with applications to the Eastman challenge problem. In 2003, the first review paper on SPM was published by S.J. Qin [86]. Recently, more achievements were reported in the area of SPM. Readers can refer to the review papers and books [77, 79, 87–90].

Finally, the major stages in a general framework for SPM strategies are summarized in Fig. 5.3 and explained in detail next.

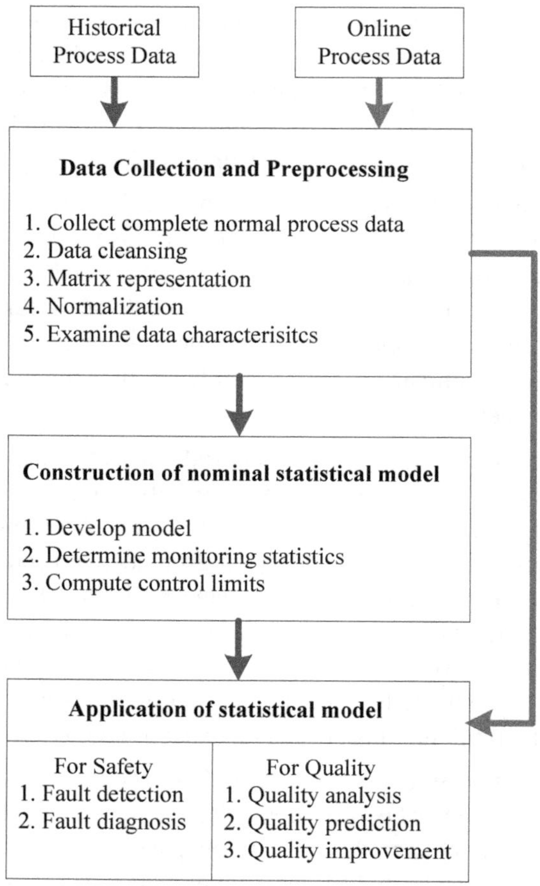


**Figure 5.3** A general framework for SPM strategies

### 5.2.1 Data Collection and Preprocessing

Data quality is critically important for the success of SPM strategies. The data used for modeling and monitoring should be able to correctly represent the real-world process. What are the basic requirements on data quality for process monitoring?

The modeling data should be complete in the statistical sense. This implies that the data should cover all in-control situations. All possible normal process variations should be taken into account in the data collection step. Only in this way are we able to distinguish between normal and abnormal process events accurately.

The data should be accurate, correct, and consistent. This requires performing data cleansing activities such as filtering noises, removing outliers, and filling in missing data.

It is desirable to deal with structured data in SPM strategies. Matrix representation is required, which can bring significant conveniences in implementing SPM strategies. A two-dimensional array is the most popular data structure in existing SPM models, where one dimension represents the data sample and the other represents the process variable. For an injection molding process, which is a typical batch process, process data are usually structured as a three-dimensional array, which should be unfolded into a two-dimensional array before multivariate statistical modeling.

Normalization is also an important issue in the data collection and preprocessing stage. In most multivariate analysis methods, different data samples are normalized to have zero mean and unity variance. This can eliminate the impact of different units or scales of process variables, so multivariate statistical models will not preoccupy those variables with large scales.

Besides, it is better to examine the data characteristics in this stage, in order to select appropriate statistical analysis methods accordingly. If the process to be monitored is stationary and the correlations among process variables are linear, basic multivariate statistical models like PCA and PLS can be directly used. Otherwise, advanced methods are required. Dynamic PCA, moving-window PCA, or independent component analysis (ICA) can be used for nonstationary processes; nonlinear PCA methods, which combine PCA with nonlinear modeling techniques like artificial neural networks (ANN), are preferable for nonlinear processes, to name a few.

### 5.2.2 Construction of Nominal Statistical Model

The core idea of process monitoring is to distinguish abnormal process events from normal ones. Developing accurate models to describe normal process behaviors is the most critical part. This raises an interesting question: how to define "normal behavior"?

In a statistical sense, a normal process is assumed to be subject to a certain distribution, which can reflect information on "common cause" variations in a process. "Common cause" means the variation is natural, random, and unavoidable. In contrast to "common cause" variation is "special cause" variation, which may be caused by process malfunctions, failure sensors, or unknown disturbances. A process is said to be normal or in statistical control when all variations are due to common causes. Otherwise, it is abnormal or out of control statistically.

The most common assumption is that, while a process is in control, process measurements appear to follow a normal distribution and are statistically independent. Under this normal distribution assumption, for univariate process monitoring methods like Shewhart, EWMA, or CUSUM charts, nominal statistical models will have some statistical features, such as mean, range, standard deviation, or autocorrelation structure based on which statistical control limits can be set up, and hypothesis tests can be proposed for online monitoring. For SPM strategies using methods based on multivariate projection, like PCA, a cross-correlation structure is an emphasis, and multivariate statistics such as Hotelling's $T^2$ statistic and the $Q$ statistic will be used to propose hypothesis tests for online monitoring.

Once the nominal statistical model is constructed and the statistics to be monitored are chosen, the next step is to determine the statistical control limits. Often, a confidence interval of 99% or 95% is used for process monitoring. Figure 5.4 shows three

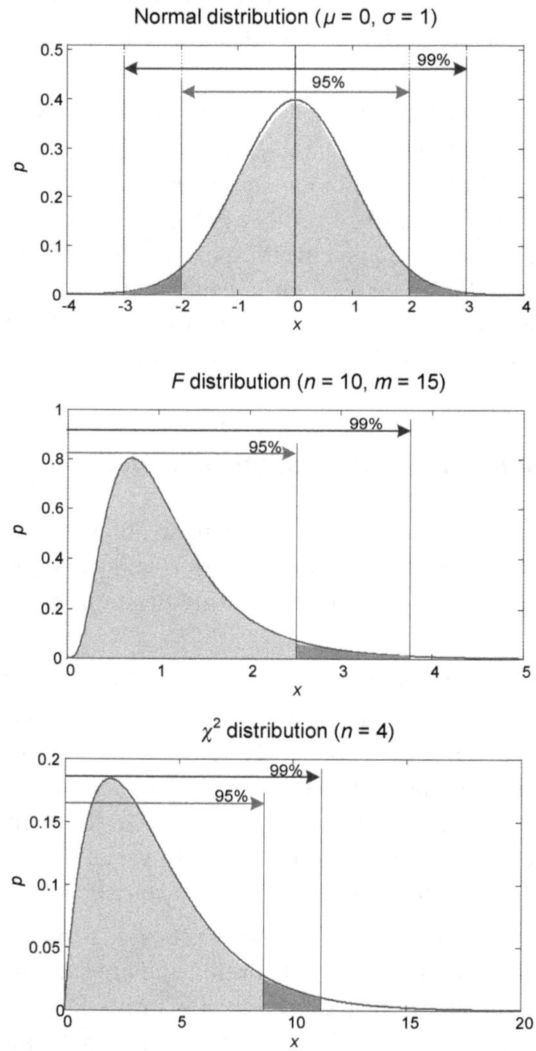


**Figure 5.4** Control limits of three statistical distributions

statistical distributions (normal distribution, $F$ distribution, and $\chi^2$ distribution) widely used in statistical process monitoring. Take a normally distributed statistic as an example, the 99% confidence interval corresponds to a $\pm 3\sigma$ (three sigma) control limit. The area bracketed by the $\pm 3\sigma$ control limits will on average contain 99.73% of all data points on the control chart, as long as the process is and remains in statistical control. The false alarm rate is only about 0.27%.

### 5.2.3 Application of Statistical Models

Applications of the developed statistical models are as varied as the different expected objectives of the process to be monitored. Roughly, there are two focuses in monitoring of industrial processes: fault detection and diagnosis, and quality prediction and analysis.

The main objective of fault detection and diagnosis is to ensure process safety. It focuses on the change of process behaviors. Only process measurements are required for statistical modeling. Based on the developed model and the corresponding statistics, we can judge whether or not an abnormal event happens in the process by comparing the monitoring statistics with their statistical control limits. After a fault has been detected by the control chart, fault diagnosis is triggered to find out the root cause of the detected fault. Typically, in data-based SPM strategies, fault diagnosis methods such as contribution plot [97] can isolate a group of affected variables due to the fault. But it should be noted that, accurate fault diagnosis results can hardly be achieved without process knowledge. Timely and accurate fault detection and diagnosis is effective to reduce (or even avoid) the occurrence of fatal events, which is the primary task of process monitoring.

The main objective of quality prediction and analysis is to achieve online quality inspection and ultimately to improve quality consistency. It focuses on analysis of product quality variations during manufacturing, in the hope that the causes of quality variation can be found out and eliminated for process improvement. Both online process measurements and offline quality measurements are required for statistical modeling. The model should be able to describe the relationship between process fluctuation and quality variation accurately. Since online quality measurements are hardly obtainable with affordable effort in most industrial processes, statistical inference models are often used for online quality prediction. This creates the foundation for online quality inspection and closed-loop quality control. For online monitoring, multivariate statistics in both process and quality subspaces are used to judge whether or not an abnormal event happens in the process that causes the exceeding of product quality from the control limits.

## 5.3 Multivariate Statistical Analysis Methods for SPM

Principal component analysis (PCA) and partial least squares (PLS) are the most popular multivariate statistical approaches for process monitoring, which will be extensively studied and applied to an injection molding process.

### 5.3.1 Principal Component Analysis and Partial Least Squares

Given a two-dimensional data matrix $X \in R^{n\times m}$, where $n$ is the number of samples and $m$ is the number of process variables, PCA decomposes $X$ as

$$X = TP^T = \sum_{j=1}^{m} \mathbf{t}_j \mathbf{p}_j^T = \mathbf{t}_1\mathbf{p}_1^T + \mathbf{t}_2\mathbf{p}_2^T + \cdots + \mathbf{t}_m\mathbf{p}_m^T \tag{5.1}$$

where $\mathbf{t}_j \in R^{n\times 1}$ is the latent variable vector, also called the principal component (PC) vector or score vector; $\mathbf{p}_j \in R^{m\times 1}$ is the loading vector that can project the original process data into the score space and contain the variables' correlation information; $T$ and $P$ are the score matrix and loading matrix, respectively. In the decomposition, scores are made to be orthogonal to each other, which means $\mathbf{t}_i^T\mathbf{t}_j = 0$ when $i \neq j$. At the same time, the loading vectors are orthogonal to each other, satisfying $\mathbf{p}_i^T\mathbf{p}_j = 0$ when $i \neq j$ and $\mathbf{p}_i^T\mathbf{p}_j = 1$ when $i = j$. Therefore,

$$\mathbf{t}_j = X\mathbf{p}_j \tag{5.2}$$

$$T = XP \tag{5.3}$$

Algebraically, $\|\mathbf{t}_j\|$ is equal to the $j$th largest eigenvalue $\lambda_j$ of the covariance matrix $\Sigma = \frac{1}{n-1}X^T X$, and $\mathbf{p}_j$ is the corresponding eigenvector. Note that $\|\mathbf{t}_1\| > \|\mathbf{t}_2\| > \cdots > \|\mathbf{t}_m\|$ because of $\lambda_1 > \lambda_2 > \cdots > \lambda_m$. It is easy to understand that the first several PCs contain the most variance information of $X$, and the last few PCs may contain only noises. This indicates that the majority of variance information can be extracted by retaining the first few orthogonal PCs, and the dimension of original variables can be largely reduced accordingly.

By retaining only the first $A$ PCs, $X$ can be approximated by

$$\hat{X} = \overline{T}\overline{P}^T = \sum_{j=1}^{A} \mathbf{t}_j \mathbf{p}_j^T \tag{5.4}$$

$$\overline{T} = X\overline{P} \tag{5.5}$$

$$\hat{X} = X\overline{P}\overline{P}^T \tag{5.6}$$

where $\bar{T} \in R^{n\times A}$ is the score matrix in PC subspace, $\bar{P} \in R^{m\times A}$ is the corresponding loading matrix, and $\hat{X} \in R^{n\times m}$ is the reconstruction of the original data matrix $X$. Equation 5.4 interprets the PCA model from the aspect of matrix decomposition, while Eq. 5.6 is from the aspect of data reconstruction. The loading matrix $\bar{P}$ is the core, which can be considered as the parameter matrix of a PCA model.

Then,

$$X = \hat{X} + E \tag{5.7}$$

$$E = \tilde{T}\tilde{P}^T = \sum_{j=A+1}^{m} \mathbf{t}_j \mathbf{p}_j^T \tag{5.8}$$

where $E \in R^{n\times m}$ is the residual matrix, $\tilde{T} \in R^{n\times(m-A)}$ is the score matrix in residual subspace, and $\tilde{P} \in R^{m\times(m-A)}$ is the corresponding loading matrix. From Eqs. 5.4, 5.5, and 5.8, it is clear that the PC subspace spanned by $\bar{P}$ and the residual subspace spanned by $\tilde{P}$ are orthogonal to each other.

Several algorithms, such as singular value decomposition (SVD) and nonlinear iterative partial least squares (NIPALS), can be used to calculate the loading matrix [91]. There are many methods to determine the number of retained PCs, among which the cross-validation method [92] is the most popular.

Different from PCA, PLS works on two data matrices, the process data matrix $X \in R^{n\times m_x}$ and the product quality data matrix $Y \in R^{n\times m_y}$, where $n$ is the number of samples, $m_x$ is the number of process variables, and $m_y$ is the number of quality variables. PLS extracts the variation of $X$ and gives as much prediction to $Y$ as possible at the same time.

The PLS model can be formulated as

$$X = \bar{T}\bar{P}^T + E = \sum_{j=1}^{A} \mathbf{t}_j \mathbf{p}_j^T + E \tag{5.9}$$

$$Y = \bar{U}\bar{Q}^T + F = \sum_{j=1}^{A} \mathbf{u}_j \mathbf{q}_j^T + F \tag{5.10}$$

$$\mathbf{u}_j = b_j \mathbf{t}_j + \varepsilon \tag{5.11}$$

Equations 5.9 and 5.10 describe the inner projection structures in $X$ and $Y$, where $\bar{T}$ and $\bar{P}$ are the score matrix and loading matrix decomposed from $X$; $\bar{U}$ and $\bar{Q}$ are the score matrix and loading matrix decomposed from $Y$; and $E$ and $F$ are the residual matrices. Equation 5.11 describes the outer projection structure between $X$ and $Y$, where $b_j = \mathbf{t}_j^T \mathbf{u}_j / (\mathbf{t}_j^T \mathbf{t}_j)$ is the regression coefficient between the latent variable $\mathbf{t}_j$ derived from $X$ and the latent variable $\mathbf{u}_j$ derived from $Y$.

The PLS model given by Eqs. 5.9–5.11 can be written in a compact way as

$$Y = X\Theta^T + F^* \tag{5.12}$$

where $\Theta$ is the regression parameter matrix and $F^*$ is the residual matrix.

The detailed properties, the calculation methods of PLS (e.g., NIPALS), and the methods to choose the number of latent variables (cross-validation, jackknife, and so on) can be found in the references [93–95]. In addition, Dayal and MacGregor [96] have developed another algorithm called Kernel PLS to compute the model parameters.

### 5.3.2 PCA/PLS-Based Statistical Process Monitoring

After PCA modeling, the loading matrix $\overline{P}$ is obtained based on the historical process data. When online process monitoring is conducted, a new sample $\mathbf{x}_{new} = [x_{new,1}, \cdots, x_{new,m}]$ is measured and projected onto the score subspace using the loading matrix $\overline{P}$ to get its score vector $\mathbf{t}_{new}$ and the residual vector $\mathbf{e}_{new}$ by

$$\mathbf{t}_{new} = \mathbf{x}_{new}\overline{P} \tag{5.13}$$

$$\hat{\mathbf{x}}_{new} = \mathbf{t}_{new}\overline{P}^T = \mathbf{x}_{new}\overline{P}\overline{P}^T \tag{5.14}$$

$$\mathbf{e}_{new} = \mathbf{x}_{new} - \hat{\mathbf{x}}_{new} = \mathbf{x}_{new} \cdot (\mathbf{I} - \overline{P}\overline{P}^T) \tag{5.15}$$

where $\mathbf{I}$ is an $m \times m$ identity matrix.

For process monitoring, two multivariate statistics are calculated: the Hotelling $T^2$ statistic and the SPE (or $Q$) statistic.

The Hotelling $T^2$ summarizes the systematic variation information extracted in score variables:

$$T^2 = \mathbf{t}_{new} S^{-1} \mathbf{t}_{new}^T = \sum_{i=1}^{A} \frac{t_{new,i}^2}{\lambda_i} \tag{5.16}$$

where $\mathbf{t}_{new} = [t_{new,1}, \cdots, t_{new,A}]$, $t_{new,a}$ ($a$ = 1,...,$A$) is the $a$th score variable, $S = diag(\lambda_1, \cdots, \lambda_A)$, $\lambda_a$ is the $a$th largest eigenvalue of the covariance matrix $\Sigma = \frac{1}{n-1} X^T X$, and $X$ is the normal history data used in PCA modeling.

Taking the assumption that scores of normal process data obey a multivariate normal distribution, the control limits of $T^2$ can be calculated using $F$ distribution:

$$T^2 \sim \frac{A(n-1)}{n-A} F_{A,n-A,\alpha} \tag{5.17}$$

where $\alpha$ is the significance level, and $F_{A,n-A,\alpha}$ is the critical value of $F$ distribution with significance level of $\alpha$ and degrees of freedom of $A$ and $n$–$A$, whose value can be found in a statistical table.

The SPE statistic measures the projection of a sample vector onto the residual subspace:

$$\text{SPE} = \mathbf{e}_{new}\mathbf{e}_{new}^T = \sum_{j=1}^{m}(x_{new,j} - \hat{x}_{new,j})^2 \tag{5.18}$$

where $x_{new,j}$ is the $j$th element in $\mathbf{x}_{new}$, and $\hat{x}_{new,j}$ is the corresponding reconstructed value using the developed PCA model. The SPE statistic measures variability that breaks the normal process correlation, which often indicates an abnormal situation.

With the assumption that residuals are normally distributed when the process is fault free, the control limits of SPE can be derived as [91]:

$$\text{SPE}_\alpha = \theta_1\left(\frac{C_\alpha\sqrt{2\theta_2 h_0^2}}{\theta_1} + 1 + \frac{\theta_2 h_0(h_0 - 1)}{\theta_1^2}\right)^{\frac{1}{h_0}} \tag{5.19}$$

where $\theta_i = \sum_{j=A+1}^{m} \lambda_j^i \quad (i = 1,2,3)$, $h_0 = 1 - \frac{2\theta_1\theta_3}{3\theta_2^2}$, and $C_\alpha$ is the critical value of the normal distribution under the significance level of $\alpha$.

The Hotelling $T^2$ statistic measures the distance between $\mathbf{t}_{new}$ and the origin of historical normal scores in $\bar{T}$. If a sample exceeds the $T^2$ limit, it does not break the correlation structure but shifts away from the normal PC subspace. In process monitoring, the values of these two statistics are plotted in the Hotelling $T^2$ and SPE control charts and compared with the corresponding control limits. If any of them exceeds the control limits, the process is considered to be out of control, and a fault is alarmed.

Like PCA, the above $T^2$ and SPE statistics are also commonly applied for PLS-based process monitoring. But, more often, PLS is used for online quality prediction, which will be introduced in Chapter 7.

After a fault has been detected by the $T^2$ or SPE control chart, it is desirable to find out the cause of the fault. A contribution plot [97] is widely applied, which shows the contribution of each process variable to the detected fault using a bar chart. Variables with large contributions need to be investigated since they are affected by the fault significantly. It should be noted that a contribution plot cannot diagnose the fault cause definitely. Process knowledge is always required to assist the contribution plot for fault identification.

The contribution of $x_{new,j}$ to the $a$th PC score $t_{new,a}$ can be derived as [98],

$$C_{t_{new,a},x_{new,j}} = x_{new,j}p_{a,j} \Big/ t_{new,a} \tag{5.20}$$

where $p_{a,j}$ is the $a$th element in the $j$th column in the loading matrix $P$. The confidence limits for contribution plots can be used to identify a set of faulty variables [99].

For SPE, the contribution of $x_{new,j}$ is given as

$$C_{SPE,x_{new,j}} = sign(x_{new,j} - \hat{x}_{new,j}) \cdot \frac{(x_{new,j} - \hat{x}_{new,j})^2}{\text{SPE}} \tag{5.21}$$

### 5.3.3 Multiway PCA/PLS

PCA and PLS can only deal with two-dimensional data matrices. For an injection molding process that is a typical batch process, process data are usually represented by a three-dimensional data matrix $\underline{X} \in R^{I \times K \times J}$, where $I$ is the number of normal batches, $J$ is the number of process variables, and $K$ is the number of total sampling intervals in a batch. To apply multivariate statistical methods like PCA or PLS for process monitoring, a three-dimensional data matrix should be transformed into a two-dimensional data matrix. Figure 5.5 shows two meaningful ways of matrix unfolding.

Figure 5.5(a) is called batch-wise unfolding. It keeps the dimension of batches, and merges variable and time dimensions, where $\tilde{X}^k$ is the time-slice data matrix of the $k$th sampling time. Each row of the unfolded two-dimensional matrix $X \in R^{I \times JK}$ contains all data within a batch. Different from batch-wise unfolding, variable-wise unfolding, as shown in Fig. 5.5(b), keeps the dimension of variables and merges the other two dimensions, generating a two-dimensional matrix $X \in R^{KI \times J}$. Each sampling point of each batch is considered as an object. Batch-wise unfolding is more popular than variable-wise unfolding because batch variation is the main concern in batch process monitoring.

Based on batch-wise unfolding, multiway PCA (MPCA) and multiway PLS (MPLS) were proposed [100, 102]. The basic idea of MPCA is to perform PCA on the unfolded data matrix $X \in R^{I \times JK}$, supposing $X$ has been normalized, while MPLS does a regression between the unfolded process data matrix $X \in R^{I \times J_x K}$ and quality data matrix $Y \in R^{I \times J_y}$, where $J_x$ is the number of process variables and $J_y$ is the number of quality variables.

MPCA is essentially similar to PCA. The only difference lies in that MPCA is able to deal with three-dimensional process data, which makes PCA available for batch process monitoring. The main principle of MPCA can be expressed as the following three steps, which are shown in Fig. 5.6:

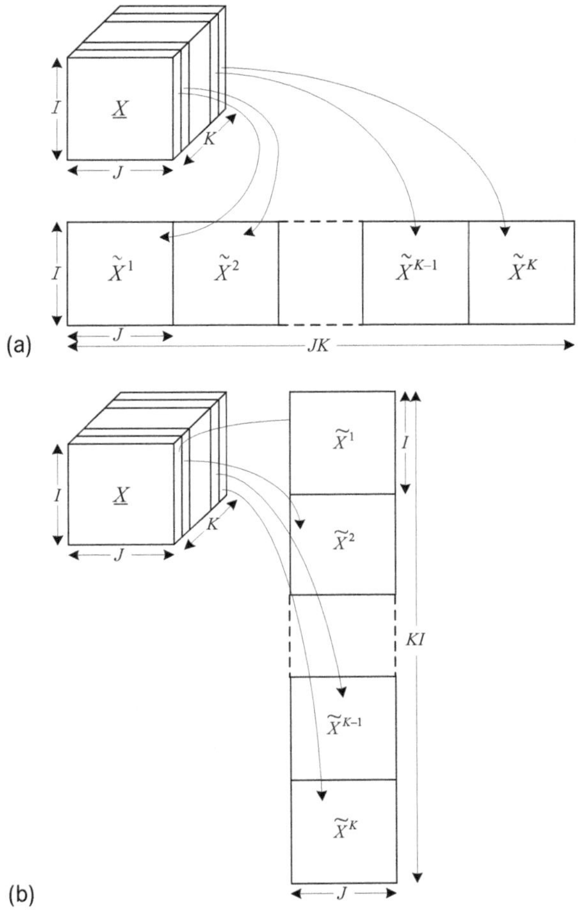


**Figure 5.5** Three-dimensional matrix unfolding:
(a) batch-wise unfolding; (b) variable-wise unfolding

1. Unfold the three-dimensional data matrix $\underline{X}$ into two-dimensional data matrix $X$ using batch-wise unfolding.
2. Calculate the score vector $\mathbf{t}_a \in R^{I\times 1}$ and corresponding loading vector $\mathbf{p}_a \in R^{JK\times 1}$ ($a$ = 1,..., $A$; $A$ is the number of retained PC).
3. Transfer the one-dimensional loading vector $\mathbf{p}_a$ into a two-dimensional matrix, $P_a \in R^{J\times K}$.

The MPCA model is defined as

$$\underline{X} = T_A \otimes P_A + \underline{E} = \sum_{a=1}^{A} \mathbf{t}_a \otimes P_a + \underline{E} = \sum_{a=1}^{A} \mathbf{t}_a \mathbf{p}_a^T + E \tag{5.22}$$

$$\hat{\underline{X}} = T_A \otimes P_A$$

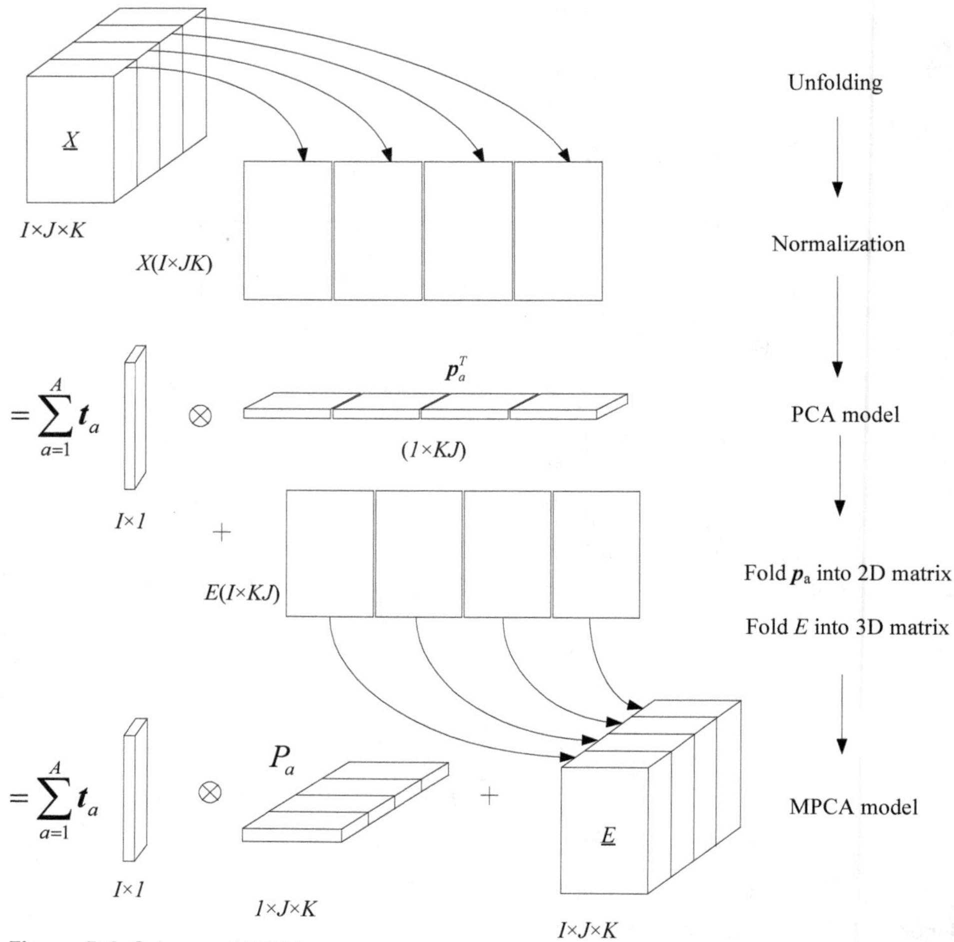


**Figure 5.6** Scheme of MPCA

where the score matrix $T_A$ has the dimension of $I \times A$, i.e., $T_A = [\mathbf{t}_1, \ldots, \mathbf{t}_a, \ldots \mathbf{t}_A]$ ($a = 1,\ldots,A$; $\mathbf{t}_a \in R^{I\times 1}$) or $T_A = [\mathbf{t}_1^T, \ldots, \mathbf{t}_i^T, \ldots \mathbf{t}_I^T]^T$ ($i = 1,\ldots,I$; $\mathbf{t}_i \in R^{1\times A}$); the loading matrix $P_A$ has the dimension of $A \times J \times K$; $\underline{E}$ is the three-dimensional residual matrix; and $\otimes$ denotes a augmented multiplication operation.

The MPCA model can include both cross-correlation between the process variables and inherent autocorrelation in each process variable. The vector $\mathbf{t}_i \in R^{1\times A}$ ($i = 1,\ldots,I$) summarizes the variation information in normal batch operations, which can be considered to obey a normal distribution. This normal distribution assumption is the basis for applying MPCA to statistical analysis and online monitoring of batch processes.

When the quality data are available at the same time, MPLS is used to build a regression model between the three-dimensional process data $\underline{X} \in R^{I\times J_x \times K}$ and the two-dimensional end-product quality data $Y \in R^{I\times J_y}$ in a batch process, by unfolding

$\underline{X}$ into $X$ just as MPCA does, and then constructing the regression relationship between $X$ and $Y$. The same details as MPCA will not be repeated here anymore.

### 5.3.4 Multiway PCA/PLS-Based Batch Process Monitoring

As mentioned previously, MPCA model can reflect cross-correlation among process variables as well as inherent autocorrelation in each variable. When an unusual event happens and leads to a change of process variable trajectory or correlation, it can be detected by monitoring the multivariate statistics $T^2$ and SPE. Due to its different model structure, for MPCA model, $T^2$ and SPE can be divided into two types: offline and online. Different methods of computing the statistical control limits will be adopted.

Because $\mathbf{t}_i$ ($i = 1,...,I$) can be regarded as obeying a multidimensional normal distribution, the $T^2$ statistic of the $i$th normal batch in the training dataset can be formulated as

$$T^2{}_i = \frac{I}{(I-1)^2}\mathbf{t}_i S^{-1}\mathbf{t}_i^T \sim \mathrm{B}_{A/2,(I-A-1)/2} \quad (5.23)$$

where $S \in R^{A\times A}$ is the covariance matrix of $T_A \in R^{I\times A}$, B means a binomial distribution with degrees $A/2$ and $(I- A- 1)/2$. Given the significance level $\alpha$ (0.05 or 0.01), the control limits of $\mathrm{B}_{A/2,\,(I-A-1)/2}$ can be induced according to $F$ distribution:

$$\mathrm{B}_{A/2,(I-A-1)/2,\alpha} = \frac{A/(I-A-1)F_{A,I-A-1,\alpha}}{1+A/(I-A-1)F_{A,I-A-1,\alpha}} \quad (5.24)$$

Another statistic to be monitored, SPE, can be defined as

$$\mathrm{SPE}_i = \sum_{j=1}^{KJ} E(i,j)^2 \quad (5.25)$$

The control limits for the SPE statistic can be calculated by Eq. 5.19.

In Eq. 5.22, the loading matrix of the MPCA model is $P_A \in R^{A\times J\times K}$, or $\mathbf{p}_a{}^T \in R^{1\times JK}$, $a = 1,...,A$. When using the obtained MPCA model for online monitoring of a new batch of data, we need to collect all process data in one batch operation to form a data matrix $X_{new} \in R^{J\times K}$, whose unfolding form is $\mathbf{x}_{new} \in R^{1\times JK}$. Then, the score vector $\mathbf{t}_{new} \in R^{1\times A}$, the reconstructed original data $\hat{\mathbf{x}}_{new} \in R^{1\times JK}$, and the residual vector $\mathbf{e}_{new} \in R^{1\times JK}$ can be obtained by

$$\begin{aligned} \mathbf{t}_{new} &= \mathbf{x}_{new} \otimes P_A = \mathbf{x}_{new}[\mathbf{p}_1,\cdots,\mathbf{p}_A] \\ \hat{\mathbf{x}}_{new} &= \mathbf{t}_{new} \otimes P_A{}^T = \mathbf{t}_{new}[\mathbf{p}_1,\cdots,\mathbf{p}_A]^T \\ \mathbf{e}_{new} &= \mathbf{x}_{new} - \hat{\mathbf{x}}_{new} \end{aligned} \quad (5.26)$$

According to the nature of batch processing, $X_{new}$ is only available at the end of each batch operation. This raises a question of how to implement online process monitoring in middle of a batch operation. Nomikos and MacGregor [102] have presented several solutions. For an instance, at the sampling time $k$ in a batch operation, $X_{new}$ or $\mathbf{x}_{new}$ can be obtained by estimating the future measurements from $k$ to the batch end, and then Eq. 5.26 can be changed as

$$\begin{aligned}
\mathbf{t}_{new,k} &= \mathbf{x}_{new,k}[\mathbf{p}_1,\cdots,\mathbf{p}_A] \\
\mathbf{x}_{new,k} &= [x_{new,1,1},\cdots,x_{new,J,1},\cdots,x_{new,1,k},\cdots,x_{new,J,k}, \\
&\qquad \hat{x}_{new,1,k+1},\cdots,\hat{x}_{new,J,k+1},\cdots,\hat{x}_{new,1,K},\cdots,\hat{x}_{new,J,K}] \\
\hat{\mathbf{x}}_{new,k} &= \mathbf{t}_{new,k}[\mathbf{p}_1,\cdots,\mathbf{p}_A]^T \\
\mathbf{e}_{new,k} &= \mathbf{x}_{new,k} - \hat{\mathbf{x}}_{new,k}
\end{aligned} \tag{5.27}$$

where $\hat{x}_{new,j,k_f}$ ($j$ = 1,...,$J$; $k_f$ = $k$+1, ..., $K$) is the prediction of future measurement $x_{new,j,k_f}$.

Accordingly, the online statistics to be monitored can be written as

$$T^2{}_{new,k} = \frac{I(I-A)}{A(I^2-1)}\mathbf{t}_{new,k}S^{-1}\mathbf{t}^T_{new,k} \sim F_{A,I-A+1,\alpha} \tag{5.28}$$

$$\text{SPE}_{new,k} = \mathbf{e}_{new,k}\mathbf{e}_{new,k}{}^T \tag{5.29}$$

For details on the prediction methods for $\hat{x}_{new,j,k_f}$ and the method for computing the control limits of $T^2{}_{new,k}$ and $\text{SPE}_{new,k}$, the reader is referred to [100, 102].

## 5.4 Challenges in Monitoring Injection Molding Process

### 5.4.1 Multiple Operation Phases

As introduced in Chapter 1, an injection molding process contains a series of operation phases, such as filling, packing, plastication, and cooling. Obviously, process data will show different characteristics during different phases. Figure 5.7 shows the trajectories of cavity temperature, injection velocity, stroke, nozzle pressure, and screw rotation speed (RPM) of a normal batch run. The magnitudes, variances, or ranges of the five variables are quite different along the time direction. The correlations among them are also varying with time. It will be very useful to extract the phase information and model the phase characteristics for process monitoring.

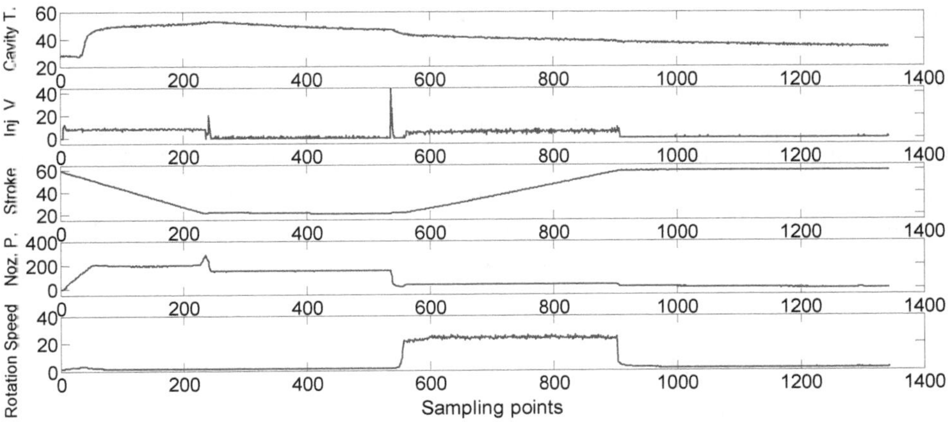


**Figure 5.7** Variable trajectories in an injection molding process

However, MPCA/MPLS cannot reveal such information because they merge everything in a batch run together and focus merely on batch-wise information.

Focusing on phase characteristics will be beneficial to fault diagnosis. Some faults happen only in some specific operation phases. For example, a check-ring fault happens in the filling phase and can eventually affect batch-end product qualities. This fault alters slightly process behaviors in the filling phase, but does not cause dramatic global changes in process variable trajectories. Therefore, it is difficult to detect a check-ring fault using MPCA/MPLS-based monitoring methods because they cannot reveal process details.

In addition, an understanding of phase characteristics can contribute to quality improvement. As mentioned in previous chapters, different control strategies are applied in different phases. In the filling phase, the major objective is to obtain a satisfactorily controlled injection velocity; in the packing phase, packing pressure is the main concern. Although the relationship between the controlled variables in each phase and the end-product quality has been studied in the past, and some advanced process control strategies have been applied in recent years, quality variation is still a troublesome issue in injection molding industry. This indicates that some unknown factors affecting the end-product quality are as yet undiscovered. If some operation phases and process variables are identified as the main causes of batch-wise quality variation, tight control of these phases and variables can definitely improve quality consistency.

### 5.4.2 Within-Batch and Batch-to-Batch Dynamics

An injection molding process can be regarded to have two time scales: a within-batch time scale and a batch-to-batch time scale. For the within-batch scale, it is obvious that an injection molding process is a nonstationary process and shows clear dynamic process behaviors. As observed from Fig. 5.7, there is no steady state as in continuous processes. Thus, it is easy to understand the within-batch dynamics in an injection molding process. Besides the within-batch dynamics, there may exist

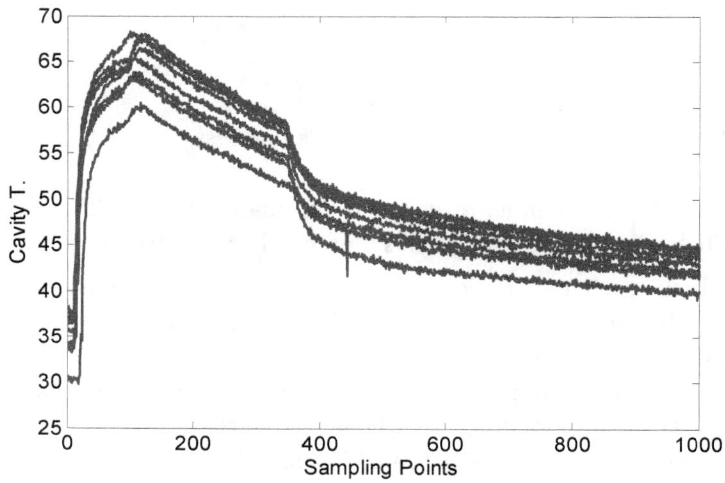


(a) Within-batch scale

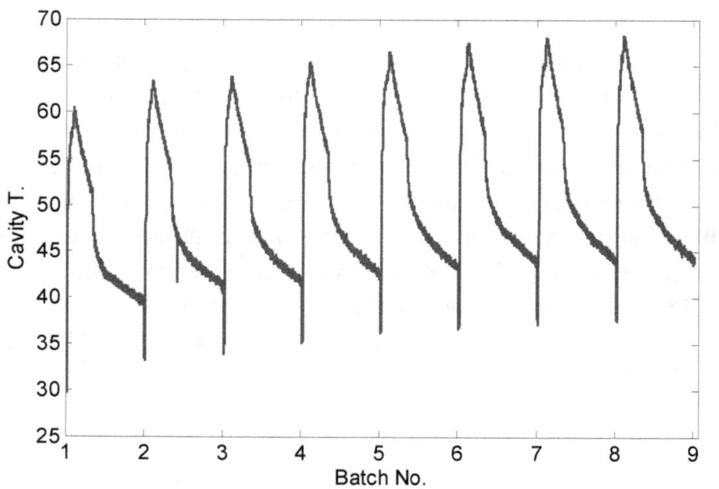


(b) Batch-to-batch scale

**Figure 5.8** Cavity temperature curves in a process startup batch-to-batch dynamics at a larger time scale. For example, in process startups or process changeovers, batch-to-batch variations can be observed. Figure 5.8 gives the cavity temperature curves in a process startup, which show clear batch-wise dynamics as well as within-batch dynamics. When the injection molding process has both within-batch and batch-to-batch variations, traditional PCA/PLS, MPCA/MPLS, and their extensions to dynamic processes are not suitable anymore. New statistical modeling and monitoring techniques are desired for such a process with complex multiscale dynamics.

### 5.4.3 Unequal Batch Length

Automatic V/P switch-over is a mature technique, which has been widely adopted in industrial injection molding processes. According to the process mechanism, the filling phase stops when the mold is completely filled. At this time, the control strategy is switched from velocity control in the filling phase to pressure control in the packing phase. Automatic V/P switch-over improves process automation, but it leads to unequal batch length issues. The filling time is not fixed, but rather depends on the injection velocity and other factors. Obviously, a lower injection velocity requires a longer filling time, resulting in a longer batch with more process data than a batch operation with a higher injection velocity, as illustrated in Fig. 5.9. This creates difficulty for MPCA/MPLS-based SPM methods because they require the modeling data having equal length to be arranged in a three-way matrix.

There are several methods to deal with the unequal batch length problem, such as cutting the longer batches or filling the shorter batches. All of these methods assume

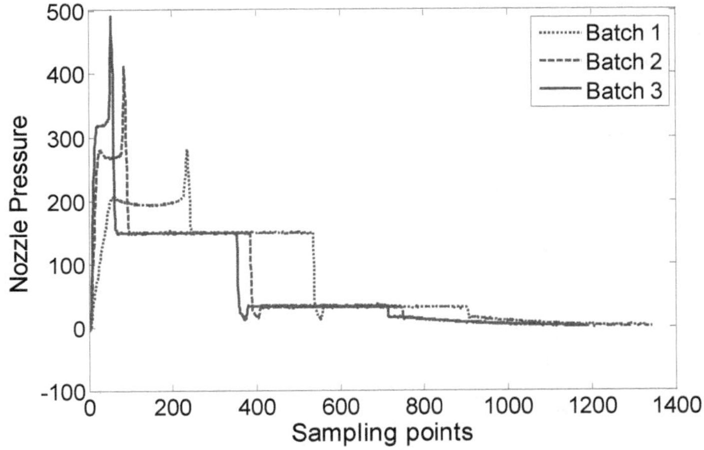


**Figure 5.9** Illustration of unequal batch length in injection molding process

that the trajectories overlap in a common time span. But in the injection molding process, and many other multiphase batch processes, unequal-length phenomena may occur in one or more specific operation phases, resulting in the total batch duration and the durations of operation phases both being unequal. The existing methods cannot handle this problem effectively.

## References


[72] X.Z. Wang and C. McGreavy, *Data Mining and Knowledge Discovery for Process Monitoring and Control*, Springer, London, 1999

[73] E.L. Russell, L.H. Chiang, and R.D. Braatz, *Data-driven Methods for Fault Detection and Diagnosis in Chemical Processes (Advances in Industrial Control)*, Springer, London, 2000

[74] V. Venkatasubramanian, R. Rengaswamy, K. Yin, and S.N. Kavuri, *Comput. Chem. Eng.*, **27**, 293–311, 2003(a)

[75] V. Venkatasubramanian, R. Raghunathan, and S.N. Kavuri, *Comput. Chem. Eng.*, **27**, 313–326, 2003(b)

[76] V. Venkatasubramanian, R. Raghunathan, S.N. Kavuri, and K. Yin, *Comput. Chem. Eng.*, **27**, 327–346, 2003(c)

[77] A. Das, J. Maiti, and R.N. Banerjee, *Int. J. Qual. Reliab. Manage.*, **29**, 720–752, 2012

[78] Z.Q. Ge, Z.H. Song, and F.R. Gao, *Ind. Eng. Chem. Res.*, **52**, 3543–3562, 2013

[79] S. Yin, S.X. Ding, and X.C. Xie, *IEEE Trans. Ind. Electron.*, **61**, 6418–6428, 2014

[80] D.C. Montgomery, *Introduction to Statistical Quality Control* (6th ed.), John Wiley & Sons, 2009

[81] D.O. Kazmer, S. Westerdale, and D. Hazen, *Int. Polym. Process.*, **23**, 447–458, 2008

[82] B.M. Wise, D.J. Veltkamp, B. Davis, N.L. Ricker, and B.R. Kowalski, *Waste Management '88 Proc.*, 811–818, 1988

[83] J.F. MacGregor, *AIChE Meet.*, San Francisco, CA, 1989

[84] T. Kourti and J.F. MacGregor, *Chemom. Intell. Lab. Syst.*, **28**, 3–21, 1995

[85] M. Kano, K. Nagao, S. Hasebe, I. Hashimoto, H. Ohno, R. Strauss, and B. Bakshi, *Comput. Chem. Eng.*, **26**, 161–174, 2002

[86] S.J. Qin, "Statistical process monitoring: basics and beyond," *J. Chemom.*, **17**, 480–502, 2003

[87] M. Kano and Y. Nakagawa, *Comput. Chem. Eng.*, **32**(1), 12–24, 2008

[88] Y. Yao and F.R. Gao, *Annu. Rev. Control*, **33**, 172–183, 2009

[89] Y.W. Zhang and Y. Zhang, *IETE Tech. Rev.*, **27**, 406–420, 2010

[90] U. Kruger and L. Xie, *Advances in Statistical Monitoring of Complex Multivariate Processes: With Applications in Industrial Process Control*, Wiley, Hoboken, 2012

[91] J. Jackson, *A User's Guide to Principal Components*, Wiley, Hoboken, 1991

[92] S. Wold, *Technometrics*, **20**(4), 397-405, 1978

[93] S. Wold, A. Ruhe, H. Wold, and W. Dunn III, *SIAM J. Sci. Stat. Comput.*, **5**, 735-743, 1984

[94] P. Geladi and B.R. Kowalshi, *Analytica Chimica Acta*, **185(**1), 1-17, 1986

[95] A. Hoskuldsson, *J. Chemom.*, **2**, 211-228, 1988

[96] B. Dayal and J.F. MacGregor, *J. Chemom.*, **11**, 73-85, 1997

[97] P. Miller, R.E. Swanson, and C.E. Heckler, *Appl. Math. Comput. Sci.*, **8**(4), 775-792, 1992

[98] J. Westerhuis, S. Gurden, and A. Smilde, *Chemom. Intell. Lab. Syst.*, **51**, 95-114, 2000

[99] A. Conlin, E. Martin, and A. Morris, *J. Chemom.*, **14**, 725-736, 2000

[100] P. Nomikos and J.F. MacGregor, *AIChE J.*, **40**, 1361-1375, 1994

[101] P. Nomikos and J.F. MacGregor, *Chemom. Intell. Lab. Syst.*, **30**, 97-108, 1995(a)

[102] P. Nomikos and J.F. MacGregor, *Technometrics*, 41-59, 1995(b)

# 6 Phase-Based SPM Strategies

## 6.1 Introduction

Multiplicity of the operation phase is an inherent nature of many batch processes; each phase has its own underlying characteristics, and process can exhibit significantly different behaviors over different operation phases. For example, an injection molding process can be roughly divided into three operation phases, injection phase, packing phase, and cooling phase; a fermentation process can be divided into lag phase, exponential growth phase, and stationary phase. Apparently, each operation phase has different process variable trajectories, running modes, and correlation structures. So, statistical modeling and online monitoring of multiphase batch processes should include not only the overall operation performance, but the specific phase characteristics.

For better monitoring performance, a batch process should be divided into blocks or segments according to the changes in process characteristics. Such a block or segment is called a "modeling phase," representing a process characteristic over a certain process duration, which may be different from the actual "operation phase." The effectiveness of a multiphase model is doubtful without a proper phase division. There are three major different ways to divide a batch process into several phases [103].

The first way is based on expert knowledge. A process can be divided into segments according to different processing units and distinguishable operation phases inside each unit [104, 105] The techniques of multiblock modeling [106, 107] and many other research works [108, 109] assume that the phase division points are known beforehand. A problem of this kind of phase division is that the process knowledge may not be available in many cases, or the known prior knowledge may not be sufficient to divide a batch processes into several phases reasonably.

The second way is based on process analysis. Gollmer and Posten [110] used the dynamic time warping (DTW) technique to compare the cycles with a prototype cycle whose phase division points are known. The comparison results give suggestions on the phase division of the new batches. Muthuswamy and Srinivasan [111] identified the division points using the process variable features described in the form of multivariate rules. Ündey and Cinar [112] used an indicator variable that contains significant landmarks to detect the completion of each phase. Facco et al. [113] also divided phases based on some easily detectable landmark events occurring in a batch in their phase PLS model building. Doan and Srinivasan [114] did phase division based on singular points in some known key variables, which are also a kind of landmark. These methods work well only when certain required process features are known. Besides the above methods, Kosanovich et al. [115] pointed out that the changes in the process variance information explained by the principal components (PCs) can indicate the process phase division point during a batch. However, in their work, the proposed idea only gives some rough indications of the phase division. They did not give an algorithm to identify phases precisely.

To overcome the shortcomings of the above methods, Lu et al. [116, 117] developed another way for phase division, according to the changes of process variables' correlation along time in each batch. In each phase, the variables' correlation structures are similar, although the process may vary in time. The division results of modeling phases may not be exactly the same as the operation phases. An alternative method developed by Camacho et al. [103, 118, 119] detects the phase division points based on the PCA model prediction abilities. The data-based automatic phase division methods show the advantage since they do not rely on prior process knowledge and can be performed on different batch processes. For online monitoring, the division of phases based on process correlation structures can provide more accurate PCA/PLS models because the process is nearly time invariant in each phase identified in this way.

Considering that an injection molding process can be characterized as a multiphase batch process, two phase-based process analysis and statistical modeling and monitoring strategies are introduced in this chapter. A general phase division algorithm and phase-based PCA modeling and monitoring strategy are presented in Section 6.2, followed by a case study on an injection molding process in Section 6.3. An improved phase-based modeling and monitoring strategy to solve unequal batch length issues will be introduced in Section 6.4. The application to an unequal-length injection molding process is presented in Section 6.5.

## 6.2 Phase-Division-Based Sub-PCA Modeling and Monitoring

### 6.2.1 Overview

Given a batch process with $J$ process variables and $K$ sampling points in each batch run, a data matrix $\overline{X} \in R^{J \times K}$ can be obtained for each batch. A set of $I$ number of normal batch runs, hence, results in a three-way data matrix, $\underline{X} \in R^{I \times J \times K}$, as illustrated in Fig. 6.1, where the horizontal data slice is $\overline{X}_i \in R^{J \times K}$ $(i = 1,..., I)$ and the vertical data slice is $\tilde{X}_k \in R^{I \times J}$ $(k = 1,..., K)$. Here $\tilde{X}_k$ is the basic unit in the following phase-based modeling methods, which will be decomposed by PCA to obtain the corresponding loading matrix at each sampling point $k$.

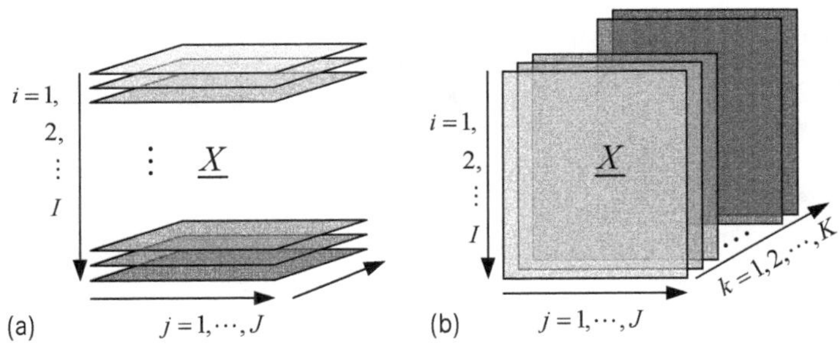


**Figure 6.1** Batch process data

According to the introduction of PCA in Section 5.3, the loading matrix represents some sort of process correlation. The phase-based PCA modeling begins with analysis of the loading matrix at each sampling interval. The time-slice matrices $\tilde{X}_k$ should have a similar loading matrix within each "operation" phase. Different "operation" phases result in different loading matrices, reflecting that the process correlation changes over different phases. Likewise, changes in the loading matrices, reflecting changes in the underlying process behavior, can be used to determine the "operation" phases. A *k-means* clustering [120] will be adopted and modified to divide the $K$ number of loading matrices. The clustering results, associated with process time span or indicator variables, can be used to define process "operation" phases. Sub-PCA models will be developed for different phases, based on which a phase-based online monitoring strategy is proposed. The scheme of the phase-based PCA modeling is illustrated in Fig. 6.2.

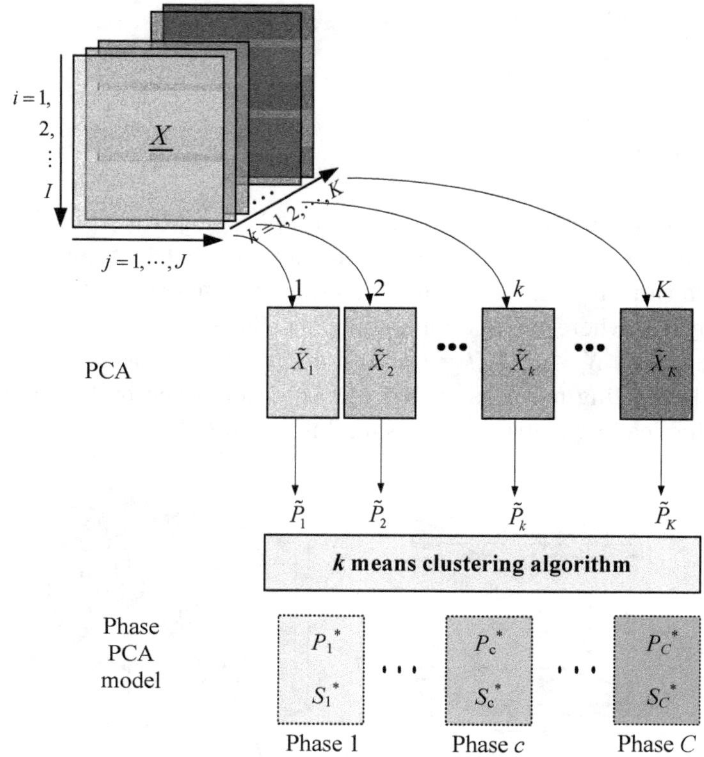


**Figure 6.2** Scheme of the phase-based sub-PCA modeling

### 6.2.2 Data Normalization

Data normalization is necessary and important for data-based modeling and monitoring. The most common normalization method is to center and scale the process variable's measurements to have zero mean and unity variance.

Mathematically, for a data matrix $X \in R^{n \times m}$, where $n$ is the number of samples and $m$ is the number of variables, zero-mean centering can be done by removing the mean value as

$$\tilde{x}_{i,j} = x_{i,j} - \overline{x}_j (i = 1, \cdots, n; j = 1, \cdots, m) \tag{6.1}$$

where $x_{ij}$ is the $i$th measurement of the $j$th variable, and $\overline{x}_j = \frac{1}{n}\sum_{i=1}^{n} x_{i,j}$. Centering does not alter the correlation of variables.

Unity-variance scaling transforms all of the variable's variances to be 1 as

$$\tilde{x}_{i,j} = x_{i,j} / s_j (i = 1, \cdots, n; j = 1, \cdots, m) \tag{6.2}$$

where the standard deviation of $x_j$ is calculated by $s_j = \sqrt{\frac{1}{n-1}\sum_{i=1}^{n}(x_{i,j} - \bar{x}_j)^2}$.

In a batch process, the data matrix is $\underline{X} \in R^{I \times J \times K}$, and measurements of the $j$th process variable in the $i$th batch run are denoted as $x_{i,j,k}$ ($k$ = 1,...,$K$), which forms a trajectory. The mean values of $x_{i,j,k}$ over batches can be regarded as the mean trajectory, computed by

$$\bar{x}_{j,k} = \frac{1}{I}\sum_{i=1}^{I} x_{i,j,k} \tag{6.3}$$

Subtracting the mean trajectory can highlight batch-to-batch variances. Similarly, a trajectory of standard deviation can be obtained by

$$s_{j,k} = \sqrt{\frac{1}{I-1}\sum_{i=1}^{I}(x_{i,j,k} - \bar{x}_{j,k})^2} \tag{6.4}$$

Data normalization is formulated as

$$\tilde{x}_{i,j,k} = \frac{x_{i,j,k} - \bar{x}_{j,k}}{s_{j,k}} \tag{6.5}$$

Equation 6.5 is called the NM normalization in this chapter because it was used in the MPCA-based process monitoring strategy first proposed by Nomikos and MacGregor [100]. For simplicity, the time-slice matrices $\tilde{X}_k$ ($k$ = 1,..., $K$) consist of measurements after normalization in the following sections.

### 6.2.3 Phase Recognition and Division

Phase recognition and division are based on normal batch data. A normal batch implies that the process follows the predetermined sequence, process variables track the desired trajectories, and the end product falls within the acceptable region. For normal batches, batch-to-batch variations are caused only by common causes, and process measurements at the same sampling time over different batches can be considered to be subject to normal distributions statistically. Then, PCA can be performed on these time-slice matrices as

$$\tilde{X}_k = \tilde{T}_k(\tilde{P}_k)^T \quad (k = 1,..., K) \tag{6.6}$$

where $\tilde{T}_k \in R^{I \times J}$ is the time-slice score matrix, and $\tilde{P}_k \in R^{J \times J}$ is the time-slice loading matrix.

In a matrix $\tilde{P}_k$, the first column, $\mathbf{p}_{1,k}$, indicates the projection direction for the first principal component (PC) that summarizes the greatest amount of variance information in $\tilde{X}_k$; the second column, $\mathbf{p}_{2,k}$, is the projection direction for the second PC that should be orthogonal to $\mathbf{p}_{1,k}$, and so on. The last few columns of $\tilde{P}_k$ usually represent the projection directions of process noises when correlations exist among process variables. The importance and ordering of $\mathbf{p}_{j,k}$ ($j$ = 1,2,...$J$) are determined by the corresponding eigenvalue $\lambda_{j,k}$ of the covariance matrix for $\tilde{X}_k$.

After performing PCA on time-slice matrices, $\tilde{X}_k$ ($k$ = 1,..., $K$), $K$ loading matrices $\tilde{P}_k$ are obtained. In order to automatically divide a process into phases and identify the phase characteristics, a clustering algorithm is required, which is given as follows:

**1. Preprocessing of $\tilde{P}_k$**

The first few columns in $\tilde{P}_k$ indicate the projection directions of the major batch-to-batch variances, which show good consistency for $\tilde{X}_k$ at different sampling intervals in the same operation phase, and the last few columns represent the correlation of process noises, which are quite different along sampling intervals even in the same operation phase. To ensure the robustness of the clustering algorithm, it is better to transform $\tilde{P}_k$ into a weighted form as follows:

$$\breve{P}_k = \tilde{P}_k \cdot \text{diag}(g_{1,k}, \cdots, g_{J,k}) = [\breve{\mathbf{p}}_{1,k}\ \breve{\mathbf{p}}_{2,k}\ \cdots\ \breve{\mathbf{p}}_{j,k}] \tag{6.7}$$

where the weighting factor is calculated by $g_{j,k} = \lambda_{j,k} / \sum_{j=1}^{J} \lambda_{j,k}$

These weighted loading matrices form a characteristic data set, $\{\breve{P}_k\}$ ($k$ = 1,..., $K$), for the original time slice matrices, $\{\tilde{X}_k\}$ ($k$ = 1,..., $K$).

**2. Distance Measure**

The Euclidean distance, the most popular metric, is used to calculate the similarity between two patterns. Thus, the distance between two loading matrices $\breve{P}_1$ and $\breve{P}_2$ is defined as

$$\text{dist}(\breve{P}_1, \breve{P}_2) = \left( \sum_{j=1}^{J} (\breve{\mathbf{p}}_{j,1} - \breve{\mathbf{p}}_{j,1})^T (\breve{\mathbf{p}}_{j,1} - \breve{\mathbf{p}}_{j,2}) \right)^{1/2} \tag{6.8}$$

Obviously, the distance between a loading matrix $\breve{P}_k$ and the zero matrix $\mathbf{0} \in R^{J \times J}$ is $\text{dist}(\breve{P}_k, 0) = \sum_{j=1}^{J} \left\| \breve{\mathbf{p}}_{j,k} \right\| = 1$.

### 3. *k*-Means Clustering

A variant *k*-means algorithm [120] is adopted for dividing the $K$ number of patterns $\breve{P}_k$. The optimal number of final clusters is determined by minimizing the local squared error (for patterns within each cluster) and the global squared error (for all of the patterns), where a threshold of the minimal distance between the centers of two clusters or the maximal radius of a cluster should be properly specified. This algorithm transforms modeling accuracy and complexity into the specification of the threshold. A large threshold results in few clusters, but less accurate modeling. A step is added in the clustering algorithm to eliminate singular clusters that catch few patterns in the iterative clustering procedure, to enhance the robustness and reliability of the division algorithm.

The clustering algorithm is summarized below:

**Input**

- Weighted time-slice loading matrices, $\breve{P}_k$ ($k$ = 1,..., $K$)
- Threshold of the minimal distance between two cluster's centers, $\theta$
- Termination precision, i.e., a small positive number, $\varepsilon$

**Output**

- Number of clusters, $C$
- Centers of clusters, $\{W_1, W_2, \cdots, W_C\}$
- Membership function, $m(k): \breve{P}_k \rightarrow \{1, 2, \cdots C\}$

**Steps**

1. Determine the initial cluster number, $C_0$, and the initial cluster centers, $\{W_{i,c}\}$ $(c = 1, 2, \cdots, C_0$, $i$ is index of iteration). A common method is randomly selecting $C_0$ centers from $\{\breve{P}_k\}$ and $C_0 = (K/3 \sim K/2)$.
2. If the distance of two centers in $\{W_{i,c}\}$ is smaller than the threshold $\theta$, eliminate one of them from $\{W_{i,c}\}$.
3. For each $\breve{P}_k$ $(k = 1, 2, \cdots, K)$, calculate the distances between $\breve{P}_k$ and $W_{i,c}$, i.e., $\text{dist}(\breve{P}_k, W_{i,c})$ $(c = 1, 2, \cdots, C_0)$; find the minimal distance, e.g., $\text{dist}(\breve{P}_k, W_{i,c^*})$, and then let $m(k) = c^*$.
4. Eliminate the singular centers that catch no loading matrices within $I_{num}$ (e.g., $I_{num} = 5$) number of iterations.
5. Update the center number, $C_{i+1}$, and calculate the new centers, $W_{i+1,c}$ $(c = 1, 2, \cdots, C_{i+1})$.
6. Calculate the distances between $W_{i+1,c}$ and $W_{i,c}$ $(c = 1, 2, \cdots, C_{i+1})$. If the average distance is less than $\varepsilon$, terminate the iteration procedure; otherwise, return to Step 3 before updating $W_{i,c}$ by $W_{i+1,c}$.

**4. Phase Recognition**

The $k$-means clustering algorithm can group optimally $\breve{P}_k$ $(k=1,\cdots,K)$ into $C$ number of clusters, representing $C$ kinds of pattern features. Since these patterns are extracted along the sampling time of a batch process, the clustering result can be directly associated with the operation time, which makes the division of the patterns very interpretable. Normally, each cluster should contain a series of successive samples. The exception may be with a process that has the same underlying characteristics for several disjointed periods of operation time; this would result in a cluster with samples disjointed in time. In most cases, process phases can be determined based on the clustering result associated with the operation time.

The number of clusters may be different from the actual operation phases. For example, a process with two or more actual operation phases of similar correlation may be clustered together, resulting in a single representative loading matrix. On the other hand, a long phase of operation having significant changes in the correlation may be divided into several suboperation phases. The above phase division method emphasizes the changes of process correlation rather than the physical operation. In this chapter, $C$ denotes the number of "operation" phases obtained by the clustering algorithm based on the process correlation characteristic.

Misclassification may occur at the beginning and end of each phase because the $k$-means clustering algorithm is a hard-division method in dealing with patterns between two neighboring clusters. Such a possible misclassification has little influence in the sub-PCA model development, but it may lead to false alarms (type I error) and missing alarms (type II errors) in online monitoring due to batch variation. Alternative methods may be used to resolve this problem. One is to relax the monitoring conditions at the beginning and the end of each phase; the other is to associate one or more characteristic process variables with the phase division, rather than using the process time.

### 6.2.4 Phase PCA Modeling

Define $P_c^* \in R^{J\times J}$ $(c=1,2,\cdots,C)$ as the representative loading matrix for the $c$th phase as

$$\begin{aligned} P_c^* &= \mathrm{Min}\Big(\sum_{\mathrm{k}} \mathrm{dist}(\tilde{P}_k, P_c^*)\Big) \\ &= \frac{1}{n_{stage\_c}} \sum_{k} \tilde{P}_k \\ &(c=1,2,\cdots,C; k=1,2,\cdots,n_{stage\_c}) \end{aligned} \tag{6.9}$$

where $n_{stage\_c}$ is the number of process data belonging to the phase $c$. Note that $\tilde{P}_k$ rather than $P_k$ is used in Eq. 6.9.

Similarly, define $S_c^* \in R^{J \times J}$ ($c$ = 1,2,..., $C$) as the representative singular-value diagonal matrix for each phase, which will be used for determination of the control limits later:

$$S_c^* = \frac{1}{n_{stage\_c}} \sum_{k=1}^{n_{stage\_c}} \tilde{S}_k = \text{diag}(\lambda_{1,c}^*, \lambda_{2,c}^*, \cdots, \lambda_{J,c}^*)(c = 1, 2, \cdots, C) \tag{6.10}$$

where $\tilde{S}_k = \text{diag}(\lambda_{1,k}, \lambda_{1,k}, \cdots, \lambda_{J,k})$ is the singular-value diagonal matrix for the time-slice matrix $\tilde{X}_k$ in phase $c$.

The number of retained principal components, $A_c^*$, for each phase could be determined by the cumulative explained variance rate, defined as follows:

$$\text{Cumulative explained variance rate} = \frac{\sum_{j=1}^{A_c^*} \lambda_{j,c}^*}{\text{trace}(S_c^*)} \geq 80\% \tag{6.11}$$

Then $P_c^*$ is divided into two parts, $\bar{P}_c^* \in R^{J \times A_c^*}$ and $\tilde{\bar{P}}_c^* \in R^{J \times (J - A_c^*)}$, for the principal component subspace and residual space, respectively. Similarly, $S_c^*$ is divided into $\bar{S}_c^* \in R^{J \times A_c^*}$ and $\tilde{\bar{S}}_c^* \in R^{J \times (J - A_c^*)}$. Then, a sub-PCA model for phase $c$ is defined as

$$\begin{aligned}
\tilde{\bar{T}}_k &= \tilde{X}_k \bar{P}_c^* \\
\hat{\tilde{X}}_k &= \tilde{\bar{T}}_k \bar{P}_c^* \\
\tilde{E}_k &= \tilde{X}_k - \hat{\tilde{X}}_k = \tilde{X}_k (\mathbf{I} - \bar{P}_c^* (\bar{P}_c^*)^T)
\end{aligned} \tag{6.12}$$

Note that $\tilde{\bar{T}}_k \in R^{I \times A_c^*}$ is the first $A_c^*$ columns of $\tilde{T}_k \in R^{I \times J}$; $\mathbf{I} \in R^{I \times I}$ is an identity matrix.

### 6.2.5 Statistics and Control Limits

Hotelling $T^2$ in the principal component subspace and the squared prediction error (SPE) in the residual subspace are used as monitoring statistics. The control limits for $T^2$ and SPE monitoring charts, which can be estimated from the reference data, should be computed in the modeling procedure.

In the principal component subspace, an augmented $T^2$ statistic [91], $T_m^2$, is adopted to describe the average variability of process variables over all batches at time interval $k$. This provides the estimation of the control limits for the principal component scores of future measurements in a new batch:

$$
\begin{aligned}
(T_m^2)_k &= I(\bar{\mathbf{t}}_k^T (\bar{S}_c^*)^{-1} \bar{\mathbf{t}}_k) \\
\bar{\mathbf{t}}_k &= \frac{1}{I} \sum_{i=1}^{I} \mathbf{t}_{i,k}
\end{aligned}
\tag{6.13}
$$

where $\mathbf{t}_{i,k} \in R^{1 \times A_c^*}$ is the $i$th row of $\tilde{\bar{T}}_k$. The $T_m^2$ statistic is subject to the $F$ distribution, so the control limits in phase $c$ can be determined by

$$
T_m^2 \sim \frac{A_c^* I(n_{stage\_c} - 1)}{I(n_{stage\_c} - 1) - A_c^*} F_{A_c^*, (I(n_{stage\_c} - 1) - A_c^*), \alpha}
\tag{6.14}
$$

In residual subspace, from the prediction error matrix, $\tilde{E}_k \in R^{I \times J}$, at the $k$th sampling interval, the following SPE vector can be obtained:

$$
\mathrm{SPE}_k = [\mathrm{SPE}_{1,k}, \cdots, \mathrm{SPE}_{i,k}, \cdots, \mathrm{SPE}_{I,k}]^T
\tag{6.15}
$$

where $\mathrm{SPE}_{i,k} = \tilde{\mathbf{e}}_{i,k}^T \tilde{\mathbf{e}}_{i,k}$, and $\tilde{\mathbf{e}}_{i,k}$ is the $i$th row of $\tilde{E}_k$.

In batch processes, the distribution of the SPE statistic at each sampling interval can be approximated by a weighted $g\chi_h^2$ distribution [102]:

$$
\begin{aligned}
\mathrm{SPE}_k &\sim g_k \chi_{h_k, \alpha}^2 \\
g_k &= v_k / 2m_k, h_k = 2(m_k)^2 / v_k
\end{aligned}
\tag{6.16}
$$

where $m_k$ represents the mean value of $\mathrm{SPE}_{i,k}$ ($i$ = 1,...,$I$) at the $k$th sampling interval, and $v_k$ is the corresponding variance. Note that the control limit of $(T_m^2)_k$ in phase $c$ is a constant value, while the control limit of $\mathrm{SPE}_k$ in phase $c$ is a trajectory.

### 6.2.6 Online Process Monitoring

A batch operation is monitored at every sampling interval $k$ with both the Hotelling $T^2$ and SPE monitoring charts. The monitoring procedure is to judge which phase the new coming data belongs to, and then call the corresponding phase PCA model to calculate the values of two indices, the Hotelling $T^2$ and SPE. The values of the two statistics for normal batch runs will be well below the control limits, while abnormal batches may have large values of the Hotelling $T^2$ and/or SPE statistics. Once an abnormal condition is detected by the monitoring charts, the contribution plot, a commonly used and effective diagnosis tool, is used to diagnose the fault cause for that phase.

Suppose the current data in current batch run, $\mathbf{x}_k$, belongs to phase $c$ according to its sample time, its principal score and residual can be obtained by:

$$
\begin{aligned}
\mathbf{t}_k &= \mathbf{x}_k \bar{P}_c^* \\
\mathbf{e}_k &= \mathbf{x}_k - \mathbf{x}_k \bar{P}_c^* (\bar{P}_c^*)^T
\end{aligned}
\tag{6.17}
$$

The $T^2$ and SPE statistics are calculated by

$$T_k^2 = \mathbf{t}_k^T (\overline{S}_c^*)^{-1} \mathbf{t}_k \tag{6.18}$$

$$\text{SPE}_k = \mathbf{e}_k^T \mathbf{e}_k \tag{6.19}$$

### 6.2.7 Summary

The main implementation steps in the offline modeling phase can be summarized as follows:

1. Collect and preprocess the historical data when the process is at normal operation conditions, forming the reference data matrix $\underline{X}$.
2. Unfold $\underline{X}$ into a series of time-slice matrices, $\tilde{X}_k$.
3. Calculate the time-slice loading matrix, $\tilde{P}_k$, and the singular value diagonal matrix, $\tilde{S}_k$, for each time-slice matrix, $\tilde{X}_k (k = 1, 2, \cdots, K)$.
4. Use the clustering algorithm to classify the loading matrices into $C$ groups representing $C$ "operation" phases and define a phase division variable based on time or characteristic variables.
5. Calculate the representative loading matrix, $P_c^*$, the representative singular value matrix, $S_c^*$ and the retained number of PCs, $A_c^*$, for each phase, then get $\overline{P}_c^*$ and $\overline{S}_c^*$.
6. Develop PCA models for each phase accordingly.
7. Compute the control limits of the two statistics for each phase.

For phase-based online process monitoring, the main steps are as follows:

1. Normalize newly available measurements for a new evolving batch run.
2. Determine which phase the new data should belong to according to the phase division variable and compute the scores and SPE values using a proper phase PCA model.
3. Posit hypotheses on the two statistics, $T^2$ and SPE. If the hypothesis tests hold, the current process data are considered to be normal. Go to Step 1 for the next new data; otherwise further analyze the process to diagnose the cause of the detected abnormality.

## 6.3 Application of Phase-Based SPM to Injection Molding

### 6.3.1 Experimental Setup

Injection molding has achieved high degree of automation. After the process conditions are properly set, the process repeats itself to produce molded parts at a high rate. The process is, however, susceptible to the production of off-specification products due to various process malfunctions, drifting of process conditions, changes in materials, and unknown disturbances. Abrupt, gross faults in the key process variables can be easily and reliably detected by the conventional control charts. Slow drift or faults involving multiple process variables, however, can be hard to detect. These process faults, even if they are small and not common, can lead to production of a large quantity of bad parts if they are not detected at the early stage.

In this section, the phase-based PCA modeling and monitoring strategy introduced in Section 6.2 is applied to an injection molding machine. Process variables are listed in Table 6.1. The operation conditions are set as in Table 6.2. Sixty normal batch runs are conducted under this operation condition, and 1000 samples are reserved during each batch operation, so the modeling data matrix can be represented as $\underline{X}$ with matrix dimension 60×16×1000. Another three abnormal batch runs are intentionally conducted for method verification.

**Table 6.1** Process Variables Used in the Phase-Based SPM

| No. | Variable Description | Unit |
|---|---|---|
| 1 | Nozzle Pressure | bar |
| 2 | Stroke | mm |
| 3 | Injection Velocity | mm/sec |
| 4 | Hydraulic Pressure | bar |
| 5 | Plastication Pressure | bar |
| 6 | Back Pressure | bar |
| 7 | Cavity Pressure | bar |
| 8 | Screw Rotation Speed | RPM |
| 9 | SV1 Opening | % |
| 10 | SV2 Opening | % |
| 11 | Mold Temperature | °C |
| 12 | Nozzle Temperature | °C |
| 13 | Barrel Temperature 1 | °C |
| 14 | Barrel Temperature 2 | °C |
| 15 | Barrel Temperature 3 | °C |
| 16 | Barrel Temperature 4 | °C |

**Table 6.2** Operation Condition Settings for Injection Molding Process

| Operation Variable | Settings |
|---|---|
| Injection Velocity | 24 mm/sec |
| Pack-Holding Pressure | 200 bar |
| Mold Temperature | 25°C |
| Seven-Band Barrel Temperature | 200, 200, 200, 200, 180, 160, and 120°C |
| Pack-Holding Time | 3 sec |
| Cooling Time | sec |

### 6.3.2 Result Analysis of Phase Division and Modeling

As shown in Fig. 6.3, the process is divided into four main phases and two transient phases according to the change of process correlation. The cooling phase, a long operation phase, actually consists of plastication of the plastic material in the barrel and cooling of the produced part in the mold, which can be clearly divided by the phase-based PCA modeling method. A few samples in the transient response from pack-holding phase to plastication phase and from plastication phase to cooling phase form two new phases, called the transient phases. Phase PCA models are then derived for each phase. This PCA analysis results in phase division similar to the actual phases used in the polymer processing industry, which suggests that the phase division based on the change of process correlation can indeed promote process understanding.

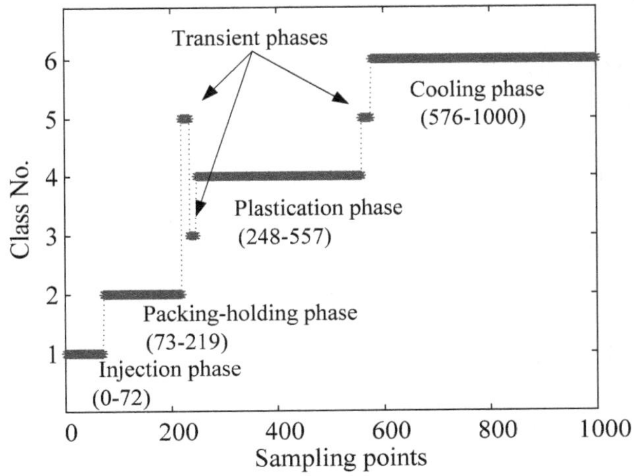


**Figure 6.3** Phase division result for the injection molding process

**Table 6.3** Sub-PCA Models in Four Operation Phases

| $\overline{P}_c^*$ | Injection | Packing | Plastication | Cooling |
|---|---|---|---|---|
| **Phase PCA models** ($A_c^* = 2$) | $\begin{bmatrix} 0.26 & -0.11 \\ 0.2 & -0.08 \\ 0.38 & -0.17 \\ 0.34 & -0.15 \\ 0 & 0 \\ 0.37 & -0.13 \\ 0.02 & -0.01 \\ 0 & 0 \\ 0.38 & -0.18 \\ 0 & 0 \\ 0.29 & -0.14 \\ 0.3 & -0.11 \\ 0.18 & 0.48 \\ 0.2 & 0.49 \\ 0.18 & 0.47 \\ 0.23 & 0.39 \end{bmatrix}$ | $\begin{bmatrix} 0.36 & -0.15 \\ 0.38 & -0.16 \\ 0.01 & -0.01 \\ 0.35 & -0.15 \\ 0 & 0 \\ 0.06 & 0 \\ 0.37 & -0.16 \\ 0 & 0 \\ 0.27 & -0.11 \\ 0 & 0 \\ 0.34 & -0.16 \\ 0.34 & -0.12 \\ 0.18 & 0.48 \\ 0.19 & 0.5 \\ 0.18 & 0.47 \\ 0.23 & 0.4 \end{bmatrix}$ | $\begin{bmatrix} 0.05 & -0.03 \\ 0.23 & -0.12 \\ 0.1 & -0.06 \\ 0.06 & -0.03 \\ 0.44 & -0.23 \\ 0 & 0 \\ 0.02 & -0.01 \\ 0.41 & -0.22 \\ 0.22 & -0.13 \\ 0.45 & -0.25 \\ 0.23 & -0.15 \\ 0.08 & -0.02 \\ 0.22 & 0.46 \\ 0.22 & 0.47 \\ 0.22 & 0.45 \\ 0.29 & 0.37 \end{bmatrix}$ | $\begin{bmatrix} 0.02 & 0.03 \\ 0 & 0 \\ 0 & 0 \\ 0.04 & 0.04 \\ 0 & 0 \\ 0.14 & 0.15 \\ 0 & 0 \\ 0 & 0 \\ 0.55 & 0.68 \\ 0 & 0 \\ 0.2 & 0.28 \\ 0.1 & 0.09 \\ 0.35 & -0.38 \\ 0.38 & -0.38 \\ 0.36 & -0.33 \\ 0.47 & -0.14 \end{bmatrix}$ |
| **Explained variance percentage** | 90.02 | 90.51 | 88.53 | 84.82 |

The loading matrices of the four main phase PCA models are listed in Table 6.3. The loading plots, which are obtained by plotting the second loading vector against the first one, indicate the correlation structure of process variables. Variables, located in the same cluster, have high correlation with each other; variables in different clusters have weak correlation [121]. As illustrated in Fig. 6.4, process variables (except barrel temperatures) form different clusters in different phases, indicating that these variables have different correlation structures at different phases. The variables located in the circles have small values in the loading vectors, indicating that they are nonsignificant variables and have no contributions to the phase PCA model. For example, in the injection phase, the key objective is to control the injection velocity, the variations in pressure variables are of no interest so that they show very little significance to the phase correlation structure of process variables. All barrel temperatures lie in the rectangles (variable numbers 13, 14, 15, and 16), forming an independent clustering, implying that they have weak relations with other process variables. It is important to point out that the variables in the diamonds change from phase to phase, indicating the varying process characteristics and the necessity for a phase-based analysis.

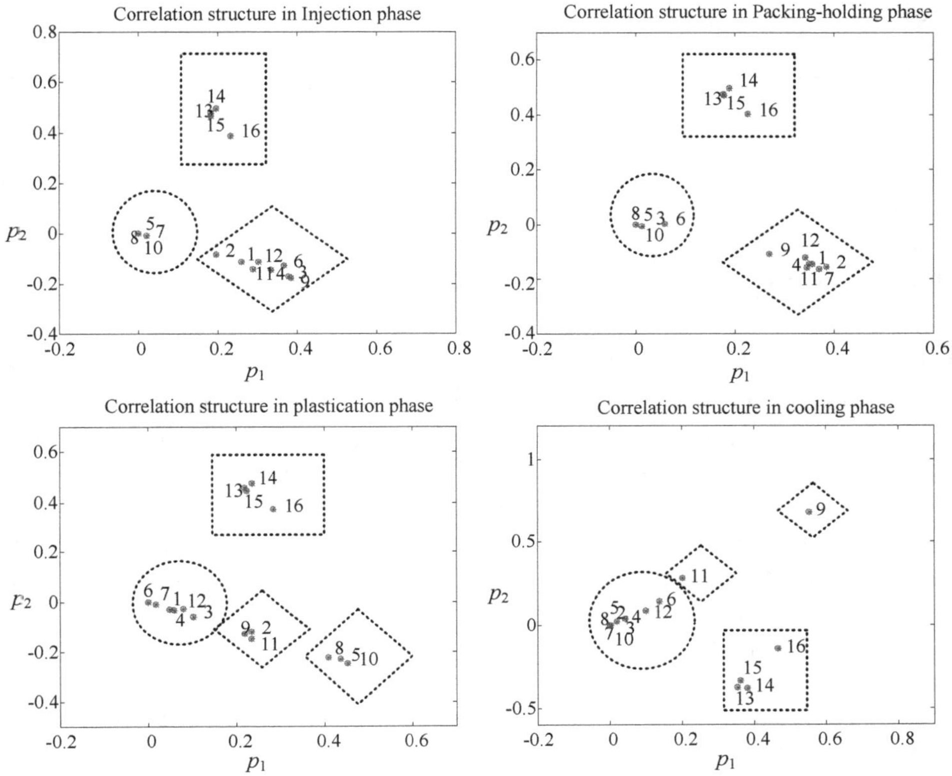


**Figure 6.4** Process correlations in the four main phases for an injection molding process

### 6.3.3 Result Analysis of Process Monitoring and Fault Diagnosis

For process monitoring and fault diagnosis, three typical faults are intentionally introduced. Fault 1 is a material disturbance by adding a few grams of polypropylene (PP) into the HDPE; the major physical properties for HDPE and PP are listed in Table 6.4. Fault 2 is a barrel temperature sensor failure, simulated by unplugging the I/O interface. Fault 3 is caused by a check-ring failure, which is a common problem in injection molding.

All faults can change the correlation structure, generating unexplained information for the phase PCA models. They can be promptly detected by the monitoring charts in the corresponding phase.

**Table 6.4** Physical Properties for HDPE and PP

| Polymers | HDPE | PP |
|---|---|---|
| **Type** | Crystalline | Crystalline |
| **Density (g/cm$^3$)** | 0.94–0.96 | 0.90 |
| **Melting Temp. (°C)** | 130–135 | 170 |
| **Thermal Decomposition Temp. (°C)** | 300 | 300 |
| **Injection Temp. (°C)** | 200–260 | 220–275 |
| **Dielectric Constant** | 2.30–2.35 | 2.3 |

### Fault 1: Material Disturbance

A small amount of PP is added to the processing of HDPE. The $T^2$ and SPE monitoring charts, as shown in Fig. 6.5, indicate that this fault can be identified soon after the start of the injection phase. In terms of the four contribution plots as shown in Fig. 6.6, contamination of PP into the HDPE results in a lower cavity temperature (No. 11) throughout the cycle because PP cools and solidifies faster than HDPE. At the same time, the viscosity of PP is higher than that of HDPE, which generates larger shear heating for PP, resulting in a higher nozzle melt temperature (No. 12). The contribution plot of the packing phase is different from the others. The cavity pressure (No. 7) has lower values, due to the faster solidification of PP. This characteristic difference among different phases can only be revealed by such a phase-based approach.

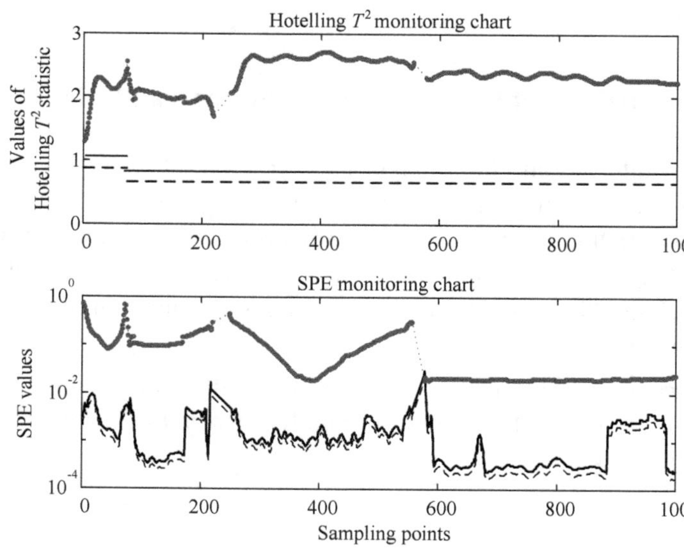


**Figure 6.5** Monitoring charts for a batch process with material disturbance; (solid line, 99% control limit; dashed line, 95% control limit; dotted line, $T^2$ or SPE)

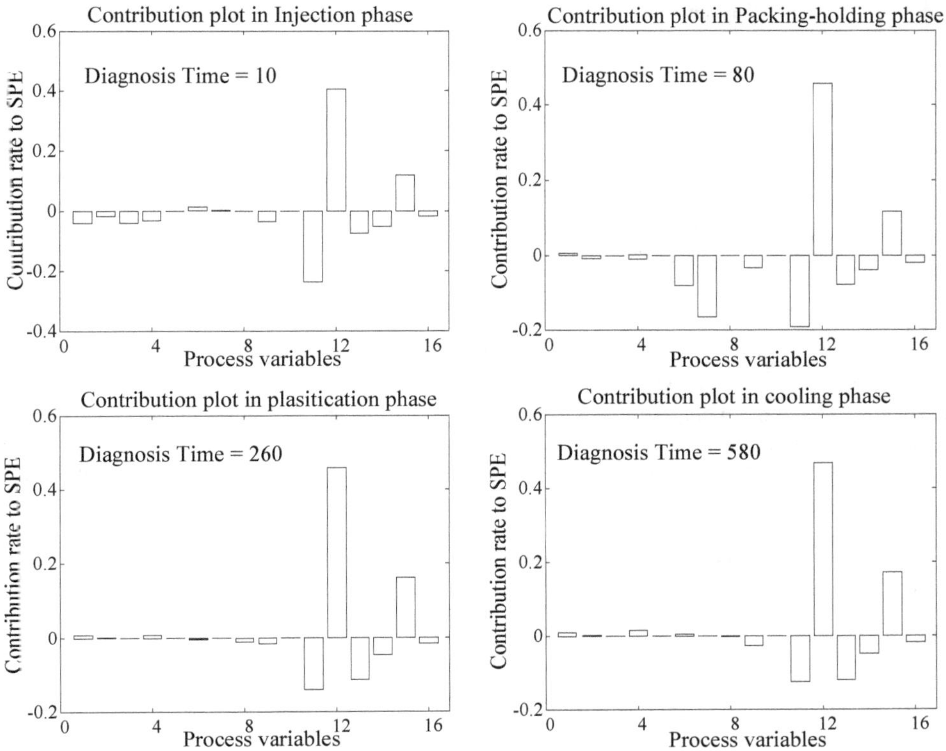


**Figure 6.6** SPE contribution plots for a batch with material disturbance

### Fault 2: Sensor Failure

When one thermocouple that measures the barrel temperature fails, the reading of this sensor drops, resulting in full heating of the corresponding zone. This causes excessive heat to be conducted to the neighboring zones, even when the heating of those zones are fully shutdown. To avoid possible hazards, the faulty batch run with sensor failure is suspended after a few seconds. The fault can be quickly picked up by the Hotelling $T^2$ and SPE monitoring charts, as shown in Fig. 6.7. The contribution plots of the four operation phases, shown in Fig. 6.8, clearly indicate the drop of the failed zone temperature (No. 14) and the increased temperatures of the neighboring zones. The contribution plots in all four phases are similar because this fault has similar impact on the four phases.

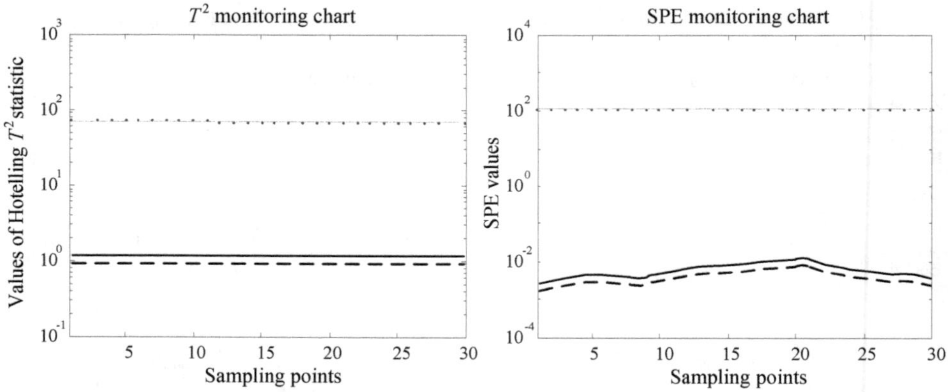


**Figure 6.7** Monitoring charts for the first phase of a batch with the sensor failure of a barrel temperature variable;
(solid line, 99% control limit; dashed line, 95% control limit; dotted line, $T^2$ or SPE)

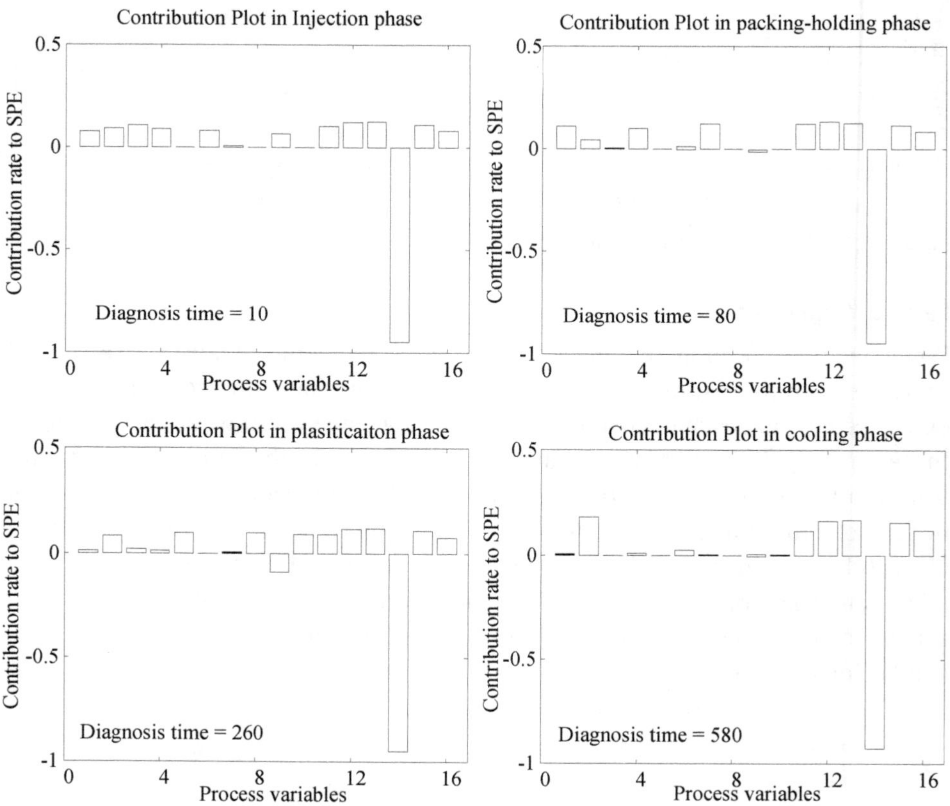


**Figure 6.8** SPE contribution plot for a batch with the sensor failure of a barrel temperature variable

### Fault 3: Check-Ring Failure

The check-ring valve, as schematically shown in Fig. 6.9, consists of three parts: a slide ring, a seat, and a main body. The seat and main body are fixed on the reciprocating screw, while the check ring slides on the main body. The check ring fits closely to the barrel. During plastication, the melting polymer pushes the check ring towards the nozzle tip, allowing the melt to flow into the nozzle reservoir, as shown in Fig. 6.9(a). When the screw moves forward in the injection phase, the check ring is pushed backward to the seat due to pressure increase in the nozzle and gradually seals the flow channel, as shown in Fig. 6.9(b).

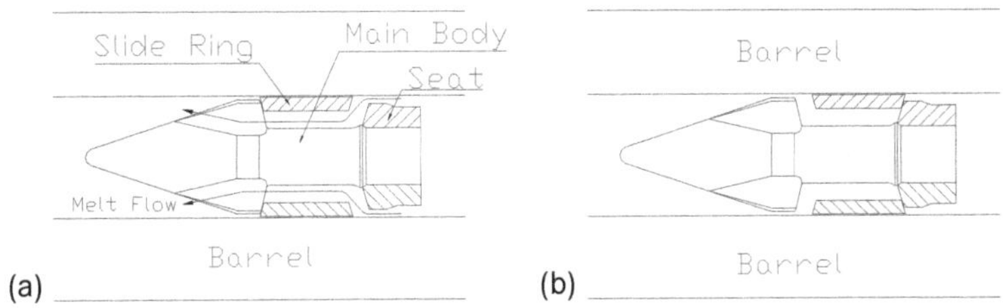


**Figure 6.9** Schematic of check-ring valve operation during injection and plastication: (a) during plastication; (b) during injection

The performance of a check-ring valve strongly affects the dimensional stability during injection molding, especially for the injection and packing phases. In some cases, the check-ring valve may be stuck and a significant amount of melt will leak to the screw channel; this phenomenon is referred to as a check-ring failure problem. When the check ring fails, less material backflows from the nozzle to the screw channel, therefore, excessive material is injected into the mold in order to compensate for the lack of material during the packing phase, which causes all of the process variable trajectories and correlations to differ from the ones observed under normal conditions.

A check-ring failure upsets the process correlations of different phases in different ways. As shown in Fig. 6.10, a check-ring fault can be readily detected by the SPE monitoring charts. The contribution plots in the first three phases, however, are different, indicating that the fault has different fault characteristics in different phases. The application of a phase-based sub-PCA method is advantageous for diagnosing such a fault.

During the injection phase, a smaller amount of material will be injected into the cavity at the same injection velocity, due to the backflow associated with the failure of the check-ring valve. The nozzle pressure (No. 1), injection pressure (No. 4), and cavity pressure (No. 7) are lower, as clearly indicated by the contribution plot of the phase. During the pack-holding phase, more material will have to be packed into the cavity to make up for the shortfall in the injection, resulting in higher stroke (No. 2), screw speed (No. 3), and pressures (No. 4, 5), as expected from the analysis of the process.

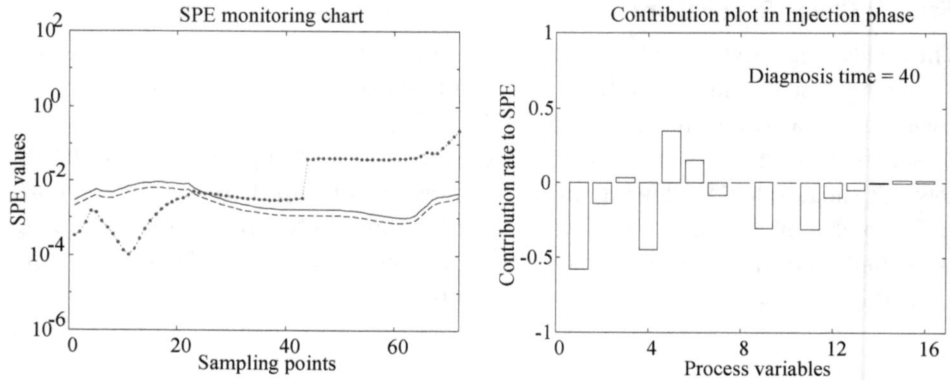


Monitoring result in Injection phase

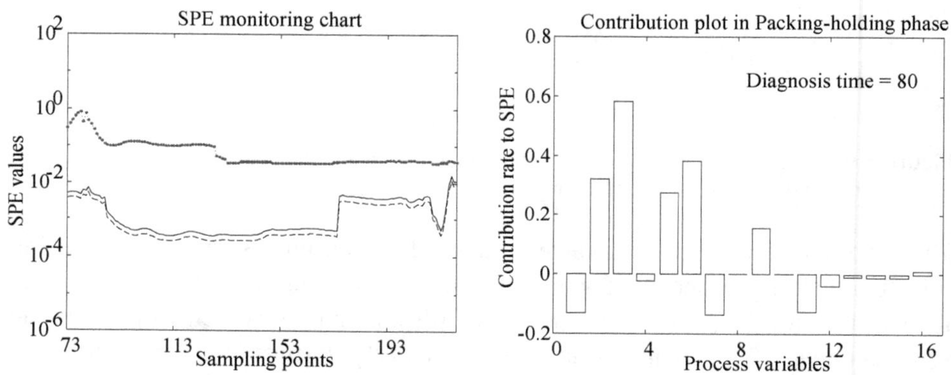


Monitoring result in Packing-Holding phase

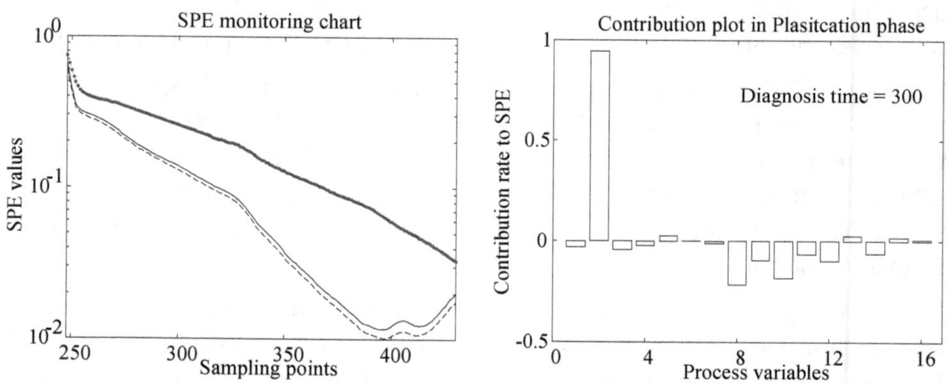


Monitoring result in Plastication phase

**Figure 6.10** Monitoring and diagnosis charts for a batch with a check-ring problem; (solid line, 99% control limit; dashed line, 95% control limit; blue dash-dotted line, online $T^2$ or SPE)

This can also be clearly seen in the contribution plot. During plastication, because a longer stroke has travelled in injection and packing, a longer plastication stroke (No. 2) has to be recovered, which is clearly seen in the contribution plot of this phase.

The above analysis accords well with the process knowledge of injection molding. For the faults that show different fault characteristics in different phases, it is desirable to analyze the contribution plots of different phases to give a reasonable cause to the fault. This suggests that the application of the phase-based PCA modeling method can indeed enhance process understanding and improve the ability of fault detection and diagnosis.

## 6.4 Improved Phase-Based SPM for Unequal-Length Batch Processes

### 6.4.1 Overview

The phase-based SPM method introduced in Section 6.2 requires that batch data should have the same sampling intervals, i.e., $\overline{X}_i \in R^{J\times K}$ $(i = 1,..., I)$ have the same data length $K$. Then, $\underline{X} \in R^{I\times J\times K}$ can be cut into $K$ time-slice matrices, $\tilde{X}_k \in R^{I\times J}$, $(k = 1,..., K)$, generating $K$ loading matrices, $\tilde{P}_k \in R^{J\times J}$, which contain process correlation information at each sampling interval $k$.

However, in many industrial batch processes, the total batch duration and the durations of various operation phases within the batch are not fixed in length because of changes in operation conditions or control objectives. For example, in injection molding processes, the duration of the injection phase varies according to the injection velocity. In such cases, as shown in Fig. 6.11, the reference batches $\overline{X}_i \in R^{J\times K_i}$ have different trajectory lengths, $K_i$. Figure 6.12 shows an example batch process consisting of an unequal-length phase (phase I) and an even-length phase (phase II). The modeling and online monitoring methods mentioned in Section 6.2 cannot be applied unless all batch runs are of the same length.

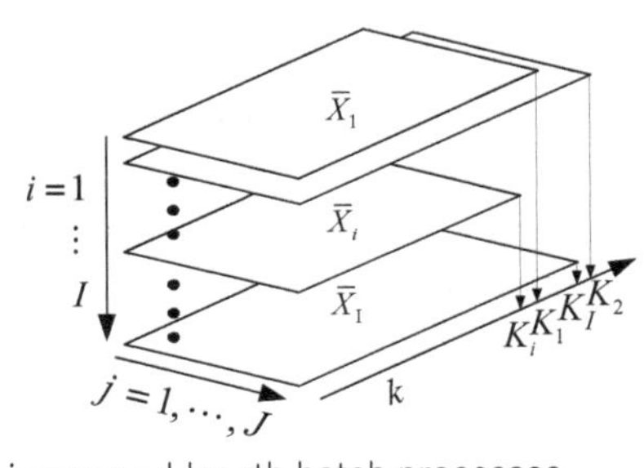


**Figure 6.11** Process data in unequal-length batch processes

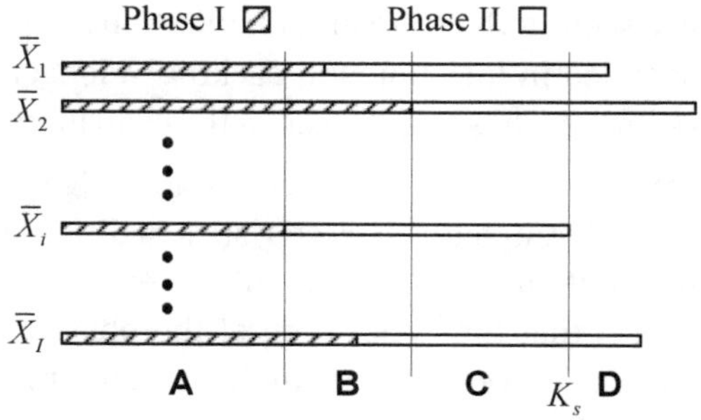


**Figure 6.12** Illustration of a two-phase unequal-length batch process

There are two simple methods that have been used to deal with the cases in which the historical successful batches have different lengths but the trajectories overlap in a common time part. One is to cut the batches to have the minimum length, provided that the unequal-length problem is not serious and the main events have occurred in the common time part. The other is to derive a model using the data from long batches and to treat the absent parts of the trajectories of the shorter batches as missing data. In a general case, unequal-length phenomena can occur in one or more specific operation phases, resulting in the total batch duration and the durations of sub operation phases both being unequal. The above two methods are not applicable to such a general case. To solve this problem, exploration of phase characteristics is still an emphasis. In this section, an improved phase-based SPM method is introduced.

### 6.4.2 Data Normalization

For unequal-length batch processes, particularly for the processes with unequal-length phases, the NM normalization method (i.e., Eq. 6.5 in Section 6.2) cannot be used directly because it cannot obtain a meaningful average trajectory, $\bar{x}_{j,k}$, and trajectory of standard deviation, $s_{j,k}$. Taking the process illustrated in Fig. 6.12 as an example, it consists of two operation phases, and without loss of generality, the first phase is assumed to have varying duration. Note that the above information on process phases might be unknown in advance for real industrial batch processes before statistical modeling. There arise two problems when applying the NM normalization to such a process: (1) How should the data in the irregular tail of span D in Fig. 6.12 be normalized? (2) Subtracting the average trajectories and dividing by the standard deviations will become meaningless, especially in span B.

Kourti [122] attempted to solve the first problem by adding zeros for the unavailable data, resulting in the difficulty that the "average trajectory" is not the true average trajectory because it has been altered by the added zeros. In the following phase-based SPM method, the average trajectory and the standard deviation trajectory in the irregular tails of unequal-length phases will be calculated from the actual available data rather than these data with added zeros. Although the confidence of the estimated mean and standard deviation is reduced statistically because fewer data are used, this can be compensated by increasing the corresponding control limits.

The second problem can be solved by separating the data according to the corresponding process phases, as illustrated in Fig. 6.13, where $L_c$ ($c = 1, 2$) stands for the data length in phase $c$. Thus, the NM normalization can only be used for data pretreatment after the phase-division information becomes available.

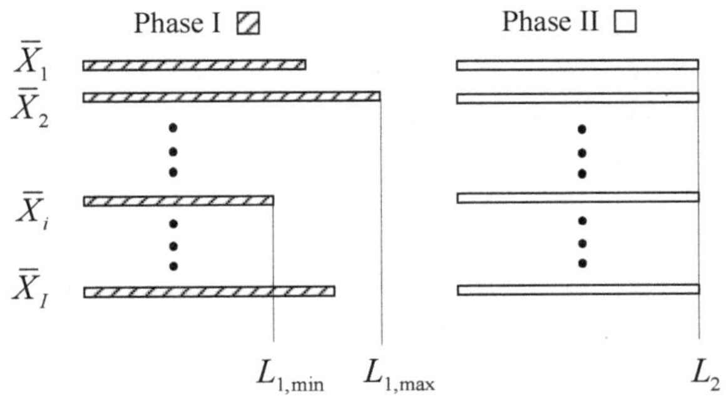


**Figure 6.13** Two separated phases of the illustration process of Fig. 6.12

For batch processes, another normalization technique was proposed by Wold et al. [123], abbreviated as the WKFH normalization here. It is conducted on the two-way matrix $X \in R^{KI \times J}$, unfolded from $\underline{X} \in R^{I \times J \times K}$ in such a way as to retain the variable direction. Each point of the trajectory of a batch is considered as an object; WKFH normalization involves subtraction of a constant, the average of each variable over all batches and over the entire batch durations, and division of each variable at each time by the overall standard deviation. Mathematically

$$
\begin{aligned}
\tilde{x}_{i,j,k} &= \frac{x_{i,j,k} - \overline{x}_j}{s_j} \\
&\text{where} \\
\overline{x}_j &= \frac{1}{KI} \sum_{k=1}^{K} \sum_{i=1}^{I} x_{i,j,k} \\
s_j &= \frac{1}{KI - 1} \sqrt{\sum_{k=1}^{K} \sum_{i=1}^{I} \left( x_{i,j,k} - \overline{x}_j \right)^2}
\end{aligned}
\tag{6.20}
$$

WKFH normalization can be applied directly to unequal-length data, although it is ill-suited for process monitoring because it focuses on the variation along the time direction, not the batch direction.

Considering the merits and deficiencies of these two normalization techniques, WKFH normalization is used for pretreatment of the unequal-length data for phase division. After the phase-division information becomes available, the NM method is used to normalize the phase data for development of the monitoring model.

### 6.4.3 Phase Recognition and Division

Assume that the reference database contains $I$ normal batches, $\overline{X}_i \in R^{J\times K_i}$ ($i$ = 1,2,..., $I$), and the shortest batch length is $K_s$. Thus, $K_s$ time-slice matrices, $\tilde{X}_k \in R^{I\times J}$ ($k$ = 1,2,..., $K_s$), can be obtained from the reference data for calculation of the time-slice loading matrices, $\tilde{P}_k$ ($k$ = 1,2,..., $K_s$). As stated in the development of the original phase-based SPM method, (1) a batch process can be divided into several "operation" phases, reflecting its inherent process correlation nature, and (2) even though the process might be time-varying, the correlation of its variables remains largely the same within the same "operation" phase. We further show here that process correlation remains essentially the same within the same phase for normal batches, notwithstanding variations in duration. That is, process correlation will not change significantly until the process enters the next phase. On the basis of this conclusion, one can find the duration of each phase by checking whether the data can be explained by the current-phase PCA model, using the corresponding SPE statistics.

For the process illustrated in Fig. 6.12, process correlation for the data in span A, the common part belonging to phase I for all reference batches, will be largely unchanging according to the aforementioned assertions. With the clustering algorithm introduced in Section 6.2.3, the time-slice PCA loading matrices in span A should form an independent and stable cluster, representing the inherent process characteristics of the first phase. A phase PCA model can be developed on the basis of the clustering result, as described later. The data in the first phase should be explained well by this first-phase PCA model, with relatively small SPE value. In contrast, for data not belonging to the first phase, this model will result in large SPE values, as the first-phase PCA model can no longer explain the correlation among variables beyond the first phase. The phase PCA model derived from the common part of the first phase can be used to find the duration of the first phase for all reference batches, given a properly selected threshold, $\theta$.

For each time-slice matrix, $\tilde{X}_k$ ($k$ = 1,2,..., $K_s$), the conventional two-way PCA can be applied directly:

$$\tilde{X}_k = \tilde{T}_k \left(\tilde{P}_k\right)^T \left(k = 1,2,\cdots,K_s\right) \tag{6.21}$$

The $K_s$ loading matrices $\tilde{P}_k \in R^{J\times J}$ ($k$ = 1,2,..., $K_s$) are generated and transformed into the weighted form, $\left\{\breve{P}_1, \breve{P}_2, \ldots, \breve{P}_{K_s}\right\}$. The $k$-means clustering algorithm is used to divide the loading matrices into $C$ groups, representing $C$ different process characteristics. Process phases are then associated with process operation time.

However, for the unequal-length case, the clustering algorithm can obtain clear and stable clusters only for the common parts of various phases, such as spans A and C in Fig. 6.12. From the clustering results, a two-way PCA model can be developed for the first phase, but not for other phases. For example, for the process illustrated in Fig. 6.14, where the difference in duration of the first phase is greater than the length of the second phase, no common part will exist for the second phase; hence, it will not be possible to derive the second-phase model using the above clustering results. For this reason, the PCA models developed for phase division will have to be derived one by one in accordance with the sequence of process "phases"; that is, determination of the process phases will have to be conducted in series, not in parallel as in Section 6.2.3.

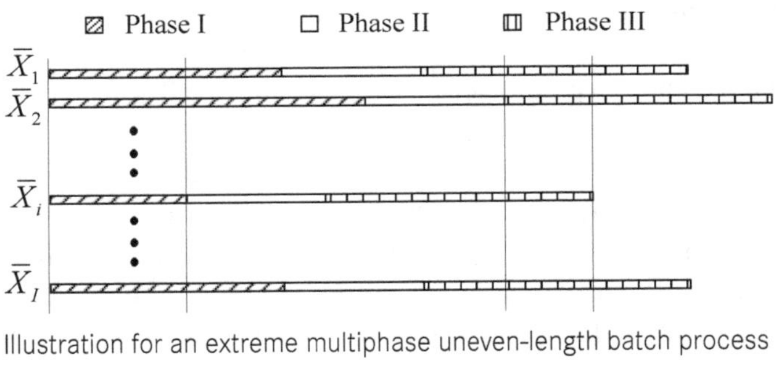


**Figure 6.14** Illustration for an extreme multiphase uneven-length batch process

The same loading matrix clustering algorithm is used as in Section 6.2, except that the output of this algorithm is the information on the first phase, namely $m(k)=1\,(k=1,\cdots,K_s)$. Then, one can define $P^*_{s1}$, the first-phase representative loading matrix, as

$$P^*_{s1}=\mathrm{Min}(\sum_k \mathrm{dist}(\tilde{P}_k, P^*_{s1}))=\frac{1}{L_1}\sum_{k=1}^{L_1}\tilde{P}_k \quad \text{for } \{k\,|\,m(k)=1, k=1,2,\ldots,K_s\} \tag{6.22}$$

where $L_1$ is the number of data in the common part belonging to the first phase (e.g., the length of span A in Fig. 6.12). In Eq. 6.22, the subscript $s$ indicates that the information is used only for phase division, the "1" stands for the first phase, and $k$ is the index for sampling intervals.

The parameter $P^*_{s1}$ can be divided into two parts, $\overline{P}^*_{s1}$ and $\tilde{\overline{P}}^*_{s1}$, for the principal component and residual subspaces, respectively. The number of retained principal components, $A^*_{s1}$, can be determined by the same method as in the standard PCA method. Then, for each data vector $\mathbf{x}_{i,k}\in R^{1\times J}$ at the $k$th sampling interval of the

$i$th batch run, the first-phase PCA loading matrix $\overline{P}_{s1}^{*}$ is used to calculate the principal component score vector $\mathbf{t}_{i,k,s1} \in R^{1\times A_1}$, the residual vector $\mathbf{e}_{i,k,s1} \in R^{1\times J}$, and the corresponding SPE value $\mathrm{SPE}_{i,k,s1}$ as

$$\begin{aligned}
\mathbf{t}_{i,k,s1} &= \mathbf{x}_{i,k}\overline{P}_{s1}^{*} \\
\mathbf{e}_{i,k,s1} &= \mathbf{x}_{i,k} - \mathbf{x}_{i,k}\overline{P}_{s1}^{*}(\overline{P}_{s1}^{*})^{T} \\
\mathrm{SPE}_{i,k,s1} &= \mathbf{e}_{i,k,s1}{}^{T}\mathbf{e}_{i,k,s1}
\end{aligned} \tag{6.23}$$

By checking at which sampling interval does the SPE statistic go beyond the predetermined threshold, $\mathrm{SPE}_c^{*}$ ($c = 1$), one can readily determine the first-phase duration for each reference batch $\tilde{X}_i$.

With the information of the first phase duration, the first-phase data of each reference batch can be removed and the remaining data form a new reference data set. The same procedure is applied to determine the duration of the second phase, and so on. The phase division can hence be conducted iteratively until all phase information becomes available. The two parameters, $\theta$ in clustering and $\mathrm{SPE}_c^{*}$ ($c = 1,\ldots,C$) in phase division, are important for the success of the proposed phase-division algorithm. The threshold $\theta$ determines the accuracy and complexity of the developed phase-based sub-PCA model. Obviously, large $\theta$ results in coarse clustering and less accurate modeling, whereas small $\theta$ can improve modeling accuracy but need more submodels and increase the modeling complexity. The parameter $\mathrm{SPE}_c^{*}$ is recommended to be assigned a value that is slightly larger than the maximum SPE values of the data belonging to the common part of phase $c$.

### 6.4.4 Sub-PCA Modeling Procedure

After phase recognition and division, a series of separated phases are derived from the batch process, as illustrated by Fig. 6.13. The durations of all phases, the longest and shortest batches (or phases), become available. The phase PCA model, Eq. 6.23, however, is derived for dividing the process into phases and determining the duration of each phase. It is ill-suited for process monitoring because of the WKFH normalization method adopted. WKFH normalization focuses on the variation in variables' trajectories, rather than the variation among the batches. For this reason, the data in each phase are renormalized by the NM method to develop phase PCA monitoring models for fault detection and diagnosis.

The time-slice loading matrices for the data in the irregular tail region are calculated from the time-slice matrices, $\tilde{X}_k$, of the available data of the longer batches. The contribution to the phase PCA monitoring model should be reduced by a weighting factor $w_k = I_{a,k} / I \left(0 < w_k \leq 1\right)$, where $I_{a,k}$ is the number of available batches for calculating the time-slice loading matrix and $I$ is the total number of reference batches.

Finally, a weighted phase loading matrix is obtained by

$$P_{mc}^{*} = \frac{\sum_{k} w_k \tilde{P}_k}{\sum_{k} w_k} \quad (k = 1, \cdots, L_c) \tag{6.24}$$

where the subscript $m$ indicates that the derived phase representative loading matrix is used for process monitoring, $c$ is the phase index, and $k$ is the data index in phase $c$. The expression $P_{mc}^{*} \in R^{J \times J}$ is divided into two parts, $\overline{P}_{mc}^{*} \in R^{J \times A_c^{*}}$ and $\tilde{\overline{P}}_{mc}^{*} \in R^{J \times (J - A_c^{*})}$, for the principal component subspace and residual space, respectively, according to the number of retained principal components in each phase PCA model, $A_c^{*}$. Similarity, the eigenvalue matrix $S_{mc}^{*}$ for phase $c$ can be obtained and divided into $\overline{S}_{mc}^{*} \in R^{J \times A_c^{*}}$ and $\tilde{\overline{S}}_{c}^{*} \in R^{J \times (J - A_c^{*})}$.

For the current data vector $\mathbf{x}_k$ in a new batch run, which is assumed to belong to phase $c$, the principal component score and SPE are calculated by the following phase PCA monitoring model:

$$\begin{aligned} \mathbf{t}_k &= \mathbf{x}_k \overline{P}_{mc}^{*} \\ T_k^2 &= \mathbf{t}_k^{T} (\overline{S}_{mc}^{*})^{-1} \mathbf{t}_k \\ \mathbf{e}_k &= \mathbf{x}_k - \mathbf{x}_k \overline{P}_{mc}^{*} (\overline{P}_{mc}^{*})^{T} \\ \mathrm{SPE}_k &= \mathbf{e}_k^{T} \mathbf{e}_k \end{aligned} \tag{6.25}$$

Determination of the control limits can follow the same procedure as in Section 6.2.

### 6.4.5 Process Monitoring Procedure

Two phase models have been developed in the previous section. The phase-division model (Eq. 6.23) is used to divide the process into phases, find the unequal-length phases, and determine the duration of each phase. With the phase-division information, that is, the length of each phase, $L_{c,i}$ ($i$ = 1,2,..., $I$; $c$ = 1,2,..., $C$) and the shortest and longest lengths of an unequal-length phase $c$, $L_{c,\min}$ and $L_{c,\max}$. The reference batch data are processed by the NM normalization to develop the phase-monitoring model (Eq. 6.25) for fault detection and diagnosis.

With the proposed modeling method, an issue in process monitoring is how to differentiate an abnormality from process phase changeovers.

In process monitoring, there are three possible results: normal conditions, phase changeover, and abnormal conditions:

1) For the normal case, the two statistics, $T^2$ and SPE, calculated by the current-phase model (Eq. 6.25), should be below their corresponding control limits.
2) For the phase changeover case, the process enters a new phase, and there is no abnormality. The current-phase model will give significant $T^2$ and SPE statistics indicating that the corresponding PCA model can no longer explain the correlation of the current data. The data, however, should be well explained by the monitoring model of the next phase, generating in-control $T^2$ and SPE statistics. Phase changeover, therefore, can readily be identified by applying the two consecutive phase-monitoring models and checking their corresponding $T^2$ and SPE values simultaneously.
3) For the abnormal case, both the current and the next phase-monitoring models should give large $T^2$ and SPE values.

As mentioned before, process time is insufficient for defining process phases because of unequal-length issue. The above knowledge is used to determine the phase to which new data belong in online process monitoring. For even-length phases, online monitoring is similar to that in the previous phase PCA method. For unequal-length phases, data are divided into two parts for online monitoring. The data before the shortest phase length, $L_{c,\min}$, can be directly monitored by the current-phase model. Significant $T^2$ or SPE statistics can be used with confidence to indicate a process abnormality. For the data between $L_{c,\min}$ and $L_{c,\max}$, out-of-control statistics represent two possibilities: a process abnormality or a phase changeover. The above-stated knowledge can be used to associate the change with a proper cause. As with any statistical modeling and monitoring method, the reference data used for the modeling are assumed to cover statistically all normal operation conditions. Any case in which the unequal-length phase of a new batch is shorter than the minimum phase length, $L_{c,\min}$, or longer than the maximum length, $L_{c,\max}$, will be considered an abnormality by the monitoring procedure.

### 6.4.6 Summary

The modeling and online monitoring strategies for unequal-length batch processes can be summarized as follows.

#### Modeling Procedure

The available information includes the number of successful historical batches, $I$, and the shortest batch length, $K_S$.

The purposes are to divide the process into phases according to the change of process correlation and to develop phase PCA monitoring models.

The main implementation steps include:

1) Normalize the reference batches by the WKFH normalization method.
2) Generate $K_s$ time-slice matrices and calculate the corresponding loading matrices.
3) Derive the phase-division PCA model (Eq. 6.23) by the clustering algorithm for the first phase.
4) Determine the duration of the first phase, $L_{1,i}$ ($i = 1,2,\ldots, I$), by checking SPE values for each reference batch, and then find the longest and shortest durations in the reference data, $L_{1,\max}$ and $L_{1,\min}$, respectively, if the first phase has a different duration.
5) Extract the first-phase data, normalize them by the NM method, derive the corresponding phase-monitoring PCA model (Eq. 6.25), and compute the corresponding control limits for the $T^2$ and SPE statistics.
6) Remove the first-phase data to form a new reference data set, and repeat steps 2–5 for the next phase.

**Monitoring Procedure**

The available information includes the phase-division information, $L_{c,i}$ ($i = 1,2,\ldots, I$; $c = 1,2,\ldots,C$), $L_{c,\max}$, and $L_{c,\min}$, the average trajectories, and the standard deviation trajectories for each phase and the phase-monitoring PCA models.

The purposes are to monitor the process conditions, detect any abnormality, and possibly provide the cause for the detected abnormality.

Case I: For data belonging to the range $\left[1, L_{c,\min}\right]$:

1) Normalize the data by the NM method using the variables' mean trajectories and the standard deviation trajectories.
2) Call the current phase-monitoring PCA model and calculate the two statistics $T^2$ and SPE.
3) If either of the two statistics exceeds its control limit, the monitoring procedure gives an alarm indicating an abnormality in the process; otherwise, if the two statistics are both within the normal operation regions, the current data are normal; return to step 1 for the next data.

Case II: For data belonging to the range $[L_{c,\min}, L_{c,\max}]$:

1) Apply the same procedures as in case I until either of the statistics is beyond the normal control limit.
2) Normalize the data by regarding them as the first data of the next phase, and use the monitoring PCA model of the next phase to calculate new $T^2$ and SPE values.
3) If both the statistics are below their control limits, the process enters the next phase; otherwise, a process abnormality has occurred. Use a contribution plot to analyze the possible cause for the fault.

# 6.5 Application of Improved Phase-Based SPM to Injection Molding

## 6.5.1 Experimental Setup

In the experimental injection molding machine, the injection phase is designed to have a varying duration as the injection velocity is changed from 22 to 26 mm/s. The maximum difference in the number of data points in the injection phase is 13, about 15% of the average injection length. The other phases are controlled to have exactly the same data length.

The material used is high-density polyethylene (HDPE). The process variables selected for modeling are listed in Table 6.5. Barrel temperatures are not selected because they have weak relations with the other process variables according to the results in Fig. 6.4. The operation conditions are set as Table 6.6, and the sampling interval is 20 ms. In total, 35 normal batch runs are conducted, where the shortest and longest batch lengths are the 1180 and 1193 samples, respectively.

**Table 6.5** Process Variables for Unequal-Length Injection Molding Process

| No. | Variable Description | Unit |
|---|---|---|
| 1 | Nozzle Temperature | °C |
| 2 | Nozzle Pressure | bar |
| 3 | Stroke | mm |
| 3 | Injection Velocity | mm/sec |
| 5 | Hydraulic Pressure | bar |
| 6 | Plastication Pressure | bar |
| 7 | Cavity Pressure | bar |
| 8 | Screw Rotation Speed | RPM |
| 9 | SV1 Opening | % |
| 10 | SV2 Opening | % |
| 11 | Mold Temperature | °C |

**Table 6.6** Operation Condition Setting for Unequal-Length Injection Molding Process

| Operation Variable | Setting Value |
|---|---|
| Injection Velocity | 22–26 mm/sec |
| Pack-Holding Pressure | 197–203 bar |
| Mold Temperature | 25°C |
| Seven-Band Barrel Temperature | 200, 200, 200, 200, 180, 160, and 120±5°C |
| Pack-Holding Time | 6 sec |
| Cooling Time | 15 sec |

Three abnormal batches, with a check-ring problem, a material disturbance, and an SV1 valve-stick fault, are introduced to test the proposed process monitoring and diagnosis scheme.

## 6.5.2 Result Analysis of Phase Division and Modeling

The reference data are first normalized by the WKFH method for phase division. Assuming that no information is available in advance on the number of phases and the length of each phase, the clustering algorithm has to divide $K_s$ ($K_s$ = 1180) loading matrices for determination of the duration of the first phase. The first column vector of each time-slice loading matrix, $\tilde{\mathbf{p}}_{1,k}(k=1,\cdots,1180)$, is plotted in Fig. 6.15, where each line stands for the elements in the first loading vectors over the batch duration. From Fig. 6.15, it is obvious that the loading vectors indeed remain similar across several operation periods and show significant differences for different periods. The corresponding clustering results are shown in Fig. 6.16. It is also easy to find out the four major clusters, corresponding to the four operation phases: injection, pack-holding, plastication, and cooling. There are a few points between the major phases, which are caused by the unequal-length injection phase.

Figure 6.17 shows the phase-division results for three test batches (batches 1–3) from which determination of the durations of their injection phases can be done easily and accurately. The injection velocities of the three test batches are 22, 24, and 26 mm/s, respectively. According to process knowledge, test batch 1 is the longest batch, and test batch 3 has the shortest injection time.

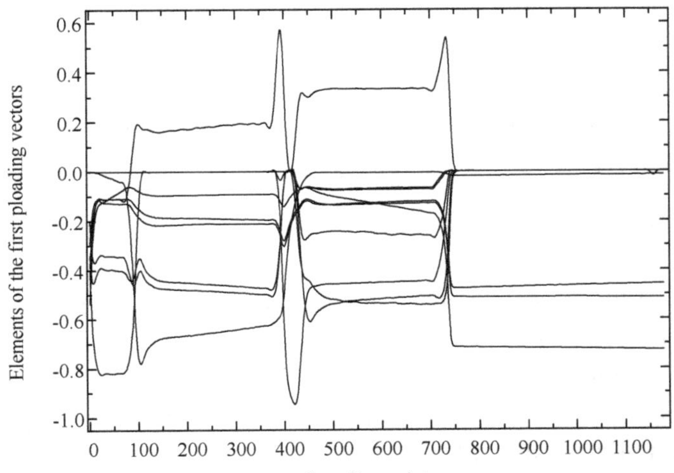


**Figure 6.15** First loading vectors of time-slice PCA loading matrices over the batch duration

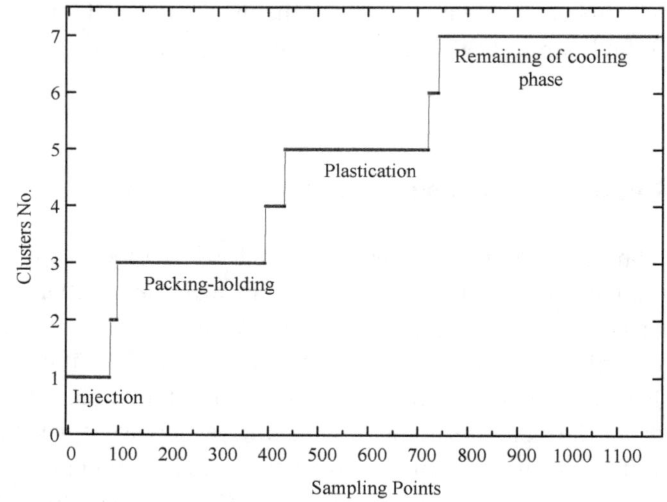


**Figure 6.16** Clustering results for determination of the first-phase PCA model

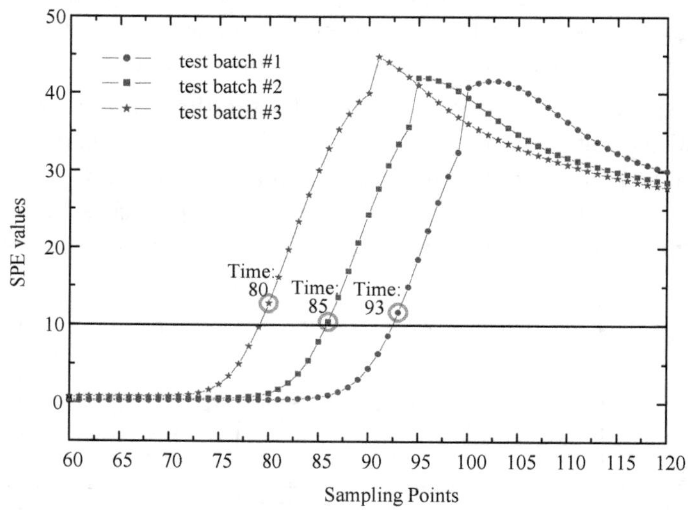


**Figure 6.17** Determination of the durations of the first phases

The SPE trajectories, obtained after application of the phase-division PCA model to these test batches, exactly agree with the true durations of the injection phase. The longest injection phase has 93 data points, whereas the shortest one has 80.

The remaining data after the removal of the injection-phase data have almost the same length and overlapping trajectories. The same phase-partition method can be applied to the remaining data to determine the durations of the pack-holding, plastication, and cooling phases. The clustering results are illustrated in Fig. 6.18.

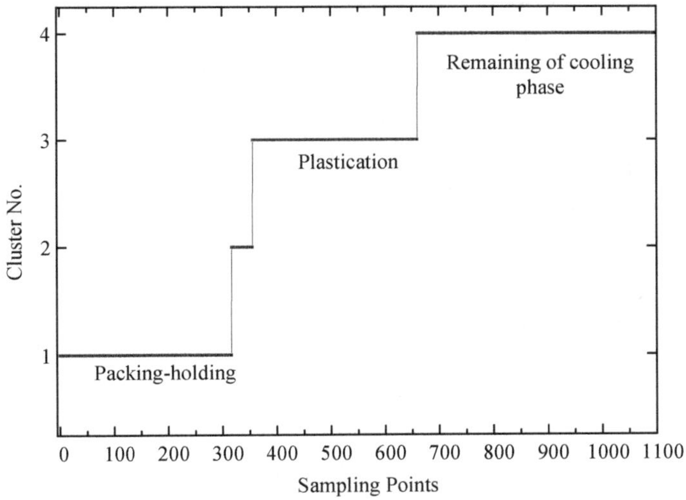


**Figure 6.18** Clustering results for the determination of the remaining phases

A short period between the pack-holding and plastication phases forms a separate phase, which was specially designed for retraction of the nozzle into the molding.

## 6.5.3 Result Analysis of Process Monitoring and Fault Diagnosis

A normal batch and three abnormal batches are selected to illustrate the performance of the online monitoring and fault diagnosis algorithm for the unequal-length injection molding process.

### 6.5.3.1 Monitoring of a Normal Batch

The SPE monitoring chart for a normal batch is shown in Fig. 6.19. For Fig. 6.19 and all remaining figures in this section, points marked with "■" are SPE values calculated by applying the injection-phase PCA model, whereas points marked with "▲" are SPE values calculated with the pack-holding phase model. The solid line represents the control limits corresponding to the significance level $\alpha = 0.01$, and the dashed line is the control limits for $\alpha = 0.05$.

After the phase-division procedure described above has been applied, the following phase information has been obtained from the experimental results: (1) The injection phase has a varying length, and the other phases have fixed operation durations; (2) The longest and shortest injection phases have 93 and 80 data points, respectively, which implies that the injection time of any normal batch should fall within the range 80–93, provided that the reference database has covered all normal possibilities.

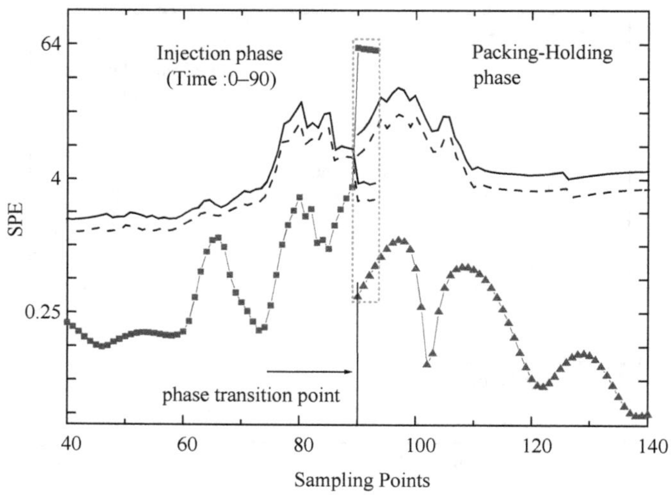


**Figure 6.19** SPE monitoring chart for a normal batch

For online monitoring, the data before sample 80, the shortest length of the injection phase, are normalized using the corresponding mean and standard deviation trajectories. The injection-phase PCA model is then applied to these data to compute the $T^2$ and SPE statistics. If both the statistics are below the control limits, the data are considered normal. Otherwise, any significantly large statistic indicates an abnormality.

For the data within the period 80–93, to differentiate a process abnormality from a phase changeover, those data that exhibit significant statistics using the injection-phase monitoring model are regarded as the beginning of the pack-holding phase and renormalized accordingly. The pack-holding phase PCA model is then applied to the normalized data to compute the $T^2$ and SPE statistics. For the case in which the process has actually entered the pack-holding phase, the evolving data should have in-control statistics according to the pack-holding phase PCA model, as illustrated in Figure 6.19.

For the data after sample 93, the longest injection period of the reference data, all normal processes should finish the first operation phase and enter the pack-holding phase. The pack-holding phase PCA model is applied to the evolving data to monitor the process conditions. For the process monitored in Fig. 6.19, the actual operation switches from the injection phase to the pack-holding phase at sampling interval 90. The continued use of the injection phase monitoring model after this transition over the period 90–93, as indicated in a dotted-line box, gives higher statistics than the corresponding control limits, even though these data are well explained by the second-phase model, the pack-holding phase monitoring model.

#### 6.5.3.2 Monitoring of Faulty Batches

Check-ring problem, material disturbance, and valve sticking are simulated in the experimental injection molding machine. Check-ring problem affects the injection and pack-holding phases; material disturbance can change process correlations overall operation phases. Valve-sticking, simulated in the middle of the pack-holding phase, will have normal process measurements in the injection phase and the early part of the packing phase. Figures 6.20–6.22 show the monitoring charts and diagnosis results for these three faults in the injection and pack-holding phases. The monitoring and diagnosis results for the other phases are not shown, as this section focuses on phase division and monitoring for an unequal-length injection molding process.

Fault mechanism of check-ring problem has been introduced in Section 6.3.3, which will result in a lower nozzle pressure (No. 2), hydraulic pressure (No. 5), and cavity pressure (No. 7). According to the correlation model, the pressures and injection velocity have a negative correlation; thus, the same pressures will result in a higher injection velocity (No. 4). The SPE monitoring chart, shown in Fig. 6.20(a), can clearly indicate the abnormality during the injection and packing phases as the SPE values are clearly beyond the control limits. The contribution plot, Fig. 6.20(b), shows that the process variables that are seriously impacted by the abnormality agree well with the fault pattern of check-ring leakage described earlier.

The abnormal batch with a material disturbance fault is carried out by adding a small amount of polypropylene (PP) to the processing of HDPE. The fault mechanism has been explained in Section 6.3.3, where the pressure variables (No. 2, 5, and 7) and the mold temperature (No. 11) are seriously affected by the fault. The monitoring chart in Fig. 6.21(a) can clearly detect the fault, and the diagnosis chart in Fig. 6.21(b) can reveal the impacts of this fault on the aforementioned process variables.

The SV1 valve-sticking fault is introduced in the middle of the packing phase, resulting in an uncontrolled nozzle pressure. Prior to this point, the process measurements in the injection phase and the early part of the packing phase should yield in-control SPE and $T^2$ statistics. When the SV1 valve fails to work in the packing phase, the nozzle pressure (No. 2) increases. The cavity pressure (No. 7) also increases following the change of nozzle pressure. The correlation between these variables is not altered by this fault, resulting in insignificant SPE statistics for the faulty data. The $T^2$ statistics, however, increase beyond the control limit because of the increased nozzle and cavity pressures, as shown in Fig. 6.22(a). The contributions of the variables to the first principal component are shown in Fig. 6.22(b), where the two pressure variables (No. 2, 7) contribute most to the significant $T^2$ statistics.

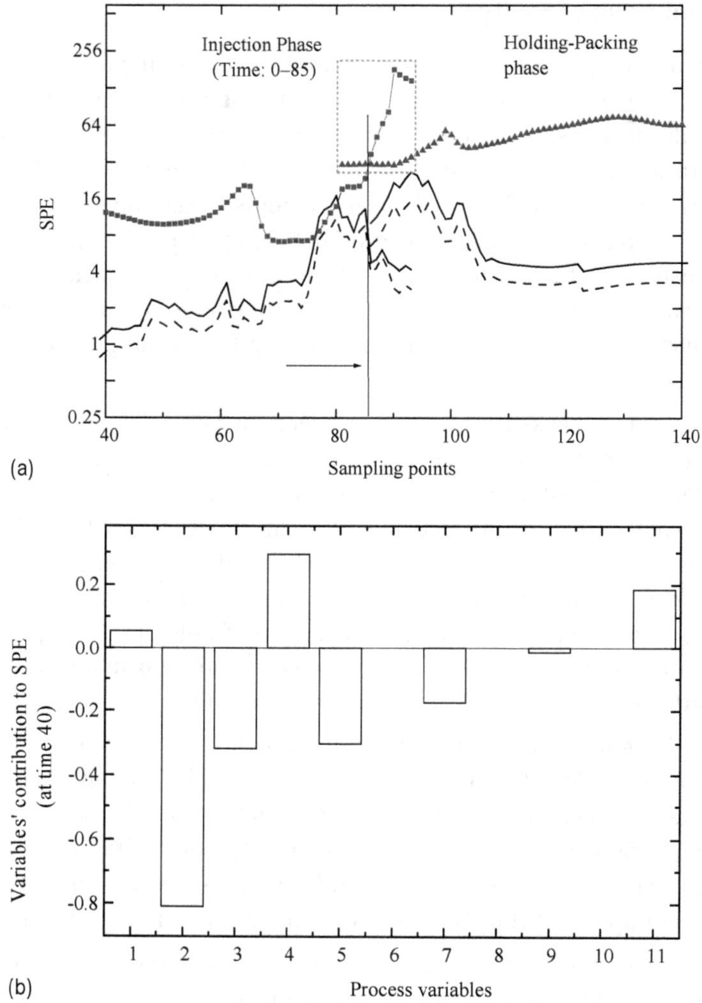


**Figure 6.20** Monitoring of an abnormal batch with check-ring problem: (a) SPE monitoring chart; (b) contribution plot

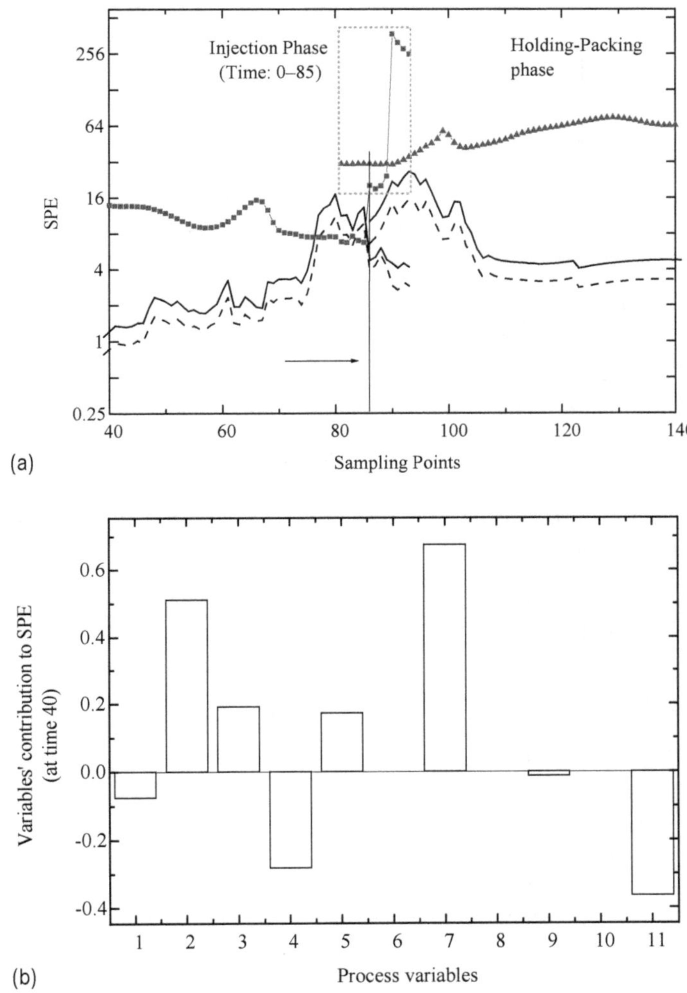


**Figure 6.21** Monitoring of an abnormal batch with material disturbance: (a) SPE monitoring chart; (b) contribution plot

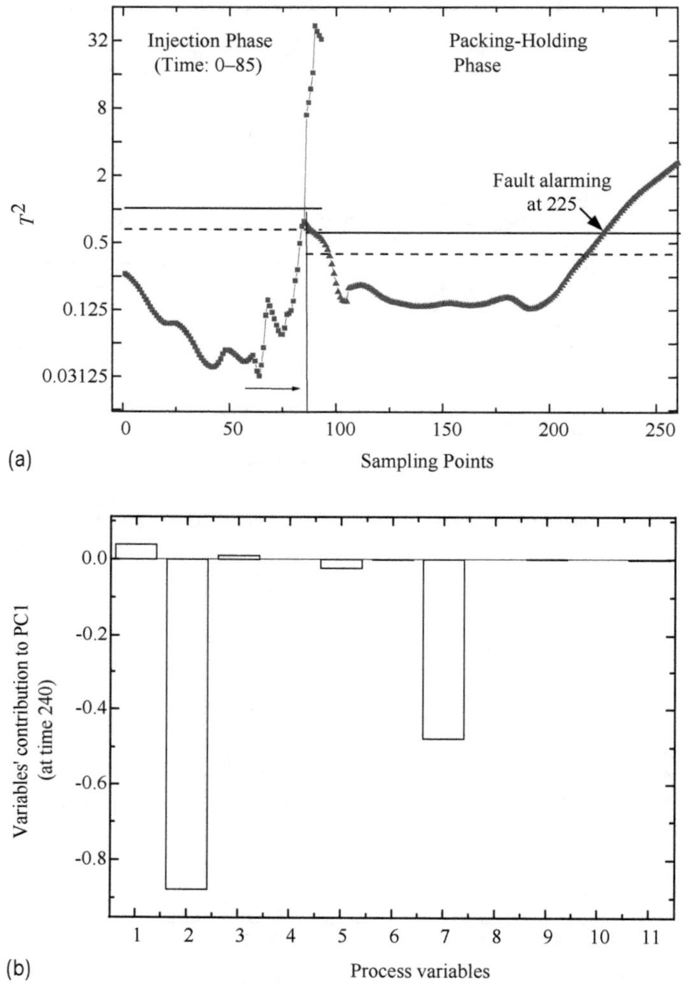


**Figure 6.22** Monitoring of an abnormal batch with a valve-stick fault: (a) $T^2$ monitoring chart; (b) contribution plot

## References

[91] J. Jackson, *A User's Guide to Principal Components*, Wiley, Hoboken, 1991

[97] P. Miller, R.E. Swanson, and C.E. Heckler, *Appl. Math. Comput. Sci.*, **8**(4), 775-792, 1992

[100] P. Nomikos and J.F. MacGregor, *AIChE J.*, **40**, 1361-1375, 1994

[102] P. Nomikos and J.F. MacGregor, *Technometrics*, 41-59, 1995(b)

[103] J. Camacho and J. Picó, *Chemom. Intell. Lab. Syst.*, **81**, 127-136, 2006(a)

[104] D. Dong and T.J. *McAvoy, AIChE J.*, **42**, 2199-2208, 1996

[105] K.A. Kosanovich, M.J. Piovoso, and K.S. Dahl, *Proc. Am. Control Conf.*, **2**, 1294-1298, 1994

[106] J.A. Westerhuis, T. Kourti, and J.F. MacGregor, *J. Chemom.*, **12**, 301-321, 1998

[107] A. Smilde, J. Westerhuis, and S. de Jong, *J. Chemom.*, **17**, 323-337, 2003

[108] S. Reinikainen and A. Höskuldsson, *Anal. Chim. Acta*, **595**, 248-256, 2007

[109] J. Liu and D. Wong, *J. Chemom.*, **22**, 385-398, 2008

[110] K. Gollmer and C. Posten, *Control Eng. Pract.*, **4**, 1287-1295, 1996

[111] K. Muthuswamy and R. Srinivasan, *J. Process Control*, **13**, 367-382, 2003

[112] C. Ündey and A. Cinar, *IEEE Control Syst. Mag.*, **22**, 40-52, 2002

[113] P. Facco, M. Olivi, C. Rebuscini, F. Bezzo, and M. Barolo, **2**, 93-98, 2007

[114] X. Doan and R. Srinivasan, *Comput. Chem. Eng.*, **32**, 230-243, 2008

[115] K. Kosanovich, K. Dahl, and M. Piovoso, *Ind. Eng. Chem. Res.*, **35**, 138-146, 1996

[116] N. Lu, F. Gao, and F. Wang, *AIChE J.*, **50**, 255-259, 2004(a)

[117] N. Lu, F. Gao, Y. Yang, and F. Wang, *Ind. Eng. Chem. Res.*, **43**, 3343-3352, 2004(b)

[118] J. Camacho and J. Picó, *J. Process Control*, **16**, 1021-1035, 2006(b)

[119] J. Camacho, J. Pico, and A. Ferrer, *J. Chemom.*, **22**, 632-643, 2008

[120] A.K. Jain, M.N. Murty, and P.J. Flynn, *ACM Comput. Surv.*, **31**, 264-323, 1999

[121] M.H. Kaspar and W.H. Ray., *AIChE J.*, **38**(10), 1593-1608, 1992

[122] T. Kourti, *J. Chemom.*, **17**, 93-109, 2003

[123] S. Wold, N. Kettaneh, H. Fridén, and A. Holmberg, *Chemom. Intell. Lab. Syst.*, **44**(1), 331-340, 1998

# 7 Phase-Based Quality Improvement Strategies

## 7.1 Introduction

The ultimate concern of batch process monitoring is to monitor the variation of end-product quality and find solutions to improve quality consistency. It is, however, difficult to monitor product quality for batch processes in an online way. The main reasons lie in that (1) end-product quality attributes are only available after a batch operation is finished, and (2) most quality attributes are difficult to measure instantly after a product is produced. Alternatively, SPM techniques are used to monitor product quality indirectly, inferring end-product quality from process behaviors, which have been introduced in Chapters 5 and 6.

For some batch processes where quality measurements are obtainable offline, it is very attractive to use offline quality information along with online process measurements for safety monitoring and quality inspection. To this end, online quality prediction is indispensable, which can reveal the relationships between online measurable process variables and offline obtainable quality attributes. Significant efforts have been made for the development of methods for online quality prediction, among which multivariate statistical modeling methods are the most popular as they are derived directly from historical data with little prior process knowledge, and have superior ability in handling high-dimensional and correlated process data.

The multiway partial least squares (MPLS) method [101] was proposed for such a purpose. But it uses process variables over the entire batch course as the inputs and requires future process measurements in the current batch for online quality prediction; thus the monitoring and quality prediction accuracy is dependent on the prediction accuracy of the future data. Furthermore, when three-way batch data are arranged in a batch-wise form in MPLS, the number of unfolded augmented variables is huge, and the complexity of autocorrelation and cross-correlation is dramatically increasing. MPLS has met great challenges when applied for batch product quality prediction. Improved PLS/MPLS modeling for quality interpretation and prediction have been reported [124], for example, by selecting the key process variables or focusing on the critical-to-quality time periods. Unfortunately, most of these methods can only be used as a post-analysis tool.

For multiphase batch processes like injection molding, which have time-specific effects of process variables on the final qualities, MPLS is inefficient in revealing time-specific relationships. To focus on the time-specific effects of process variables on the final product quality, a pathway multiblock PLS algorithm was developed by Duchesne and MacGregor [125] by introducing intermediate quality measurements to isolate the local effects of process variables to the final product quality. But they did not provide any means of dividing a batch process into the blocks, in addition, for most industrial processes, online measurements of intermediate quality are rarely available. A bootstrapping improved MPLS [126] can also isolate the local effects of process variables on the final quality based on so-called VIP (variable importance in the projection of MPLS); computation of bootstrapping-based variable selection is, however, very heavy.

This chapter introduces two phase-based quality improvement methods, which can use the inherent phase nature to improve the performances of online quality prediction models. A phase-based PLS modeling method is introduced in Section 7.2, based on which phase-based process analysis and quality prediction are developed and applied to an injection molding process. These methods can have the following advantages:

- They allow the exploration of relationships between process and quality variables over different phases, to determine the critical-to-quality phases and to identify the key variables in each phase.
- They allow earlier quality prediction in critical-to-quality phases, and avoid the prediction of future process data in a running batch.
- They have simpler model structures because a conventional two-way PLS model is directly used in each phase.

## 7.2 Phase-Based Process Analysis and End-Product Quality Prediction (Method A)

### 7.2.1 Phase-Based PLS Modeling

With the available offline quality measurements, phase division can be made according to the changes of correlation between process and quality variables. This will allow us to determine the phase-specific effects of process variables on the final product qualities. Similar to phase-based PCA modeling, a *k*-means clustering algorithm is adopted for phase division.

Consider an equal-length batch process with $J_x$ process variables measured over time intervals $k$ ($k = 1,2,\cdots,K$) and $J_y$ quality variables available at the end of each batch run; a data matrix of dimensions $J_x \times K$ for process variables and a vector of dimension $1 \times J_y$ for quality variables are generated from each batch. A set of $I$ number of historical batch data thus result in a three-way process data matrix, $\underline{X} \in R^{I \times J_x \times K}$, and a two-way quality data matrix, $Y \in R^{I \times J_y}$. Then, $\underline{X}$ is normalized using the NM normalization technique, that is, $\underline{X}$ is centered and scaled across the batches, to remove process nonlinearity along the trajectories and to focus on batch-to-batch variations. Two-way quality data matrix $Y$ should also be normalized to have zero means and unit variances for all quality variables.

Similar as in the phase-based PCA modeling, $\underline{X}$ is first cut along the time axis, resulting in $K$ time-slice matrices, $\tilde{X}_k \in R^{I \times J_x}$ ($k = 1,2,\cdots,K$); $\tilde{X}_k$ contains batch-to-batch variations of process variables at time interval k. A standard PLS algorithm is applied to $\{\tilde{X}_k, Y\}$ at each time interval, generating $K$ number of time-slice regression models:

$$\begin{aligned} \tilde{X}_k &= \tilde{T}_k(\tilde{P}_k)^T + E_k \\ Y &= \tilde{U}_k(\tilde{Q}_k)^T + F_k \end{aligned} \tag{7.1}$$

The above time-slice PLS model can be written in a compact form as

$$\begin{aligned} Y &= \tilde{X}_k \cdot \tilde{\Theta}_k + F_k^* \\ \hat{Y}_k &= \tilde{X}_k \cdot \tilde{\Theta}_k \end{aligned} \tag{7.2}$$

Note that quality prediction can be done at every time interval as shown in Eq. 7.2, but it is impractical to use time-slice PLS models for online prediction. The following phase PLS modeling is more practical and effective.

The regression parameter matrix, $\tilde{\Theta}_k \in R^{J_x \times J_y}$, containing the correlation information between process and quality variables at time interval $k$, should remain similar in the same phase and show significant differences over different phases. Changes in process correlation can be reflected by the changes in the regression parameter matrices $\tilde{\Theta}_k$ ($k = 1,\cdots,K$). The same *k-means* clustering algorithm, as detailed in Section 6.2.3, is adopted for phase division. Process phases can be determined by the clustering results associated with the process operation time.

The phase division results based on clustering of time-slice PLS regression parameter matrices could be different from that of phase-based PCA modeling, where clustering is conducted on the time-slice PCA loading matrices. It is likely that a batch process may be characterized by several quality variables, and different quality variables may have different correlations to different variables over different periods (or phases). This phase division method requires no prior process knowledge: it automatically divides a batch process into meaningful phases, allows analysis of different local effects of process variables on the product qualities. The phase division also allows the conventional PLS model to be applied directly to batch processes.

Suppose that $K$ time-slice regression parameter matrices are clustered into $C$ groups, indicating $C$ number of phases with different correlation structure between process and quality variables. Then, a representative regression parameter matrix for phase $c$ can be computed by

$$\Theta_c^* = \frac{1}{n_c}\sum_k \tilde{\Theta}_k (c = 1,2,\cdots,C;\ k = 1,2,\cdots,n_c) \tag{7.3}$$

where $n_c$ is the data length in phase $c$.

The phase PLS model for quality prediction can be formulated as

$$\hat{\mathbf{y}}_k = \mathbf{x}_k \cdot \Theta_c^*, c = 1,2,\cdots,C, k = 1,\cdots,n_c \tag{7.4}$$

where $\mathbf{x}_k$ is composed of process measurements at time $k$, belonging to phase $c$. An illustration of the proposed phase-based PLS modeling scheme is given in Fig. 7.1.

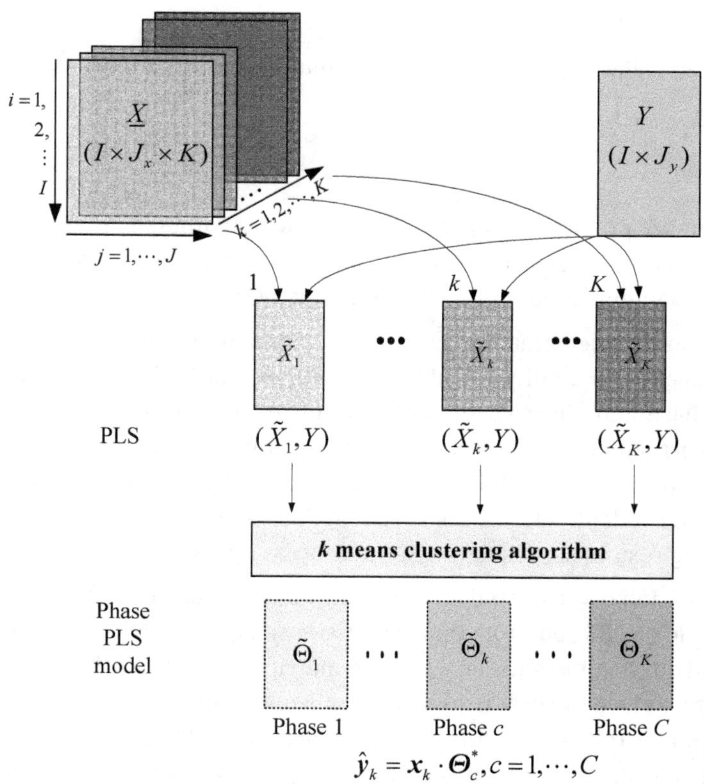


**Figure 7.1** Illustration of phase-based PLS modeling

### 7.2.2 Phase-Based Quality-Related Process Analysis

In batch processes, it is common that the product qualities are determined by some key variables in some critical operation phases, and different quality attributes may be determined by different phases. For example, in injection molding, product surface defects, such as jetting and record grooves, depend only on the injection phase, while product dimensions are mainly determined by the pack-holding phase. We define a phase that has the most important contribution to certain quality attributes as a critical-to-quality phase. A batch process may have several critical-to-quality phases if a product has several quality measures. For different quality measures, critical-to-quality phases may be different or the same but determined by different variables, depending on the process characteristics. It is, therefore, important to identify critical-to-quality phases and find out in each phase the key process variables that contribute most to the quality variations. The phase-based PLS modeling method as formulated in Section 7.2.1 can indeed provide such a platform for the phase-based quality-related analysis.

Critical-to-quality phases can be determined by measuring the goodness of fit of the phase PLS models. Phases with prediction models that can explain significant quality variations are defined as critical-to-quality phases. The multiple coefficient of determination, $R^2$, in linear regression analysis can be adopted to evaluate the fitness of each phase PLS model and to indicate the variables' contributions of these phases to the final quality variations.

The index $R^2$ can be calculated for each quality variable $y_{j_y}$ $(j_y = 1, \cdots, J_y)$ at every sampling interval $k$:

$$R^2_{j_y,k} = 1 - \frac{\sum_{i=1}^{I} (y_{i,j_y,k} - \hat{y}_{i,j_y,k})^2}{\sum_{i=1}^{I} (y_{i,j_y} - \bar{y}_{j_y})^2} \quad (j_y = 1, \cdots, J_y) \tag{7.5}$$

where $i$, $j_y$, $k$ are indices of batch run, quality variables, and sampling time, respectively; $y_{i,j_y}$ is the real measurement; $\bar{y}_{j_y}$ is the average across batches; and $\hat{y}_{i,j_y,k}$ is the quality prediction at time $k$, calculated by calling the corresponding phase PLS model. Obviously, $R^2_{j_y,k}$ ranges from 0 to 1.

For each quality variable, $y_{j_y}$, larger $R^2_{j_y,k}$ indicates better fitness of the corresponding phase PLS model; that is, the phase PLS model is more convincingly accurate and reliable for the prediction of quality variable $y_{j_y}$. The $R^2_{j_y,k}$ indices over the batch duration can therefore be used to determine the critical-to-quality phases. Like in linear regression, the *F test* can be used to test the significance of $R^2$: $F = \frac{n-m-1}{m}\left(\frac{R^2}{1-R^2}\right)$. The critical values of $R^2$ can be conversely calculated by the above expression, where the critical values of $F$ statistic with significance factors ($\alpha = 0.01$ or 0.05) can be found in statistical tables of $F$ distributions. If the average $R^2_{j_y,k}$ at phase $c$ is larger than the critical values, phase $c$ is defined as the critical phase for quality variable $y_{j_y}$.

The purposes of phase-based process analysis are twofold: to identify the critical phases for each quality variable, and to analyze the phase effects of process variables on the quality variations. Whereas the above $R^2$ index is used to find out the critical-to-quality phases, a new index will be introduced for determine the key process variables in each critical-to-quality phase.

With the phase PLS model of Eq. 7.4, prediction of quality variable $y_{j_y}$ for the reference data sets can be formulized as

$$\begin{aligned}\hat{y}_{j_y,k} &= [\mathbf{x}_{1,k}, \mathbf{x}_{2,k} \cdots \mathbf{x}_{J_x,k}] \cdot \Theta^*_{c,j_y} \\ &= \sum_{j_x=1}^{J_x} \mathbf{x}_{j_x,k} \times \Theta^*_c(j_x, j_y)\end{aligned} \tag{7.6}$$

where $\hat{y}_{j_y,k}$ is the predicted quality at time $k$, $\Theta^*_{c,j_y} = \left[\Theta^*_c(1, j_y), \cdots, \Theta^*_c(J_x, j_y)\right]^T$ and the contribution of process variable $x_{j_x}$ to $\hat{y}_{j_y}$ at time $k$ can be computed by

$$\hat{\mathbf{y}}_{j_y,k} \mid \mathbf{x}_{j_x,k} = [0, \cdots, \mathbf{x}_{j_x,k}, \cdots 0] \cdot \Theta^*_c(j_x, j_y) \tag{7.7}$$

The contribution rate of process variable $x_{j_x}$ to $\hat{y}_{j_y}$ can then be defined as

$$C_{\mathbf{y}_{j_y}\mathbf{x}_{j_x},k} = 1 - \frac{\left\| \mathbf{y}_{j_y} - \hat{\mathbf{y}}_{j_y,k} \mid \mathbf{x}_{j_x,k} \right\|}{\left\| \mathbf{y}_{j_y} \right\|} \tag{7.8}$$

The larger $C_{\mathbf{y}_{j_y}\mathbf{x}_{j_x},k}$ is, the more significant the contribution of the process variable $x_{j_x}$ is to the variation of quality variable $y_{j_y}$ at time $k$. This can therefore be used to determine the key process variables in each critical-to-quality phase.

The regression parameter matrix of each phase PLS model may be further explored for the details of the relationship between process variables and the final product qualities. The above-proposed phase-based process analysis is useful for quality improvement because it provides what process variables in which phases should be better controlled for quality improvement.

### 7.2.3 Online Quality Prediction

Quality variables in a batch process can be divided into two types: qualities determined by only one specific phase and qualities determined by more than one phase. Two online quality prediction algorithms are hence developed in this section, correspondingly.

For the first case, the PLS models of the critical-to-quality phases can be simply used without any modification for online quality prediction, mathematically represented by

$$\hat{y}_{j_y,k} = \begin{cases} \mathbf{x}_k \cdot \Theta^*_{c,j_y} & \text{if stage } c \text{ is critical to } y_{j_y} \\ \text{null} & \text{if stage } c \text{ is not critical to } y_{j_y} \end{cases} \tag{7.9}$$

$$j_y = 1,2,\cdots,J_y; c = 1,2,\cdots,C; k = 1,2,\cdots,n_{stage\_c}$$

If, according to the $R^2$ indices, the quality variable $y_{j_y}$ is only determined by phase $c$ (i.e., phase $c$ is the critical-to-$y_{j_y}$ phase), the phase-$c$ prediction model is used to estimate $y_{j_y}$ at each sampling interval of that phase, with no prediction of $y_{j_y}$ in any other phases.

The predicted quality may vary slightly with time within the same phase, and the variation is mainly caused by measurement noises and modeling errors as the phase PLS model is a linear and averaged time-slice PLS model of that phase. This variation is small, as seen in Section 7.3. Similar to the final predicted qualities in methods based on an MPLS model, the averaged value of $\hat{y}_{j_y,k}$ in phase $c$, $\hat{y}^*_{c,j_y}$, is defined as the *end-of-phase prediction.*

For the second case, the variations in quality are determined by more than one phase. Two or more critical-to-quality phases have cumulative but different effects on the product quality. This case allows a larger degree for online quality compensation; the predicted deviation in the earlier critical-to-quality phases may be compensated for in the current or future critical-to-quality phases. A strategy combining multiphase PLS models will have to be developed; stacked modeling methods [127, 128] can be adopted to weight each phase PLS model for the combination.

Without losing generality, assume that quality variable $y_{j_y}$ has two critical phases, phases $c_1$ and $c_2$. These two phases may explain different parts of quality variations, and accurate quality prediction becomes available in the last critical phase. The online quality prediction in phase $c_2$ can be formulated as

$$\hat{y}_{j_y,k} = w_1 \cdot y^*_{c_1,j_y} + w_2 \cdot \mathbf{x}_k \cdot \Theta^*_{c_2,j_y}, \quad k = 1,2,\cdots,n_{c_2} \tag{7.10}$$

where $w_c (c = 1,2)$ is the weighting factor of phase $c$; $\hat{y}^*_{c_1,j_y}$ is the quality prediction at the end of phase $c_1$; $\mathbf{x}_k$ is the current process measurement at phase $c_2$; $\Theta^*_{c_2,j_y}$ is the $j_y$th column of $\Theta^*_{c_2}$ for phase $c_2$; and $n_{c_2}$ is the length of phase $c_2$. The weighted sum of the end-of-phase predicted values is defined as the *final-phase prediction*, formulated as

$$\hat{y}_{j_y} = \sum_c w_c \hat{y}^*_{c,j_y} \tag{7.11}$$

The phase weight $w_c$ can be obtained by methods such as least squares regression (LSR) by minimizing $\sum_{i=1}^{I}(y_{i,j_y} - \sum_c w_c \hat{y}^*_{c,i,j_y})^2$ for the reference data set, where $i$, $j_y$, and $c$ are indices of batch runs, quality variables, and critical-to-$y_{j_y}$ phases. If the end-of-phase predictions in the different critical-to-quality phases are correlated, ridge regression, principal component regression (PCR), or partial least squares (PLS) can be used to calculate the weight vector in stacked regression instead of the LSR method.

## 7.3 Application of Phase PLS Model (Method A) to Injection Molding

### 7.3.1 Experimental Setup

The selected process and quality variables are listed in Table 7.1, among which the dimension qualities of weight and length can be directly measured by instruments; the surface defects, jetting, and record grooves, are quantified by a process operator expert before modeling.

An experimental design as shown in Table 7.2 is applied to excite the process, generating training data sets for modeling and analysis. Ranges of the experimental factors are determined according to the machine and mold specifications. The first 12 conditions in the design, treatments A–L, form a balanced design of experiment (DOE) with packing pressure taking three levels (low = 150 bar, middle = 300 bar, and high = 450 bar), and barrel temperature (low = 180°C and high = 220°C) and mold temperature (low = 15°C and high = 55°C) at two levels. Treatments M, N, and O are added with barrel temperature (middle = 200°C) and mold temperature (middle = 35°C), both taking the centered value and packing pressure of three levels. Because jetting and record grooves are associated with injection velocity, four additional experiments with different velocity settings (low = 8 mm/sec and high = 40 mm/sec) are conducted in conditions A, C, and D. All other process settings are shown in Table 7.3.

**Table 7.1** Selected Process and Quality Variables

| No. | Description | Unit |
|---|---|---|
| **Process Variables** | | |
| 1 | Cavity Temperature | °C |
| 2 | Nozzle Pressure | bar |
| 3 | Stroke | mm |
| 4 | Injection Velocity | mm/sec |
| 5 | Hydraulic Pressure | bar |
| 6 | Plastication Pressure | bar |
| 7 | Back Pressure | bar |
| 8 | Cavity Pressure | bar |
| 9 | SV1 | % |
| 10 | SV2 | % |
| 11 | Barrel Temperature | °C |
| 12 | Mold Temperature | °C |
| **Quality Variables** | | |
| 1 | Weight | g |
| 2 | Length | mm |
| 3 | Jetting | – |
| 4 | Record Grooves | – |

**Table 7.2** Operation Conditions for Training Data Sets by DOE

| Packing Pressure | Barrel Temperature | | | Mold Temperature |
|---|---|---|---|---|
| | 180°C | 200°C | 220°C | |
| **150** | A | - | G | 15°C |
| **300 bar** | - | M | - | |
| **450** | B | - | J | |
| **150** | C | - | I | 35°C |
| **300 bar** | - | N | - | |
| **450** | D | - | J | |
| **150** | E | - | K | 55°C |
| **300 bar** | - | O | - | |
| **450** | F | - | L | |

**Table 7.3** Process Settings for Injection Molding Process

| Parameters | Settings |
|---|---|
| Material | High-density polyethylene (HDPE) |
| Injection Velocity | 8–40 mm/sec |
| Injection Stroke | 38.5 mm |
| Pack-Holding Time | 6 sec |
| Plastication Back Pressure | 5 bar |
| Screw Rotation Speed | 80 rpm |
| Cooling Time | 15 sec |

**Table 7.4** Operation Conditions for Validation Data Sets

| No. | Packing Pressure (bar) | Barrel Temperature (°C) | Mold Temperature (°C) | Injection Velocity (mm/sec) |
|---|---|---|---|---|
| 1 | 150 | 180 | 15 | 8 |
| 2 | 275 | 190 | 30 | 20 |
| 3 | 275 | 210 | 40 | 20 |
| 4 | 300 | 200 | 35 | 24 |
| 5 | 300 | 200 | 35 | 20 |
| 6 | 150 | 180 | 15 | 40 |
| 7 | 325 | 190 | 40 | 20 |
| 8 | 325 | 210 | 30 | 20 |
| 9 | 450 | 220 | 35 | 24 |

Altogether, 33 training batch runs were collected for modeling and post batch analysis. Nine experiments with settings shown in Table 7.3 were conducted as validation data sets to evaluate the method in Section 7.2. The details of each of those nine experiments are listed in Table 7.4.

## 7.3.2 Illustration of Phase-Based Process Analysis

### 7.3.2.1 Phase Division

As different injection velocities result in different filling times in the injection phase, reference batch runs therefore have varying data length. To arrange these batch data into matrices, batches with different injection velocities are aligned to have the same filling length, taking injection stroke as an indicator variable, as the values of injection stroke decrease monotonously from 60 mm to 21.5 mm, and the total displacement is fixed at 38.5 mm at the end of the injection phase. There are finally 78 samples in the injection phase after alignment. The dimensions of the final data matrices $\underline{X}$ and $Y$ are $33 \times 12 \times 1100$ and $33 \times 4$, respectively.

According to the modeling procedure, $\underline{X}$ is first cut into 1100 time-slice matrices, $\tilde{X}_k$ ($k = 1, \cdots, 1100$). PLS is applied to $\{\tilde{X}_k, Y\}$ generating 1100 time-slice regression parameter matrices $\tilde{\Theta}_k$. The clustering algorithm divided the process into six phases, as shown in Fig. 7.2. This phase division result is slightly different from that of the phase-based PCA modeling method in Chapter 6, where the process is divided into four phases, consistent with the real operation phases: injection, pack-holding, plastication, and cooling. From Fig. 7.2, the first two phases and the last phase are in agreement with the actual operation phases of injection, pack-holding, and cooling, but the plastication operation is partitioned into two subphases, phases 4 and 5.

These two phases indeed have different effects on the final product quality and have different correlations between process and quality variables, as explained later. A short period between the pack-holding and plastication phases forms a separate phase, that is, phase 3 in Fig. 7.2. This phase corresponds to the operation of "suck-back" to retract the screw for a certain distance to prevent the melt from drooling, which has little impact on quality. This phase will be neglected in the following phase-based process analysis.

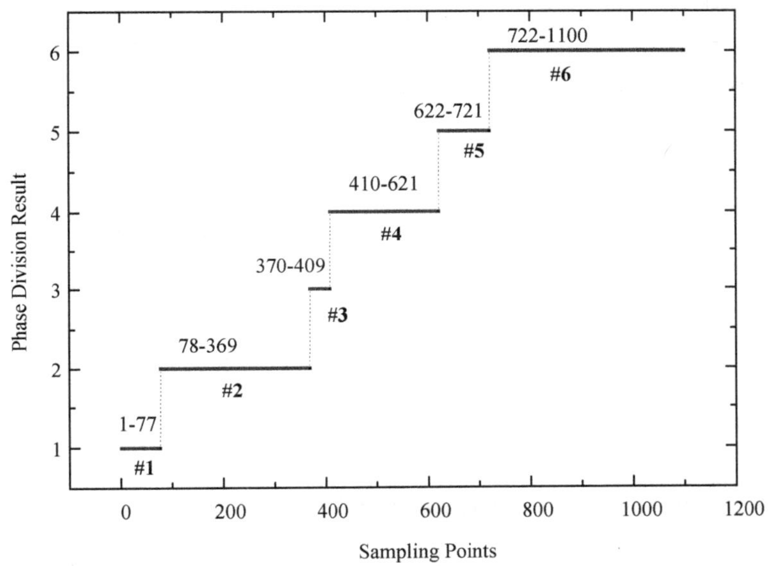


**Figure 7.2** Stage division result for injection molding process

Figure 7.3 shows the $R^2$ indices for the four quality variables over the batch duration. Obviously, the surface qualities, jetting, and record grooves, have large and stable $R^2$ in the first phase (i.e., the injection phase), while the dimension qualities, length and weight, have a significant relation to phases 2 and 4. The first phase is therefore defined as the critical-to-surface phase, and phases 2 and 4 are the critical-to-dimension phases. A detailed analysis is given in the following sections.

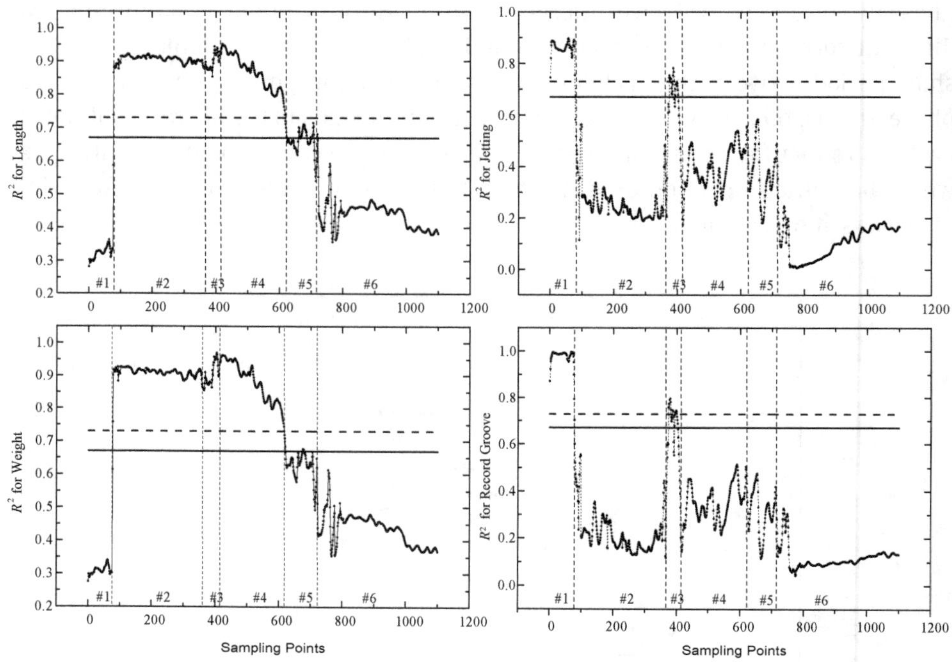


**Figure 7.3** $R^2$ plots of four quality variables in the injection molding process (solid line, 99% limit; dashed line, 95% limit)

#### 7.3.2.2 Process Analysis in the Critical-to-Surface Phase

Figure 7.4 shows the contributions of process variables to the prediction of jetting and record grooves, and the corresponding phase regression parameters that indicate the correlation pattern between process and quality variables in the first phase, respectively.

From Fig. 7.4(a), the variation of nozzle pressure (No. 2), injection velocity (No. 4), hydraulic pressure (No. 5), back pressure (No. 7), and manipulated variable SV1 (No. 9) accurately reflect the variation of jetting in the products. These process variables have positive correlations to jetting, which means that high pressures and a large injection velocity in the injection phase cause jetting in the final products. The same group of process variables also determines the variation of record grooves, as shown in Fig. 7.4(b). The correlation between these process variables and record grooves, however, are negative, indicating that low pressures and slow injection velocity are responsible for record grooves. The negative correlation between jetting and record grooves, in fact, agrees well with the process characteristics.

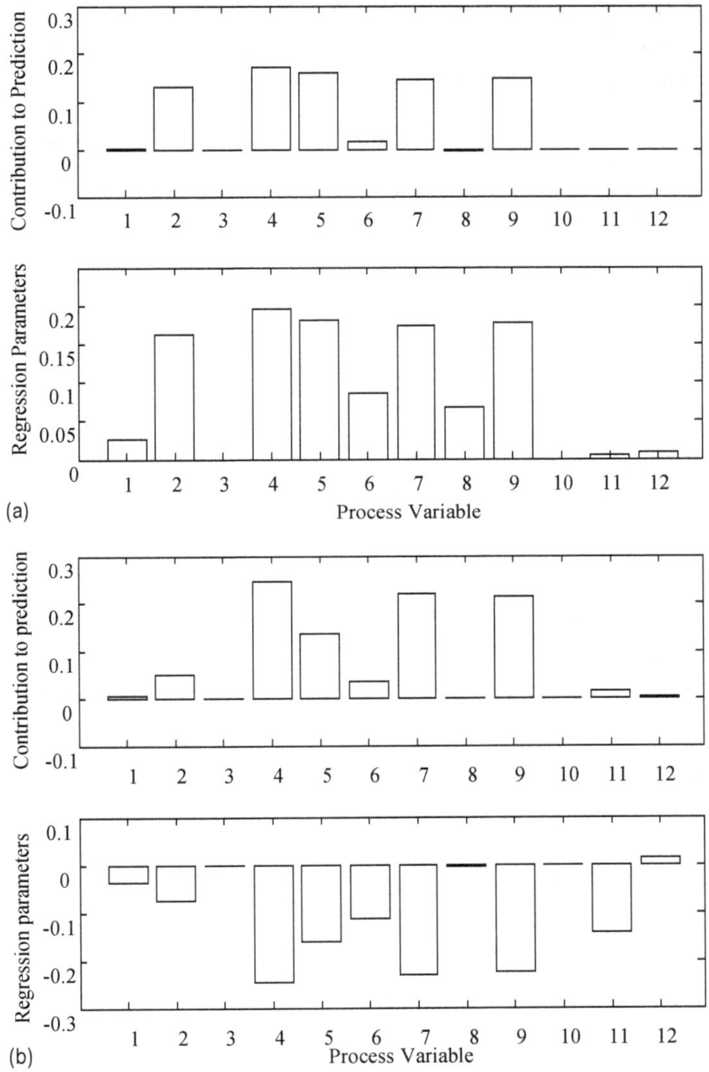


**Figure 7.4** Key process variables in the critical-to-surface stage (phase 1): (a) Jetting; (b) Record grooves

### 7.3.2.3 Process Analysis in Critical-to-Dimension Phases

In phase 2, the pack-holding phase, the dominant process variables include pressure variables (No. 2, 5, 8), temperature variables (No. 1, 11, 12), and manipulated variable SV1, as shown in Fig. 7.5. The phase regression parameters can explain how these variables affect the product dimension qualities. From Fig. 7.5, pressure variables have positive relations with the variations of length and weight, while temperature variables are negatively related to the dimension qualities. This implies that larger pressures and lower temperatures result in larger and heavier products, which again agrees well with the physics of polymer processing.

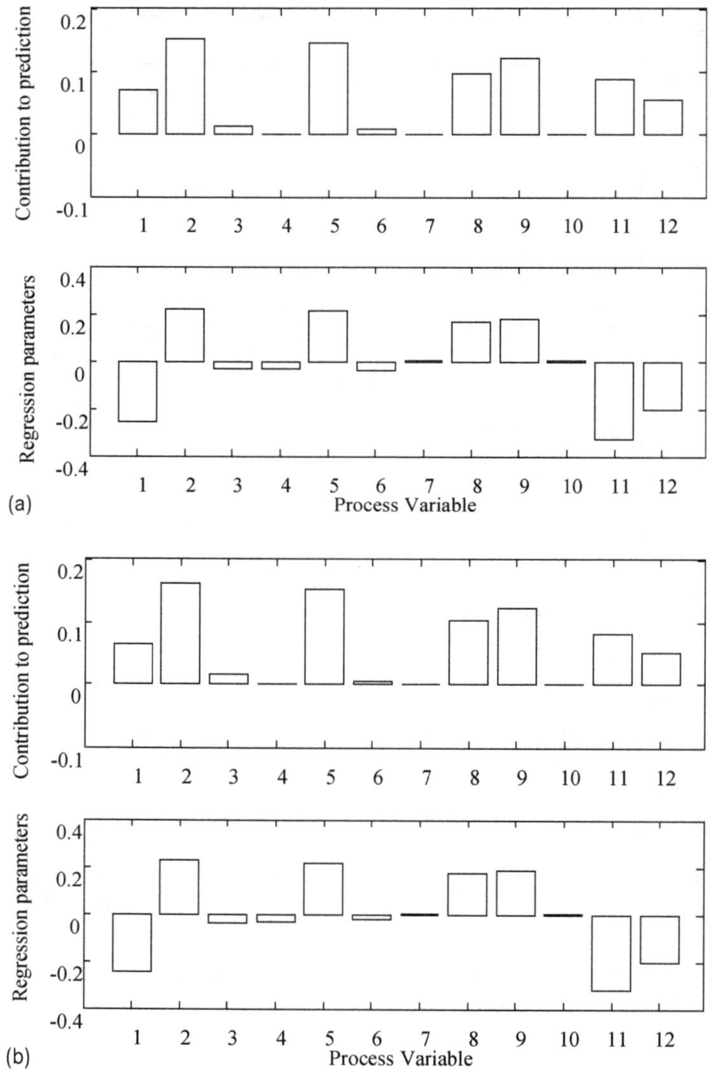


**Figure 7.5** Key process variables in critical-to-dimension stage (phase 2): (a) Length; (b) Weight

Phase 4, the earlier phase of the plastication phase, is also recognized by the algorithm as a critical-to-dimension phase according to the $R^2$ plot. Dividing the plastication operation into two subphases is reasonable in our experiments because the process shows different characteristics before and after the mold gate is frozen. From Fig. 7.6, cavity pressure (No. 8) is the most dominant variable reflecting variations in the dimension qualities.

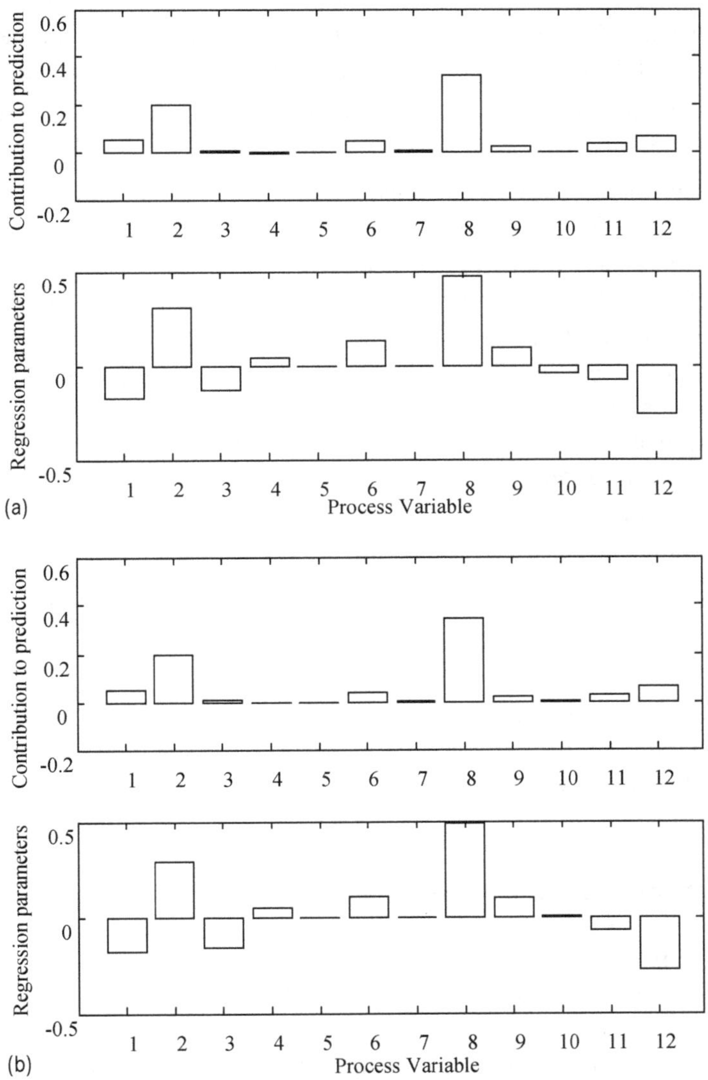


**Figure 7.6** Key process variables in the critical-to-dimension stage (phase 4): (a) Length; (b) Weight

This is because cavity pressure is mainly determined by the volume of material filled in the cavity and has a close relation with nozzle pressure (No. 2) and mold temperature (No. 12) before the mold gate is frozen in the earlier phase of plastication. The variations of the above three process variables therefore can accurately reflect the dimension variation of filled materials in the mold. The decreasing explained rate in phase 4, as shown in the $R^2$ plot, may be caused by the solidification in the filled polymer in the mold and gradual freezing of the gate.

Without use of prior process knowledge, the above analysis results in the following observations for injection molding, which agree well with the process characteristics:

- This injection molding process can be divided into six phases; different phases have different effects on the final product qualities.
- The surface qualities are determined by the injection phase (phase 1), where nozzle pressure (No. 2), injection velocity (No. 4), hydraulic pressure (No. 5), back pressure (No. 7), and manipulated variable SV1 (No. 9) are key process variables that have large contributions to the variation of surface attributes.
- The dimension qualities are mainly determined by the pack-holding phase (phase 2). Larger pressures (No. 2, 5, 7) and lower temperatures (No. 1, 11, 12) lead to heavier product. Cavity pressure (No. 8) and nozzle pressure (No. 2) at the earlier phase of the plastication phase (phase 4) can also have significant effects on dimension variations.
- The remaining period of plastication (phase 5) and the cooling phase (phase 6) have little effect on the final product qualities of concern here.
- For online quality prediction, surface qualities can be predicted by process variables in the first phase, while the dimension qualities can be predicted in the fourth phase.
- For quality improvement, tight control of injection velocity can improve surface qualities; while enhanced control of packing pressure and melt temperatures can achieve better dimension qualities.

### 7.3.3 Illustration of Phase-Based Quality Prediction

The above phase-based quality-related process analysis divides the four quality variables into two types. Surface qualities are determined only by the first phase, while the dimension qualities have close relations with phases 2 and 4. Online predictions of surface attributes are given in the first phase; predictions of weight and length are available from phase 4.

A comparison of prediction performance is conducted for the MPLS model, the phase PLS model, and the bootstrapping-based MPLS model (B-MPLS), as shown in Figs. 7.7,

7.8, and Table 7.5. Figures 7.7 and 7.8 show the quality predictions by the three methods for the training and validation data sets, respectively, and Table 7.5 summarizes the prediction performances. From Figs. 7.7, 7.8, and Table 7.5, conclusions can be drawn that the MPLS model is applicable for the prediction of quality variables that are cumulatively determined by several phases; the bootstrapping-based improved MPLS model is applicable for the prediction of quality variables that have a significant time-specific correlation to process variable trajectories; and the phase-based PLS model can give reasonable predictions for both types of quality variables.

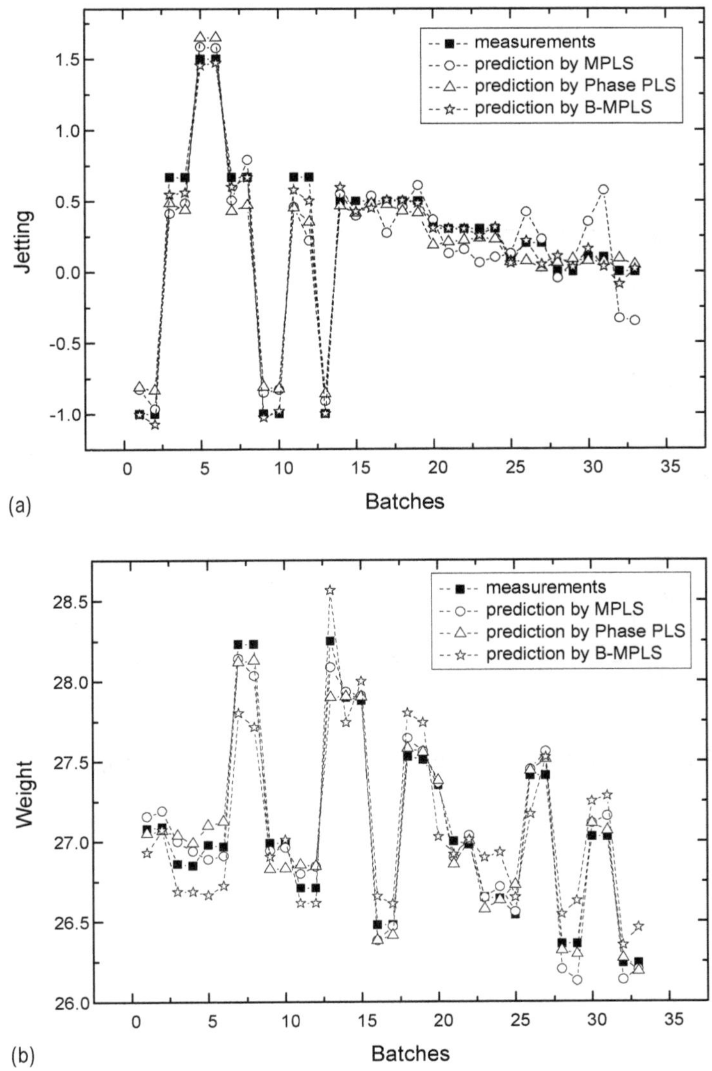


**Figure 7.7** Prediction comparison for training batch runs: (a) Jetting; (b) Weight

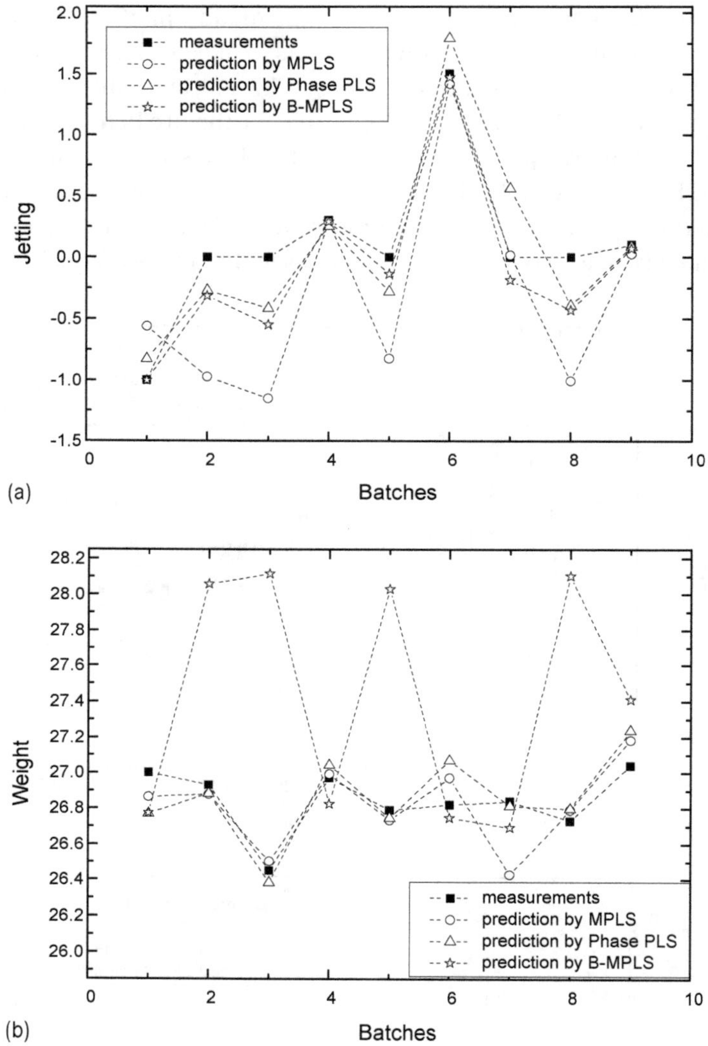


**Figure 7.8** Prediction comparison for test batch runs: (a) Jetting; (b) Weight

**Table 7.5** Prediction Performance Comparison for Both Training and Test Data Sets

| Methods | Mean Squared Error (MSE) | | | |
|---|---|---|---|---|
| | Length | Length | Jetting | R.G. |
| **Training data sets** | | | | |
| MPLS | 0.0055 | **0.0099** | 0.0394 | 0.0147 |
| Phase PLS | **0.0048** | 0.0147 | 0.0205 | 0.0019 |
| B-MPLS | 0.0213 | 0.0497 | **0.0046** | **0.0004** |
| **Test data sets** | | | | |
| MPLS | **0.0190** | 0.0272 | 0.4632 | 0.3065 |
| Proposed | 0.0254 | **0.0194** | 0.1007 | 0.0244 |
| B-MPLS | 10.45 | 0.08451 | **0.0712** | **0.0156** |

Online predictions of the product weight for test batch 4 are compared in Fig. 7.9, using the MPLS model and the phase PLS model, respectively. The offline prediction results shown in Fig. 7.8 suggest that the phase PLS model has an accuracy comparable to the MPLS model, while the online predictions of Fig. 7.9 show the superiority of the proposed method. The online weight prediction by the phase PLS model is not only more accurate but also can be made much earlier at the beginning of phase 4, rather than at the end of the batch.

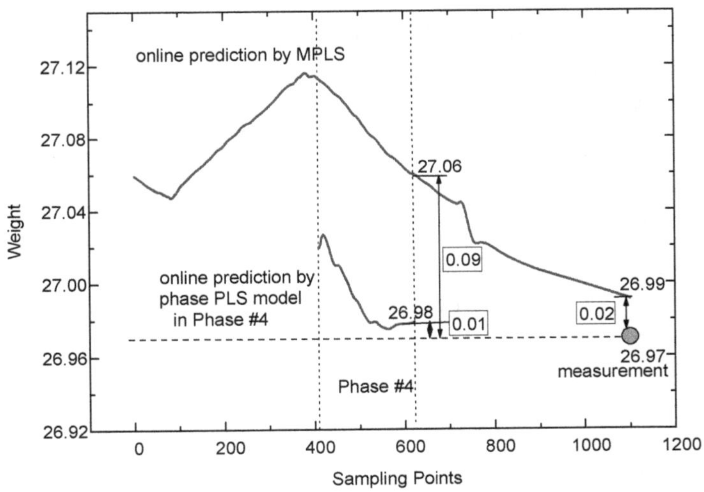


**Figure 7.9** Online weight prediction for test batch 4 by MPLS and phase PLS models

These experimental results show that the proposed phase-based process analysis and quality prediction methods can indeed be successfully applied to obtain phase information, to explore phase-specific effects of process variables, to find out the key process variables, and to give earlier and accurate online quality prediction.

# 7.4 Phase-Based Process Analysis and End-Product Quality Prediction (Method B)

The phase PLS models obtained by averaging time-slice PLS models within the same phase only focus on the variation along the batch direction without capturing the correlations along the time direction within the same phase. This may cause inaccurate phase monitoring models.

In this section, another phase-based process analysis and online quality prediction method is introduced [129]. The key difference lies in that the phase PLS model in Section 7.2 (Method A) uses batch-wise unfolding to deal with the three-dimensional batch data for phase division, modeling, and quality prediction; while the new phase PLS model in this section (Method B) is based on a variable-wise unfolding technique. Method B may have the following advantages and disadvantages compared with Method A.

Advantages:

- Using correlation analysis based on variable-wise unfolding, the identification of critical-to-quality phases and key process variables in each critical phase requires no heavy computation and has a parsimonious expression.
- In each phase, variable-wise unfolding extends data samples into time series. PLS modeling based on the augmented samples can cover the time-varying variance information and eliminate the negative influences of process variable auto- and cross-correlations on the phase PLS models.

Disadvantages:

- Method B is more complex than Method A, which is a price to pay if one wants to get better-quality prediction.
- Method B could become very convoluted if there is more than one quality variable, in which one needs to perform the analysis separately for each quality variable. Thus it may lose some correlation information among quality variables compared with Method A.

## 7.4.1 Critical Phase Identification

Let consider an equal-length multiphase batch process, whose data are arranged into a matrix $\underline{X} \in R^{I\times J\times K}$ for process variables and a vector $\mathbf{y} \in R^{I\times 1}$ for a quality variable of interest.

The phase division procedure is the same as in phase PCA modeling method in Chapter 6. After the phase division and PCA modeling for time-slice process data

$\tilde{X}_k \in R^{I\times J}$, one can get $K_c$ number of score matrices $\tilde{T}_k \in R^{I\times J}$ and loading matrices $\tilde{P}_k \in R^{J\times J}$ in phase *c*. They contain the most important process variation information along the time direction within each phase. First, $\tilde{T}_k$ is transformed into a weighted form as

$$\begin{aligned}\breve{T}_k &= \left[\tilde{T}_{k,1}\cdot g_{k,1}, \tilde{T}_{k,2}\cdot g_{k,2}, \cdots, \tilde{T}_{k,J}\cdot g_{k,J}\right] \\ &= \tilde{T}_k \cdot \mathrm{diag}(g_{k,1}, g_{k,2}, \cdots, g_{k,J})\end{aligned} \tag{7.12}$$

where $g_{k,j}$ is the same as that in Eq. 6.7.

Then, using the variable-wise data rearrangement, as illustrated in Fig. 7.10, $T^c \in R^{K_c I\times J}$ is formed for the *c*th phase by stacking up all the weighted time-slice scores $\breve{T}_k$ within the same phase. In this way, $T^c$ covers more detailed dissimilarity of the process nature varying over time within the same phase and reflects the different weights between each principal component, $T_j^c \in R^{K_c I\times 1}(j=1,\cdots,J)$. Consequently, the synthetically phase-specific process variation information closely related with the final quality can be extracted naturally by the simple sum form of the principal components:

$$\tilde{\mathbf{t}}_c^* = \sum_{j=1}^{J} T_j^c \tag{7.13}$$

where $\tilde{\mathbf{t}}_c^* \in R^{K_c I\times 1}$ is defined as the representative generalized principal component vector in phase *c*.

In addition, unlike the process variables that are collected throughout the duration of each batch run, the quality variable *y* is only measured at the end of a batch run. In order to carry out the following correlation analysis and the PLS regression, one should match the sizes of the process data and the quality data. So, the quality data are duplicated $K_c$ times, as shown in Fig. 7.10. The data is structured in this way because the quality at the end of the batch run will be acceptable if all $\tilde{X}_k$ are normal. Then we can simply perform a correlation analysis between the phase-representative principal component vector $\tilde{\mathbf{t}}_c^*$ and the restructured quality variable measurements $\mathbf{y}_c \in R^{K_c I\times 1}$ in the *c*th phase to check the critical phases. The squared simple correlation coefficient, $CP_c^2$, can be defined to represent how good the correlation relationship between predictor variables and response variable is:

$$CP_c^2(\tilde{\mathbf{t}}_c^*, \mathbf{y}_c) = \left(\frac{\mathrm{cov}(\tilde{\mathbf{t}}_c^*, \mathbf{y}_c)}{\sqrt{D(\tilde{\mathbf{t}}_c^*)\cdot D(\mathbf{y}_c)}}\right)^2 = \frac{\mathrm{cov}^2(\tilde{\mathbf{t}}_c^*, \mathbf{y}_c)}{D(\tilde{\mathbf{t}}_c^*)\cdot D(\mathbf{y}_c)} \tag{7.14}$$

where cov( ) means the covariance of two vectors, and *D*( ) denotes the variance of a vector.

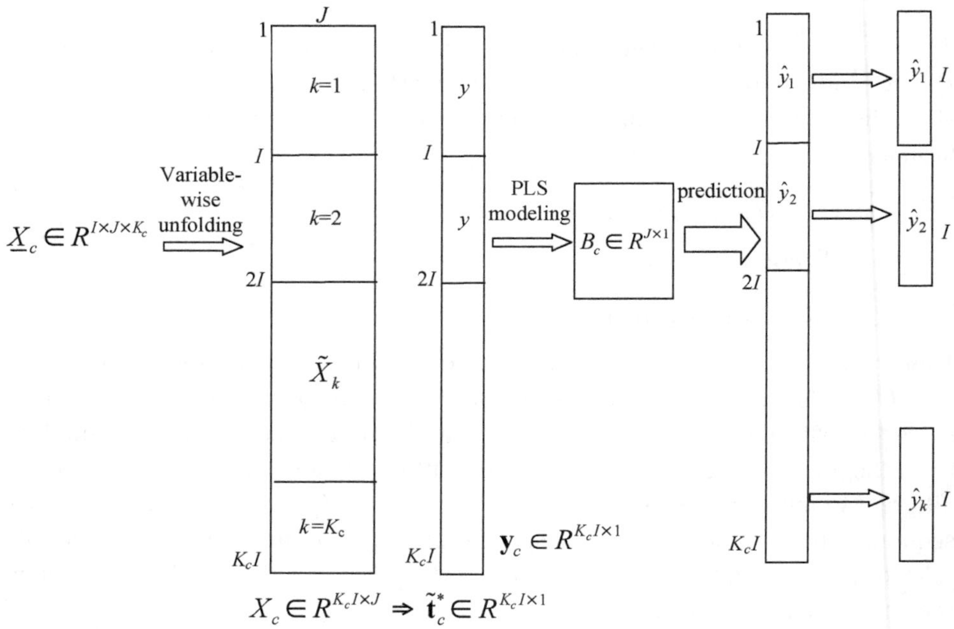


**Figure 7.10** Illustration of the phase-based PLS modeling scheme in the *c*th phase

It is easy to explain the meaning of the criterion. The PLS regression extracts the principal component scores from the process variables, which should represent most of process information and meanwhile have strong correlations with the quality variable. So in the calculation of $CP_c^2$, it is simple and sufficient to reveal the phase-specific correlation significance to quality based on the representative principal component scores instead of the original collected process measurements. Obviously, $CP_c^2$ ranges from 0 to 1 since $CP_c$ itself varies between -1 and 1.

Different phase results in different $CP_c^2$, revealing the change of specific underlying correlation relationships between process behaviors and quality over different phases. A larger $CP_c^2$ indicates a higher correlation between predictor variables and the response variable, that is, the phase region is more critical and reliable in the prediction of quality. Naturally, a critical limit should be defined as the reference standard. It is well known that the squared correlation coefficient is actually the coefficient of determination in a simple linear regression analysis. Therefore, $F$ test is used here to test the significance of the phase-specific explanation ability of process behaviors to the final quality:

$$\frac{CP_c^2}{(1-CP_c^2)/(I-2)} \sim F_\alpha(1, I-2) \tag{7.15}$$

where $I$ is the number of batches, and $\alpha$ is the significance level.

The critical limit of $CP_c^2$ can be conversely calculated by Eq. 7.15, where the critical value of the $F$ statistic with significance factors ($\alpha = 0.01$ or $0.05$) can be found in the statistical table of $F$ distributions. If the squared correlation coefficient $CP_c^2$ is larger than the critical limit, the corresponding time region is defined as the critical phase with pivotal effects on the concerned quality. The information is extremely helpful for us to further understand the different phase-specific effects of process variables on the end product quality and establish the corresponding modeling guideline. This information can also be used to improve the process, either by redesigning the process equipment or redesigning the control logic, with the aim of minimizing any variability that is present today.

### 7.4.2 Key Variable Selection Based on Variable-Wise Unfolding

It is common that the product quality is determined by the key variables of several critical phases. Degradation of the prediction performance of the quality estimation model caused by inclusion of unimportant process variables is a serious problem. It is thus important to focus only on those variables that show the highest correlation with or contribute most to the quality variations in each phase. The importance of variable selection has been often reported. The variable selection technique also can be used as a knowledge extraction tool. Generally speaking, although data compression techniques such as MPLS can handle the redundancy problem of batch process data, an acceptable prediction accuracy cannot be expected without the removal of less important variables. Because those variables have weak correlations with the quality variables, they cannot be good predictors in the regression model and only cause deterioration of the data compression performance. If those variables are removed prior to modeling, the quality prediction performance can be effectively improved because of the enhanced causal relationship between the process and quality variables. As mentioned previously, the effects of process variables on quality remain similar within one phase; they indeed have similar correlations with quality. This provides the theoretical basis for the proposed correlation analysis based on variable-wise unfolding.

After the phase division, the three-way array $X \in R^{I \times J \times K_c}$ ($K_c$ is the phase duration) is formed in each critical phase, which may be unfolded in different ways, reducing various two-dimensional matrices. Moreover, considering the underlying nature that each process variable tends to show a phase-specific ability to explain the final quality (that is, within the same phase, similar effects on quality will remain, with slight variation), it is advantageous and reasonable to conduct the index analysis using the form of phase-based variable-wise unfolding. On the one hand, by means of variable-wise unfolding of $\underline{X}_c$ for each phase, augmented samples are composed of batches and time series from the same phase. They take the characteristics of the overall phase into account, which can offer more sufficient system information and lead to more stable and comprehensive analysis results. On the other hand, the complex interactions of autocorrelations and cross-correlations between process variables often badly impact on the selection results, which accordingly induce biased regression models with worse prediction performance. Based on variable-wise unfolding of the entire phase, variable correlations between different time intervals are no longer a serious matter since only the initial $J$ process variables are considered simultaneously, which gives more explicit and reliable results. Therefore, the phase-specific implementation based on variable unfolding avoids imposing a heavy calculation burden and complexity in focusing on each time interval and alleviates the influence of process variable collinearity along time within the same phase.

Any procedure for variable selection must contain two components: a selection criterion and a search procedure. The selection criterion is the measure used to rank feature subsets. Most variable selection search algorithms have an unbalanced performance between accuracy and computational speed, which commonly suggests that the variables are selected according to the increase in statistic indices, such as $R^2$, $AdjR^2$, AIC, Mallows $C_p$, and others. In practice, these procedures often lead to over fitting because they prefer to extremely pursue the ability to model the training data by measuring the degree of fit rather than optimizing the selection of variables, specifically taking the augmented prediction aspect of the model into account. Here, the correlation contribution rate of process variable $\mathbf{x}_{c,j} \in R^{K_c I \times 1}$ to quality $\mathbf{y}_c \in R^{K_c I \times 1}$ with respect to all candidate variables can then be defined to identify quality-related variables in each phase:

$$CV_{c,j} = \frac{r(\mathbf{x}_{c,j}, \mathbf{y}_c)}{\sqrt{\sum_{j}^{J} r^2(\mathbf{x}_{c,j}, \mathbf{y}_c)}} \tag{7.16}$$

where $j$ and $c$ are indices of process variables and critical phases, respectively. Here, $\mathbf{x}_{c,j}$ is the augmented vector of the $j$th variable derived from variable-wise unfolding in the $c$th phase; $\mathbf{y}_c$ is the duplicated quality variable within the same phase. It is clear that $CV_{c,j}$ varies between −1 and 1 since it is transformed from the correlation coefficient. Also, $\pm\sqrt{1/J}$, the moderate level of contribution rate, can be easily derived supposing that all of the process variables possess an identical correlation with quality. They will be employed as important reference boundary limits of ranking the significance of variables in the following variable selection steps.

Generally, the larger $CV_{c,j}$ is, the larger the rate the $j$th process variable accounts for relative to the other candidate variables, that is, the more significance the process variable contributes to the variation of quality in the $c$th phase. Those variables with small $CV_{c,j}$ values can be initially and roughly judged that they may have weak correlations with quality. Here it should be remarked that the correlations are not adequate to interpret the models' prediction ability. Those process variables with minor absolute $CV_{c,j}$ may also improve the prediction performance if entering regression models. So besides the prior correlation analysis, it is necessary to further evaluate these variables to verify whether they should be kept in the final regression models. Thus, the mean prediction squared error is developed as

$$MSE_c = \frac{1}{K_c I} \sum_{i}^{K_c I} (y_i - \hat{y}_{c,i})^2 \tag{7.17}$$

where $c$ and $i$ are indices of phases and augmented observations, respectively, and $y_i$ $\hat{y}_{c,i}$ are the real measurement and predicted quality, respectively.

Here it should be noted that $MSE_c$ is computed by focusing on the validated batches instead of the calibrated data used in calculating the $CV_{c,j}$ index. Comparatively, the above two parameters, $CV_{c,j}$ and $MSE_c$, make full use of train and test data subsets, respectively, so that their combination ensures the generalization capability and prediction performance, overcoming the overfitting defect when only devoting one's attention to the prediction errors of the train data set. Commonly, when the number of observations is sufficient, the samples are split randomly into two parts: one used to calibrate the model (i.e., to estimate the parameter values) and the other to validate the model (to test the predictive performance). In general, it is often recommended to use more data to calibrate than to validate, usually a two-thirds to one-third split of the sample. The latter set is sometimes known as a *holdout sample*. After fitting the model using the data from the calibration subset, we can get the regression parameters and then use them to calculate the predicted values for the observations in the holdout sample. Thus we can assess more the augmented predictive accuracy of the model in the validation sample rather than the fitting ability in the calibration sample. Moreover, if the data set is too small to split the samples, an alternative is to use jackknife validation.

Generally, in previous work, the selection procedure was conducted by searching all possible feature variables one by one, such as with the backward elimination procedure, the forward selection procedure, and the step-wise regression procedure. However, they require a large quantity of calculations because the number of calculation steps grows exponentially with the amount of candidate variables. From any set of $J$ predictors, $2^J$ alternative models should be constructed, which is based on the fact that each predictor can be either included or excluded from the model. In most circumstances, therefore, it will be impossible and impractical to make a detailed examination of all possible regression models. In order to simplify the calculation and overcome the complexity of evaluating all of the possible modeling variables one by one from quantities of candidate variables, we first divide the candidate variables into different intervals and impose an order of priority on them according to the forenamed $CV_{c,j}$ criterion. Then, instead of individual variables, these intervals are searched step by step. In this way, one can greatly reduce the number of possible models to be evaluated compared with single variable selection methods, thus simplifying the analysis steps, avoiding the blindness of random step-wise selection and typically increasing reliability. Certainly, if the number of variables is not large, one can directly search all variables one by one.

The variable selection steps we suggest are listed in detail as follows:

1. Set $\pm\sqrt{1/J}$ as the two-side critical limits of the correlation relationship since they are the moderate level of the contribution coefficient $CV_{c,j}$. Retain the process variables with the $CV_{c,j}$ criterion beyond the critical values, $\pm\sqrt{1/J}$, and perform the initial PLS regression model using them.
2. Find the highest absolute value of $CV_{c,j}$ in the remaining process variables, and divide the range, from the zero value to the highest absolute value, into several minor intervals, whose number can be determined adaptively according to the specific circumstance: more intervals correspond to a more detailed variable selection analysis, whereas fewer intervals account for a reduced calculation amount but a cursory key variable searching result. Meanwhile, a descending order of priority is given to these intervals while they are approaching the zero value.
3. Find the variables contained in the first interval with highest priority, add them into the PLS regression model, and compare the squared prediction errors with the former one using the validation data set. If the prediction errors decrease, hold the variables in the new regression model and continue searching the next interval. Otherwise stop, exclude all the other variables, and use the current regression model as the final one.

By using the above procedures, which will be illustrated in detail in the later simulation, we can find a reasonable set of variables used in PLS regressions. The quality prediction ability of the regression model can be improved when one eliminates variables that probably bring only noise or idle information. The selection strategy emphasizes the fact that, when predicting quality, we should have optimized the appropriate variables to work with so that the current regression model has the least average squared prediction errors for test sets, that is, the best prediction ability. In this way, the predictor variables used in the phase regression models are refined so that one can get more stable and parsimonious PLS regression models, which show increased correlations between the predictors and the quality variable.

### 7.4.3 Phase-Based PLS Modeling Algorithm

According to the aforementioned phase-specific process behaviors, the observed variables should have a similar explanation and contribution to quality within the same phase, which provides a reasonable basis for phase-based variable identification, regression modeling, and quality prediction. Then the regression parameters, containing the information on relationships between process variables and quality, should remain similar in the same phase and show differences over different phases. Therefore, it is preferable to locate the phase-specific effects of process variables on the final product quality by phase-based regression models, demanding no prior process knowledge.

As mentioned, each of the batch data rearrangements corresponds to observing a different type of variability information. Batch-wise unfolding focuses on the variability among different batches, while variable-wise rearrangement stresses the system variances extracted along both batches and time. When working with batch-wise data, the unfolded matrix $X \in R^{I \times KJ}$ has a great number of augmented variables $KJ$ opposite the initial comparatively small quantity of observed batches. Only limited system information can be extracted from fewer batches; while also, repetitions of the same process variables at different times cause the models to be complex and confused. Another major problem in the online prediction of quality is that the dataset of the new batch has the process measurements only up to the current time point. So it is inevitable that the unknown future observations should be estimated appropriately when online quality prediction is performed. Four approaches have been proposed for complementing the future observations. However, the existing methods have been criticized for not being sensitive enough to process dynamic changes or for being time consuming. The estimated values may also distort the process information, potentially leading to worse quality prediction, since predicting the performance of regression models directly depends on the estimation accuracy of missing future values. Here, modeling based on variable-wise unfolding overcomes the above shortcomings and provides the following solution.

We can obtain a representative regression parameter matrix, $B_c \in R^{J_c \times 1}$, by performing the PLS algorithm on $\{ X_c \in R^{K_c I \times J_c}, \mathbf{y}_c \in R^{K_c I \times 1} \}$ for each critical phase (where $X_c$ is the process data matrix with the $J_c$ retained number of key variables; $\mathbf{y}_c$ is the quality vector):

$$B_c = W_c (P_c^T W_c)^{-1} Q_c^T \tag{7.18}$$

where $W_c$, $P_c$ are, respectively, the weighting matrix and loading matrix for $X_c$, and $Q_c$ is the loading matrix for $Y_c$.

Then the phase-based PLS regression model for quality prediction can be formulated as

$$\hat{y}_k = \mathbf{x}_k \cdot B_c, c = 1,2,...,C; k = 1,2,..., K_c \quad (7.19)$$

where $\mathbf{x}_k \in R^{1\times J_c}$ consists of process measurements for the selected variable at time $k$ within the $c$th phase, and $\hat{y}_k$ is the predicted value of final product quality.

In conclusion, the procedure for the new phase-based modeling method is outlined as follows, where the first two steps are the same as in Method A of Section 7.2:

1. The three-way array $\underline{X} \in R^{I\times J\times K}$ is unfolded batch-wise into $K$ number of time slices. Then they are normalized to zero mean and unit variance as well as the final quality variable vector $\mathbf{y} \in R^{I\times 1}$. Subtracting the mean of each column can approximately eliminate the main nonlinearity due to the dynamic behavior of the process and enable us to model the deviations from the mean process trajectory. And each variable is scaled to unit variance to handle different measurement units, thus giving each equal weight.
2. The PLS algorithm can be performed on these time-slice datasets $\{\tilde{X}_k, \mathbf{y}\}$ to extract the loading matrices $\tilde{P}_k$ and score matrices $\tilde{T}_k$ using Eq. 6.6, which reflect the phase-specific process property more relevant to quality variation rather than the common process behaviors. Then the phase clustering algorithm is adopted using those weighted loading matrices $\breve{P}_k$ calculated by Eq. 6.7, to partition the operation duration into different phases.
3. In each phase, the synthetic phase-representative principal score $\tilde{\mathbf{t}}_c^*$ is derived using Eq. 7.13, and the quality variables are duplicated correspondingly as $\mathbf{y}_c$. Then critical phases are identified by analyzing the phase-specific correlation between $\tilde{\mathbf{t}}_c^*$ and $\mathbf{y}_c$.
4. In each key phase, as shown in Fig. 7.10, the batch data $\underline{X}_c \in R^{I\times J\times K_c}$ ($K_c$ is the time duration in the $c$th phase) is rearranged into the form of $X_c \in R^{K_c I\times J}$. Then $J_c$ key variables closely related to quality are selected based on the corresponding correlation analysis and cross-validation for prediction square errors of the validation data set. Thus the final representative regression parameter matrix, $B_c(J_c \times 1)$, for the $c$th phase can be generated, correspondingly focusing on the predigested $\{X_c, \mathbf{y}_c\}$.
5. From the outlined procedure, two types of data unfolding, batch-wise and variable-wise, are appropriately used for different functions. Because it is known that there is a growing debate about which unfolding is better, the present work shows that both are useful as long as one knows what one wants to achieve with the two different data arrangements.

### 7.4.4 Online Quality Prediction

Without losing generality, assuming that the quality variable has two critical phases, phase 1 and phase 2, where their starting times are $K_{1s}$ and $K_{2s}$, and the retained numbers of key variables are $J_1$ and $J_2$, respectively. The current online quality prediction can then be yielded as

$$\hat{y}_k = \begin{cases} \frac{1}{k - K_{1s} + 1} \sum_{i=K_{1s}}^{k} \left( x_i(1 \times J_1) \cdot B_1(J_1 \times 1) \right) & k \in \text{the 1st critical phase} \\ w_1 \cdot \hat{y}_1^* + w_2 \cdot \frac{1}{k - K_{2s} + 1} \sum_{i=K_{2s}}^{k} \left( x_i(1 \times J_2) \cdot B_2(J_2 \times 1) \right) & k \in \text{the 2nd critical phase} \\ \text{null} & \text{others} \end{cases} \tag{7.20}$$

where $w_1, w_2$ are weights for critical phase 1 and 2, respectively, and $\hat{y}_1^*$ is the quality prediction at the end of critical phase 1. The phase weights $w_c$ are calculated directly from the previous correlation metric, $CP_c$, by a simple ratio algorithm without increasing the heavy calculation burden and complexity: $w_c = \frac{CP_c}{CP_c + CP_2}$. In this way, the explanations for quality variation throughout different phases are stacked with different significance weights.

From Eq. 7.20, we can clearly see that for every sample interval, there will be a corresponding real-time quality prediction. In detail, for critical phase 1, the real-time predicted quality $\hat{y}_k$ is indeed the average value from the starting time of phase 1 up to the current time *k*. Therefore, with the time evolvement, the end-of-phase quality prediction, $\hat{y}_1^*$, will be readily obtained when the first dominant phase completes. For critical phase 2, the online prediction at each time *k* is actually a weighted combination of $\hat{y}_1^*$ and the average value up to the current time *k* starting from the beginning of key phase 2. In the end, the quality prediction value, $\hat{y}_2^*$, obtained at the end of critical phase 2 is naturally regarded as the final quality prediction result of the whole process.

# 7.5 Application of Phase PLS Model (Method B) to Injection Molding

The same experimental data given in Section 7.3 are used in this section.

## 7.5.1 Illustration of Correlation Analysis

The weighted time-slice loading matrices calculated from the PLS analysis are fed into the clustering algorithm. The clustering result is shown in Fig. 7.11(a), clearly showing that without using any prior process knowledge, the trajectory of the injection molding is automatically divided into six phases, among which four long phases (marked with the shaded circle) agree well with the four main physical operation phases of the process, that is, the injection, pack-holding, plastication, and cooling phases, plus a few short transitional time periods. These temporary time regions, corresponding to the dynamic transient period with unstable process states, form individual durations, which have little impact on quality prediction. Dividing a batch process into "steady" and "transient" phases can not only improve quality prediction performance, but also enhance process analysis and understanding.

Without losing generality, each phase hints at different effects on the final product, as well as different correlations between process variables and product quality.

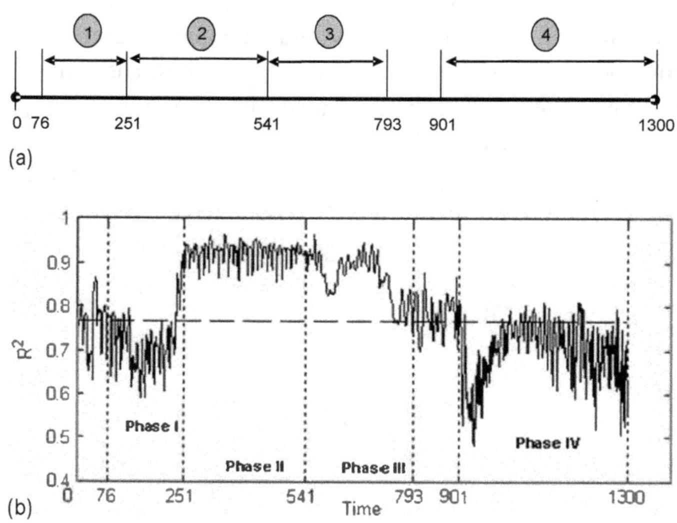


**Figure 7.11** (a) Phase division result for injection molding process;
(b) fitness of regression models
(solid line, the real-time $R^2$ values; dashed line, 99% critical limit)

**Table 7.6** Critical phase analysis result using *CP* metric for the four main phases

| Main phase No. | Duration | *CP* metric | Critical value ($\alpha = 0.01$) | Estimation result |
|---|---|---|---|---|
| 1 | 77–251 | 0.2557 | 0.5052 | Noncritical |
| 2 | 252–541 | 0.6147 | 0.5052 | Critical |
| 3 | 542–793 | 0.5349 | 0.5052 | Critical |
| 4 | 902–1300 | 0.3983 | 0.5052 | Noncritical |

Combined with the phase clustering result, Table 7.6 lists the analysis results of the *CP* metric over different phase operations. Obviously, it displays that in phases 2 and 3 the *CP* values are above the critical point derived from Eq. 7.15 and can be inferred as critical phases. To further affirm the reasonability of the above analyses, the goodness of fit of the phase-specific regression model is evaluated, where all the input process variables are retained, using multiple coefficients of determination, $R_k^2$, shown in Fig. 7.11(b). The larger the $R_k^2$, the better the fitness of the corresponding phase-specific PLS model. Normally, the regression model in the critical-to-quality phase is more accurate and reliable for the prediction of the quality variable, so it should have larger $R_k^2$. From the plot, we can clearly see that larger and more stable fitness appears in two phases, phase 2 and 3, especially the second phase, indicating the same identification result of critical phases as in Table 7.6. This demonstrates from another aspect that it is reasonable to determine the critical phases according to the correlation analysis based on the *CP* index.

In phases 2 and 3 (the pack-holding and plastication phases), the *CV* values can be calculated for each process variable based on the variable-wise arrangement in each phase, as illustrated in Fig. 7.12. Here, for comparison, the phase regression coefficients including all process variables are also plotted in Fig. 7.13, which can numerically explain how these variables will affect the product quality. Similar to the status shown in Fig. 7.12, the pressure variables have a positive relation with the quality, while the temperature variables are negatively correlated with the quality. It means that higher pressures and lower temperatures result in larger product length. Also, a longer screw displacement results in more material being crammed into the cavity, which obviously engenders a longer product.

Now taking phase 3 (plastication phase) as an example, we shall explain the variable selection guideline further. In Fig. 7.13(b), we place the two-side horizontal critical limits $\pm\sqrt{1/J}$ so that we can get the variables beyond the limits, including pressure variables (nozzle pressure, cavity pressure) and displacement variable (stroke), which are defined as the variables most correlated with quality. Moreover, they are all positively related with quality. Using the three variables, we carry out the initial PLS regression and calculate the MSE value with Eq. 7.17. Sequentially, we search the remaining variables and find the one with the highest absolute value of *CV*,

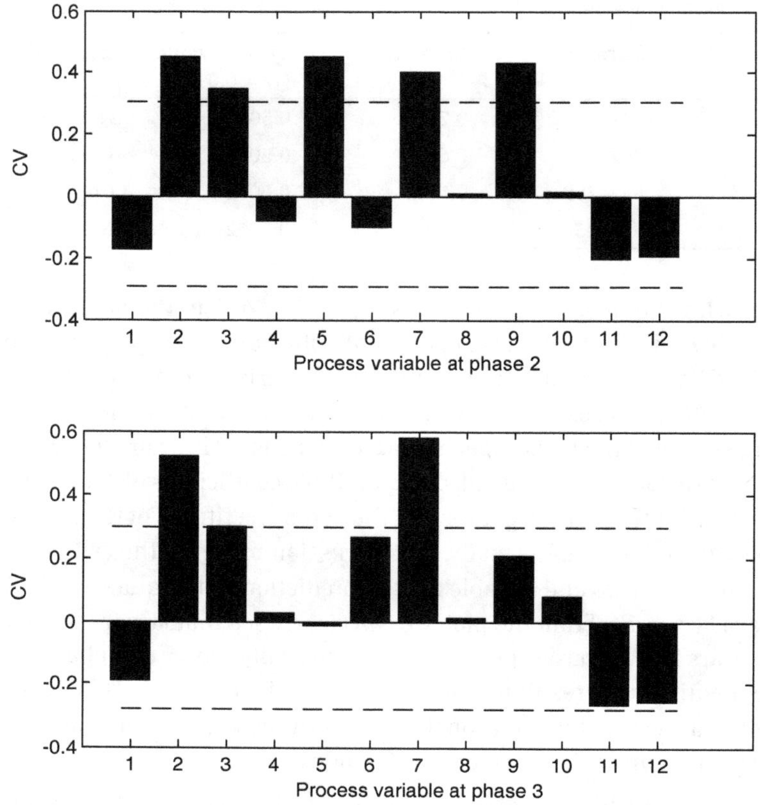


**Figure 7.12** *CV* criterion of key variable selection (dashed line, $\pm\sqrt{1/J}$, the moderate level of *CV* metric)

from which to the zero value the range is divided into ten equal spaces with nine related lower interval limits. In each space we find the variables outside the relevant interval limit, add them into the regression model, and compute the corresponding MSE value using the current prediction model. For comparison, Table 7.7(a) lists the cases we deal with in the variable selection procedures and the corresponding MSE values for the test data. It shows that the appropriate variables used in the regressions are variables No. 1, 2, 3, 6, 7, 9, 11, and 12. If we use more process variables or fewer, we get worse results, which implies that it deserves extensive computations to find a good set of variables employed in the PLS regressions.

Similarly, in phase 2 (the pack-holding phase), the initial correlation analysis shown in Fig. 7.13(a) reveals that the dominant process variables may cover the pressure variables (nozzle pressure, hydraulic pressure, and cavity pressure), the temperature variables (cavity temperature, barrel temperature, and mold temperature), the displacement variable (stroke), and the manipulated variable (SV1).

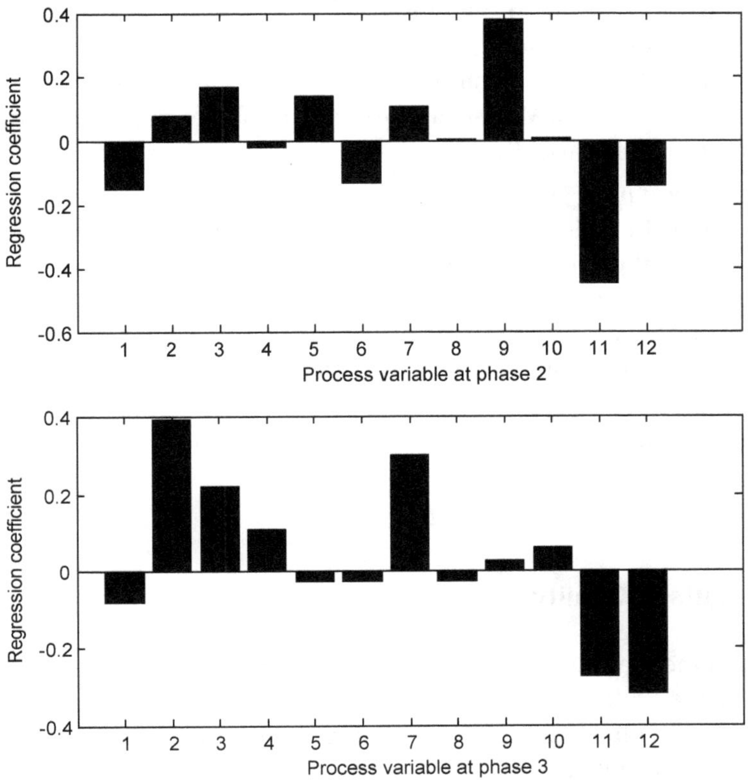


**Figure 7.13** Regression parameters for each process variable in the critical phases

**Table 7.7** Variable Selection in PLS Regression in Phase 2 (a) and Phase 3 (b)

(a)

| No. | Variables in regression model | MSE |
|---|---|---|
| 1 | 2, 3, 5, 7, 9 | 0.038287 |
| 2 | 1, 2, 3, 5, 7, 9, 11, 12 | 0.0059491 |
| 3 | 1, 2, 3, 5, 6, 7, 9, 11, 12 | 0.0045318 |
| 4 | 1, 2, 3, 4, 5, 6, 7, 9, 11, 12 | 0.0038673 |
| 5 | all | 0.0038756 |

(b)

| No. | Variables in regression model | MSE |
|---|---|---|
| 1 | 2, 3, 7 | 0.084991 |
| 2 | 2, 3, 6, 7, 11, 12 | 0.035243 |
| 3 | 2, 3, 6, 7, 9, 11, 12 | 0.034587 |
| 4 | 1, 2, 3, 6, 7, 9, 11, 12 | 0.033589 |
| 5 | all | 0.037243 |

A further variable selection result presented in Table 7.7(b) denotes that all of the process variables except SRS and SV2 should be kept in the regression, which demonstrates that it is better that the variable correlation analysis and variable selection strategy based on MSE should be closely combined to provide more reliable and concise modeling variables.

Without the use of prior process knowledge, the above phase-based analysis agrees well with the real physical process, which can be useful for quality improvement as it suggests what process variables in the critical phases should be more emphasized and better controlled for quality improvement. If we can extract enough information during a particular period that is critical for product quality prediction, it will be possible to weight that period more heavily and thus track faster any quality deviations from the average over that period. Accurate online quality prediction is available to reflect the quality status and perform process adjustments in time.

### 7.5.2 Results of Quality Prediction

According to the previous process analysis and understanding, phases 2 and 3 are indicated as the critical phases. The product length has a close relation with both phases. So when online predicting, it is necessary to conduct the quality prediction by stacking the cumulative effects over phases. The online quality prediction is performed at each sampling time of critical phase 2 and 3, respectively, as shown in Fig. 7.14(a) for a test batch. From the figure, it is clear that the real-time predicted quality will gradually approach the real quality measurement over time. The maximum online predicted error rates, shown in Fig. 7.14(b), are less than 0.06% in phase 2 and 0.07% in phase 3, which are well-accepted prediction precisions in industry. The accurate quality prediction can be successfully applied to demonstrate the phase-specific effects of process variables on quality, explore process running information, and evaluate product quality performance in advance.

Moreover, a comparison of offline prediction performance is conducted for training and testing batches using the first phase-based sub-PLS (Method A) and the conventional MPLS method, respectively. Here, multiple coefficients of determination, $R^2$, can be used to quantitatively measure a model's prediction performance. Generally, $R^2$ ranges between 0 and 1. The larger the $R^2$, the better the fitness of the prediction model. Usually, the $R^2$ of a calibration set is larger than that of a validation set because calibration models can easily lead to overfitting of the data. We can conclude that our inferential model is reliable if it works well for a new sample, so the $R^2$ of training and testing batches should be effectively combined to quantify the model's fitting capability and generalization performance.

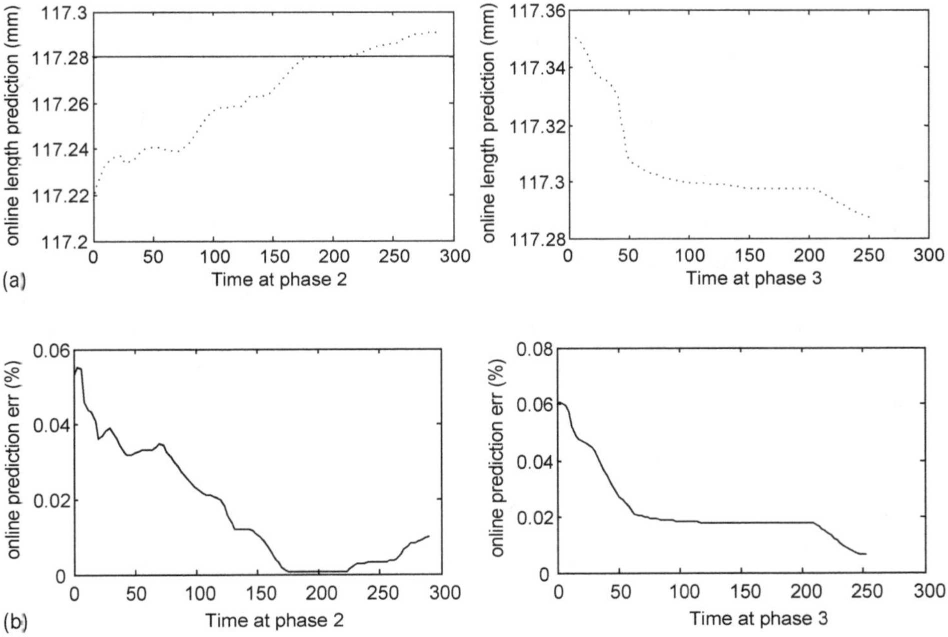


**Figure 7.14** Online quality prediction result for a test batch in phase 2 and phase 3.
(a) online predicted quality
(solid line, the real quality measurement; dashed line, online prediction value);
(b) online predicted error rate (%)

The $R^2$ of the training batches are respectively 0.8566, 0.8041, and 0.9969 for the three methods, where the fitting ability of the MPLS model is the best. The $R^2$ of the testing batches are respectively 0.8420, 0.7210, and −8.8993, where the $R^2$ for the MPLS method has gone beyond its normal measurement scope, indicating a worse generalization ability for quality prediction. From the $R^2$ values for the training batches and the testing batches, it is obvious that our proposed method yields better quality prediction performance.

The comparison results are shown in Figs. 7.15 and 7.16, respectively, which can visualize the comparison of prediction performance between the three methods. From Fig. 7.15, it can be observed that MPLS model is superior for fitting the training batches where the predicted quality is well fitted with the real quality measurement. In comparison, from Fig. 7.16, it can be seen that for test batches 3 and 7, the trained MPLS regression model fails to give a reliable enough quality prediction result. The comparison may illustrate the disadvantages of the conventional MPLS algorithm. Although MPLS is well known as an effective data compression and feature extraction technique, one cannot expect it to pick up pivotal information from a great amount of redundant candidate data information.

Thus, the conventional MPLS may not clearly distinguish the system information from normal stochastic noises, which have little relation to the final quality. Therefore, the overfitting to commonly caused process variations may spoil the models' ability to capture the useful predictor information, and thus the favorable adaptability to new batches with normal stochastic dynamics is lost. The phase PLS model (Method A) also compromises the reliability of the predicted quality since the phase-based regression relationships are extracted by only focusing on each individual time slice, which may not be able to catch stable-enough prediction information. By comparison, the superiority of the new phase-PLS model in Method B over the other two methods is obvious for both training and testing batches, which gives a satisfying overall prediction trend, demonstrating the models' fitness ability and prediction adaptability.

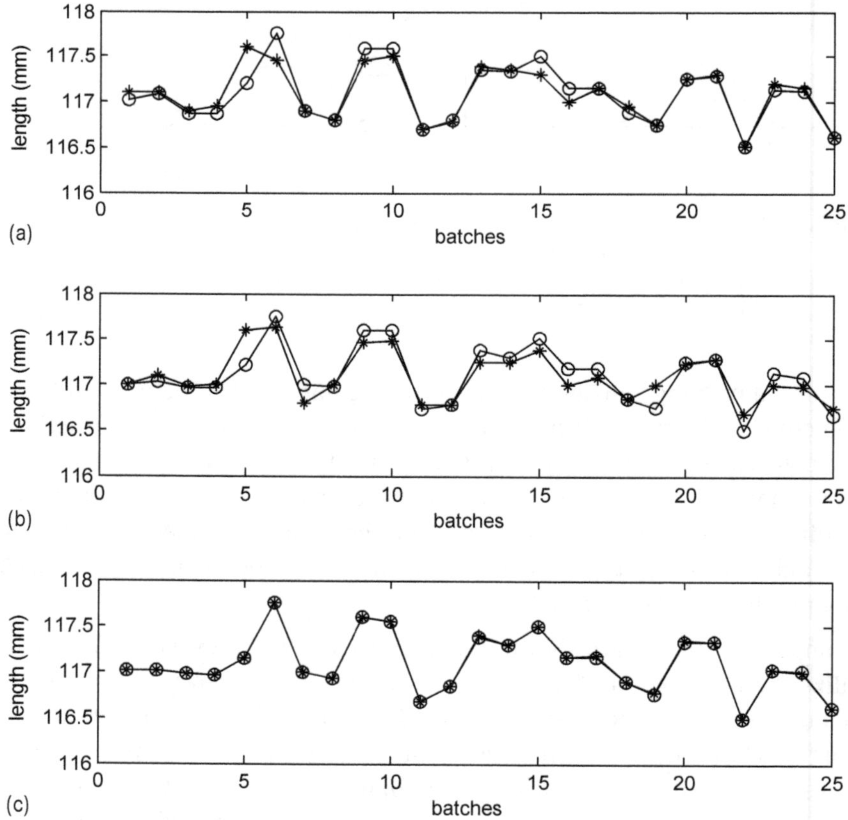


**Figure 7.15** Offline quality prediction results for the reference batches using
(a) Phase PLS in Method B,
(b) Phase PLS in Method A, and
(c) MPLS method
(-o-, real measurements; —∗—, predicted values)

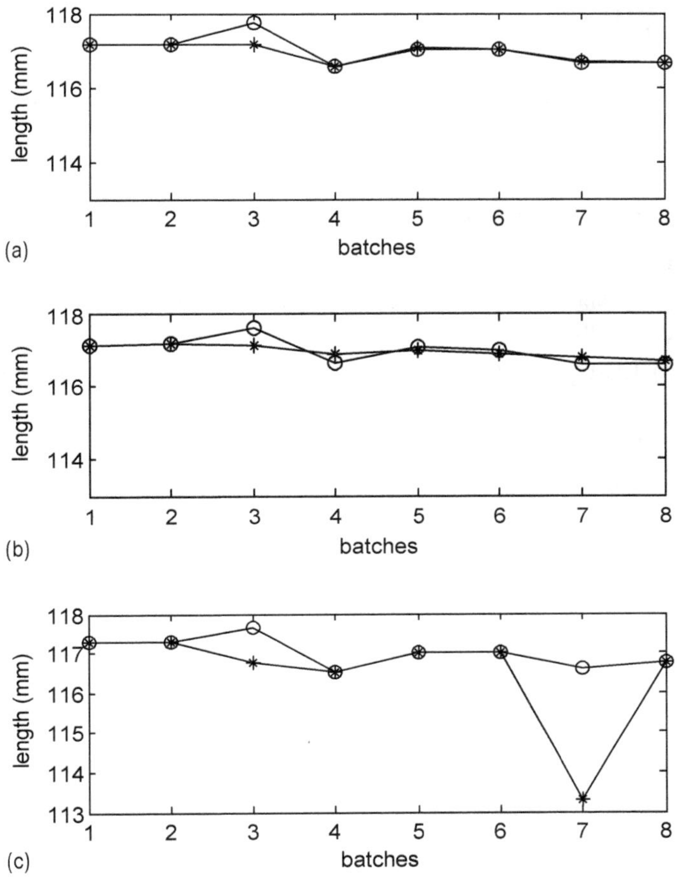


**Figure 7.16** Offline quality prediction results for the test batches using
(a) Phase PLS in Method B,
(b) Phase PLS in Method A, and
(c) MPLS method
(-o-, real measurements; —∗—, predicted values)

## References


[101] P. Nomikos and J.F. MacGregor, *Chemom. Intell. Lab. Syst.*, **30**, 97–108, 1995(a)

[124] N.Y. Lu and F.R. Gao, *Ind. Eng. Chem. Res.*, **44**, 3547–3555, 2005

[125] C. Duchesne and J.F. MacGregor, *Chemom. Intell. Lab. Syst.*, ***51***, 125–137, 2000

[126] Y.H. Chu, Y.H. Lee, and C. Han, *Ind. Eng. Chem. Res.*, **43**, 2680–2690, 2004

[127] L. Breiman, *Mach. Learn.*, **24**, 49–64, 1996

[128] J. Zhang, E.B. Martin, A.J. Morris, and C. Kiparssider, *Comput. Chem. Eng.*, **21**, 1025–1030, 1997

[129] C.H. Zhao, F.L. Wang, Z.Z. Mao, N.Y. Lu, and M.X. Jia, *Ind. Eng. Chem. Res.*, **47**(3), 825–834, 2008

# 8 In-Mold Capacitive Transducer for Injection Molding Process

The measurement or monitoring of polymer melt status inside an injection mold is important to the online quality control of an injection molding process. Many efforts have been made in this area. Most measurements of the material status inside an injection mold have been limited to that of cavity pressure [130], mold surface temperature [131], mold temperature [131], and mold heat flux [131]. Limited progress has been made with respect to the monitoring of the melt-front position. Yokoi et al. [132–135] might be the first who succeeded in observing the melt flow in a mold by visualization through glass inserts fitted on the mold wall. This technique was modified by Bress and Dowling [136] for the observation of the large part surface side. This kind of visual technique using glass inserts allows for direct observation of the melt flow status, typically using a high-speed camera. This kind of device provides good information for research, but it has limited usage for industrial applications because glass cannot withstand the high pressure required in the applications. Furthermore, the flow and heat transfer along the glass surfaces may be different from those along the mold metal surfaces.

An ultrasonic pulse-echo technique was developed by Wang et al. [137] for a simple box mold, based on the attenuation difference between the air and polymer melt, to detect the flow front of melt inside the mold. Multiple such sensors were installed at selected locations inside the mold for detection of flow positions. When the melt front reaches any probe location, it could be reflected in the sudden changes in the pulse-echo magnitude. However, this method can only measure the melt front at the locations where the ultrasonic sensors are installed, giving discrete and limited information on the mold filling. This type of detection of flow position is not very different from the detection methods by temperature and pressure transducers by Gao et al. [130, 131] and the optical fiber sensors by Thomas et al. [138, 139].

To eliminate the limitations of the existing technologies, a capacitive transducer was proposed and developed by Chen et al. [140], based on the dielectric property difference between polymer and air. In addition to the proposed sensor being able to online and continuously measure the melt-front position in a mold cavity, it can also be applied to detections of V/P transfer, gate freezing-off, and indications of overpack and solidification rate [141]. Combined with data-based regression technology, it may also be used for online check-ring fault detection and part weight estimation [142]. The details are elaborated as follows.

## 8.1 Fundamentals of Capacitive Transducers

Capacitance is a useful physical phenomenon for sensor design. It can be successfully applied to measure distance, area, volume, pressure, force, humidity, and so on [143–149]. The principle of the capacitive sensors is briefly reviewed as follows.

Taking a parallel-plate capacitor, for example, the capacitance of the capacitor, $C_0$, can be derived as

$$C_0 = \frac{\varepsilon \cdot A}{d} \tag{8.1}$$

where $\varepsilon$ is the dielectric constant (permittivity) of the medium between the plates, which are with a surface area $A$ separated by a distance $d$, as shown in Fig. 8.1. Assuming the plate length is $l$ and the width is $w$, the fringing effects can be ignored for cases when $l \gg w$.

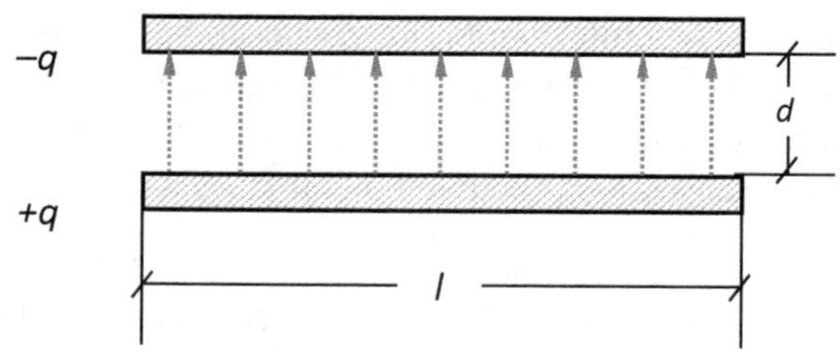


**Figure 8.1** Schematic of a parallel-plate capacitor

Equation 8.1 establishes a relationship between the dielectric constant, the plate area, and the distance. Varying any of them will linearly change the capacitance. The principle of capacitive sensors, therefore, is based on the measurement of the capacitance change, which can further predict the system change based on either of the above aspects. Accordingly, the capacitive sensors can be categorized into the following three types.

The first and also the most widely used type of capacitive sensor is designed based on the area change. As illustrated in Fig. 8.2, if the bottom plate is fixed while the upper one moves by a displacement of $\Delta l$, the capacitance will vary as

$$C = C_0 \frac{l - \Delta l}{l} \tag{8.2}$$

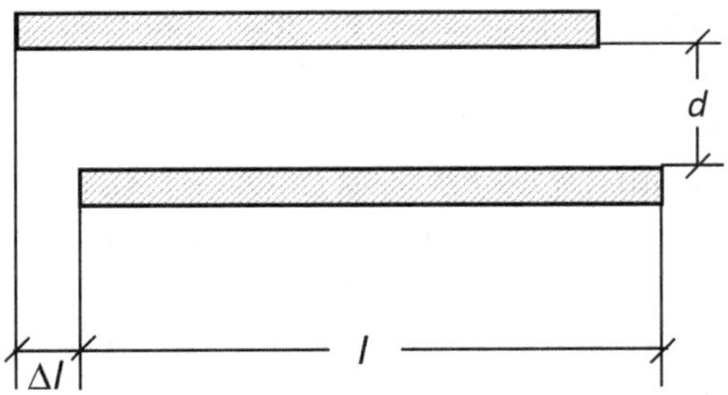


**Figure 8.2** Schematic of the displacement capacitive sensor

The displacement, $\Delta l$, is thus proportional to the capacitance change with the following relation:

$$\frac{\Delta C}{C_0}=\frac{\Delta l}{l} \tag{8.3}$$

Based on this linear relation, the displacement or position sensors can thus be built.

The second type of capacitive sensor is built according to the distance change. Under the case when the dielectric layer is deformed over its entire surface with a distance $\Delta d$, such that the plates remain parallel, as shown in Fig. 8.3, the capacitance will change as

$$C=\frac{\varepsilon \cdot A}{d+\Delta d} \tag{8.4}$$

From Eqs. 8.1 and 8.4, and for $\Delta d \ll d$, one obtains

$$\frac{C}{C_0}\approx 1-\frac{\Delta d}{d} \tag{8.5}$$

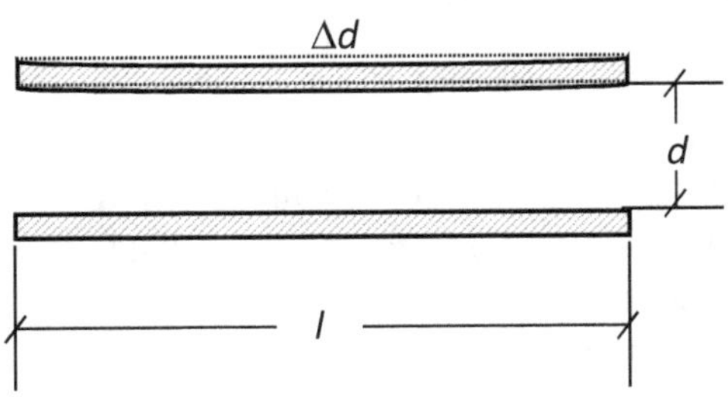


**Figure 8.3** Schematic of the pressure capacitive sensor

The capacitance keeps an approximately linear relationship with the distance change. Capacitive pressure sensors [144] adapted from the above principle have been developed with success for industrial applications.

The third type of capacitive sensor utilizes the dielectric constant change. Development of a relative humidity (RH) sensor has been reported [147]. The sensing mechanism is based on the fact that the dielectric constant of the insulating polymer, polyimide (PI), is linearly related to the ambient relative humidity. The capacitance of a parallel plate device using PI as the medium is, therefore, a linear function of ambient RH. Another successful application of this type of capacitive sensor is to measure the liquid level [148]. Basically, the mechanism can be roughly explained by the illustration shown in Fig. 8.4.

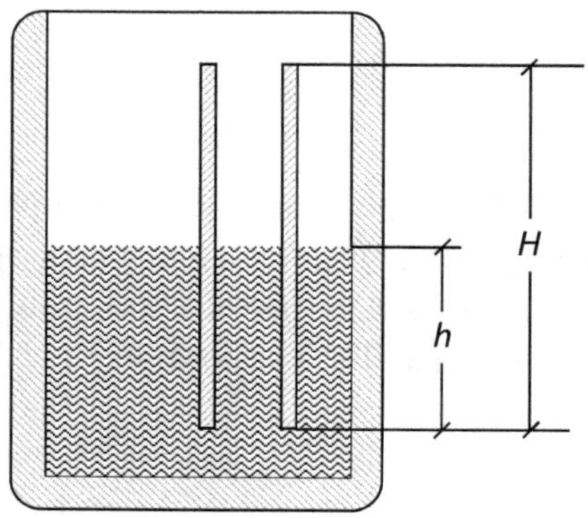


**Figure 8.4** Schematic of the water-level capacitive sensor

As shown in the figure, the sensor is immersed in a water tank. When the level increases, water fills more space between the plates, thus increasing the sensor's capacitance. The total capacitance is

$$C = C_1 + C_2 = \frac{\varepsilon_0 kwh + \varepsilon_0 w(H-h)}{d} \tag{8.6}$$

where $C_1$ is the capacitance of the water-free portion of the sensor, $C_2$ the capacitance of the water-filled portion, $\varepsilon_0$ the dielectric constant of air, $k$ the relative dielectric constant of water, $w$ the width of the plate, and $d$ the distance between two plates. In Eq. 8.6, all of the variables are constant except for the water level, $h$. The capacitance is, therefore, a linear function of the water level.

## 8.2 Dielectric Properties of Polymers

To develop a capacitive transducer for injection molding, it is essential to understand the dielectric properties of polymers [150–155]. In contrast to metals, general polymers are poor electron conductors. However, they are excellent insulators and thus can be used as the medium of capacitors. Using polymer as the medium, the capacitance of a given parallel-plate capacitor with spacing $d$ and with area $A$ is

$$C = \frac{k\varepsilon_0 A}{d} \tag{8.7}$$

where $\varepsilon_0$ is the dielectric constant of the air, and $k$ is the relative dielectric constant, defined as the ratio of the capacity of a condenser made with the particular polymer to the capacity of the same condenser with air as the dielectric. It is dimensionless and dependent on the material, temperature, and frequency. The relative dielectric constants of some polymers are listed in Table 8.1 for reference.

**Table 8.1** Relative Dielectric Constants of Various Polymers [151]

| Polymer | Relative dielectric constant (800 Hz) |
|---|---|
| Expanded Polystyrene | 1.05 |
| Polytetrafluorethylene | 2.05 |
| Polyethylene (density dependent) | 2.3–2.4 |
| Polystyrene | 2.5 |
| Polypropylene | 2.3 |
| Polyphenylene ether | 2.7 |
| Polycarbonate | 3.0 |
| ABS | 4.6 |
| Polyamide 6 (moisture content dependent) | 3.7–7.0 |
| Polyamide 66 (moisture content dependent) | 3.6–5.0 |

Figures 8.5 and 8.6 present the relative dielectric constants of selected polymers as a function of temperature and frequency, respectively [151]. These relations are important for developing the capacitive transducer for injection molding. Figure 8.5 indicates that the temperature relationship of the dielectric constant is material dependent. For certain polymers, the dielectric constant does not change much with temperature, while for other polymers, for example, PA and PVC, the dielectric constants increase greatly when the temperature rises across the glass transition temperature. It has been reported that even greater increases in the dielectric constants, for example, PVC from 3.5 to about 15, and PA from 4.0 to about 50, can be observed [155].

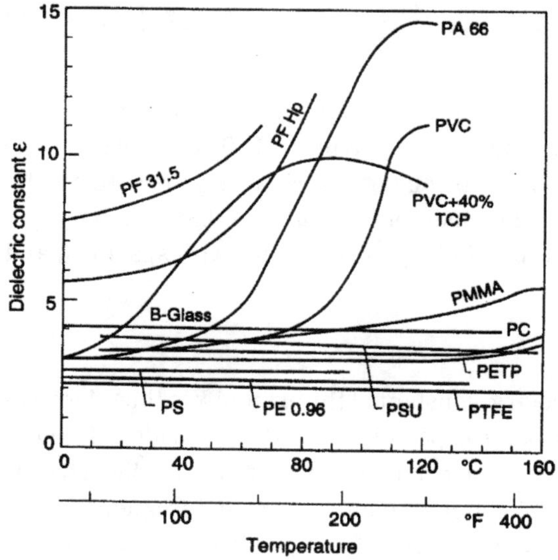


**Figure 8.5** Dielectric constant as a function of temperature [151]

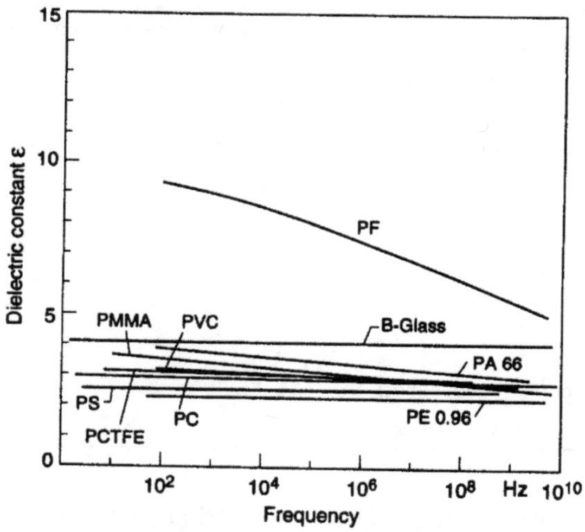


**Figure 8.6** Dielectric constant as a function of frequency [151]

The large and sudden change of the dielectric constant is attributed to changes in the polar functional group around the glass transition temperature. Studies in the literature [150, 151, 155] show that the relations of polymer dielectric constants to the molecular polarization can be categorized into three types: electronic, atomic, and orientation. The first two types of polarization exist for both the polar and nonpolar polymers and are hardly affected by temperature. The orientation polarization due to dipole moment, which is influenced by temperature, is unique for polar polymers.

This change thus can be used to detect the glass transition temperature of polar polymers, which may be applied to the study of the cooling stage in injection molding.

The dielectric constants of nonpolar polymers are less temperature sensitive. Even for polar polymers, the dielectric constants hardly change at their molding temperatures, which are typically much higher than their glass transition temperatures. Furthermore, the duration of the filling phase is typically short compared with that of the entire cycle. These ensure that the capacitive transducer developed for the measurement of melt flow during filling will not be strongly influenced by temperature variation.

## 8.3 Principle and Preliminary Tests of Capacitive Transducer in Injection Mold

The capacitive transducer developed for the melt flow measurement in an injection mold is based on a principle similar to the water-level capacitive sensor described previously. If two metal plates can be separately installed in the two halves of the mold and isolated from each other, they can constitute two electrodes of a capacitor. The medium of the capacitor is air before mold filling, and it is gradually replaced by polymer melt during filling, until completely occupied by the melt at the end of filling. The dielectric constant of the medium between the capacitor plates, therefore, is continually varying during the filling stage. This results in the variation of the capacitance. Figure 8.7 gives such a schematic illustration of the melt flow during the filling stage.

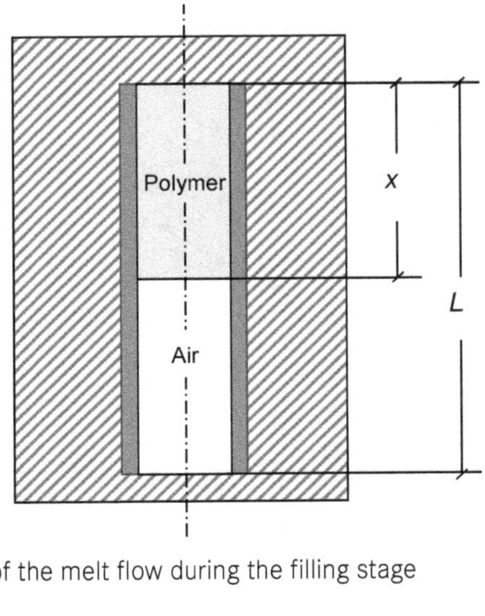


**Figure 8.7** Illustration of the melt flow during the filling stage

With a composite medium, the capacitance during filling is

$$C = C_0\left(1 - x/L + k\,x/L\right) \tag{8.8}$$

where $C_0$ is the capacitance before the start of filling, $k$ the relative dielectric constant of the filled polymer, $x$ the average melt flow length, and $L$ the total length of the plate. The relative melt flow length, $x/L$, can be predicted by the measured capacitance as

$$x/L = \left(C/C_0 - 1\right)/\left(k - 1\right) \tag{8.9}$$

The relative dielectric constant, $k$, can be obtained by measuring the capacitance at the end of filling, at which $x/L = 1$, and

$$k = C_{\text{end}}/C_0 \tag{8.10}$$

As the injection molding is a cyclic process, it is easy to derive the $k$ value from experiments.

A preliminary test [226] was conducted to experimentally verify the relationship between capacitance and polymer fill length. A capacitor was designed using two aluminum plates, each with dimensions of 159 mm × 15 mm, as the electrodes. The distance between the two plates was 4 mm. An ABS plate, with a uniform thickness of 4 mm, was inserted into the electrode gap with a different length in each run of the test to simulate a varying medium. Figure 8.8 shows the influence of the increasing polymer length on the capacitance, which was measured using a capacitance meter with a resolution of 0.1 pf. The data suggested that the capacitance output is highly linear with the polymer length, and this type of capacitive transducer may be applied to the melt flow detection.

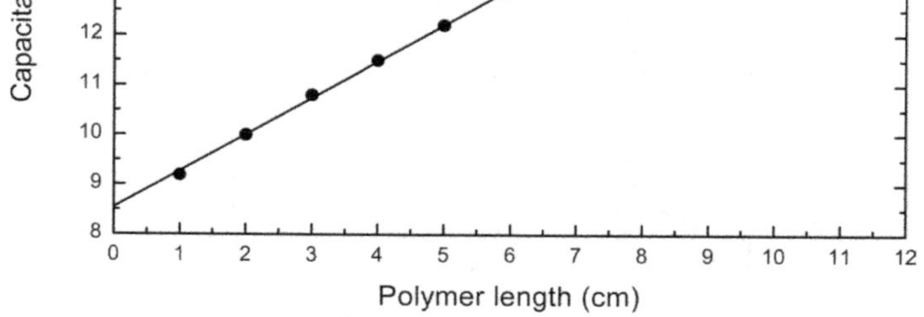


**Figure 8.8** Test of relationship between capacitance and polymer length

Offline tests were also conducted to study the temperature influence on capacitance. The dielectric constant of a nonpolar material is affected by temperature only through the density change with temperature. It, therefore, can be expected that the temperature effect on the dielectric constant, consequently the capacitance, of a nonpolar material is small, except around the melting temperature, $T_m$. This is clearly verified by the experiments on HDPE, as shown in Figs. 8.9 and 8.10. A capacitor was specifically designed using two aluminum plates as electrodes and HDPE as the medium. After the capacitor was heated to 200°C in an oven, it was taken out and cooled under the atmosphere. The capacitance was measured together with the temperature, as shown in Fig. 8.9.

The capacitance change ratio of two neighboring sampling times, as presented by

$$\text{Ratio} = \frac{C_{n-1} - C_n}{C_n} \times 100\% \tag{8.11}$$

is drawn in Fig. 8.10, from which a distinct drop can be clearly observed around the melt temperature of HDPE (about 130°C). It indicates that the polymer phase change may also be revealed by the capacitance change.

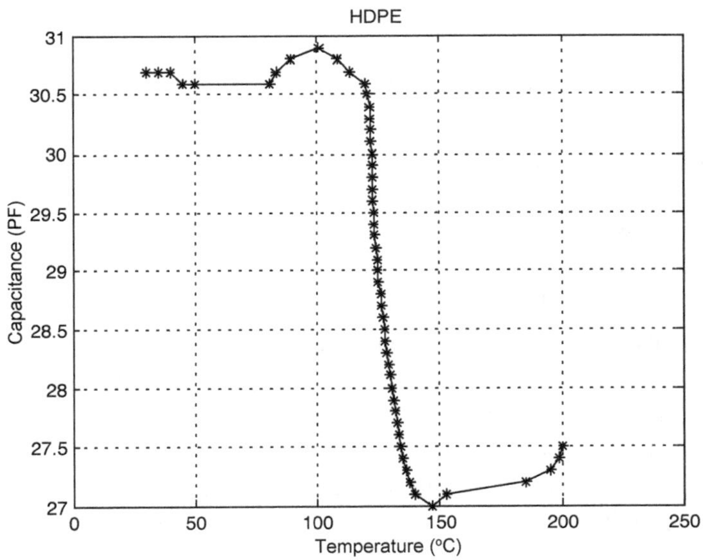


**Figure 8.9** The relation of capacitance and temperature of HDPE

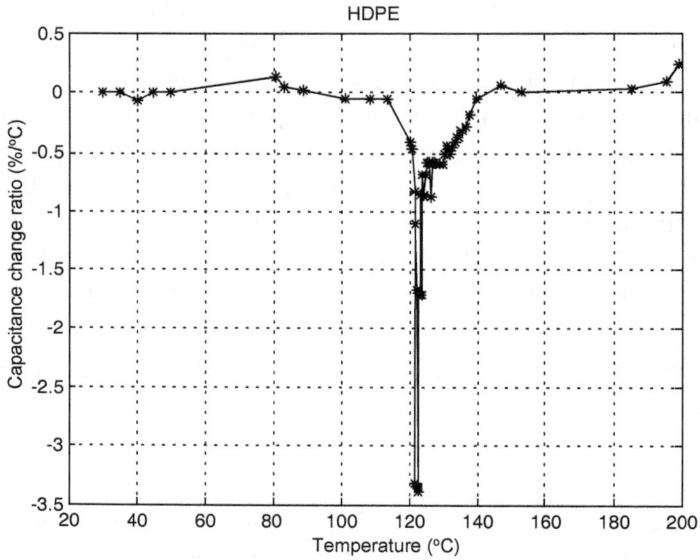


**Figure 8.10** The capacitance change ratio with temperature of HDPE

# 8.4 Design of In-Mold Capacitive Transducer

## 8.4.1 Mold Base Design

A plate mold is designed for the application of the capacitive transducer. As shown in Fig. 8.11, the basic shape of the mold insert is a 150 mm × 100 mm rectangle. The nominal thickness of the designed parts is 2 mm.

Normally two plates are used for a parallel plate capacitor. In the mold design, only one plate is specifically installed to reduce the mold complexity. As shown in Fig. 8.12, a 136 mm × 35 mm plate made of stainless steel is installed at the stationary plate part of the mold as one electrode. The moving mold base, also made of stainless steel, acts as the opposing electrode. The gap between two electrodes is the same as the designed part thickness, 2 mm. For electrode systems with a large area to distance, fringe fields will be very small and usually can be neglected [149]. Such an electrode design can thus be described by a parallel capacitor model, in which the capacitance is proportional to the polymer length.

To constitute a capacitor, the electrodes must be isolated from each other. Electrical insulators, therefore, are required to separate the specifically installed electrode from the remaining part of the mold. Five pieces of insulator are fitted to surround the electrode: one is at the back, and the others are along the four sides, as shown in Fig. 8.13.

**Figure 8.11** Basic shape of the mold insert

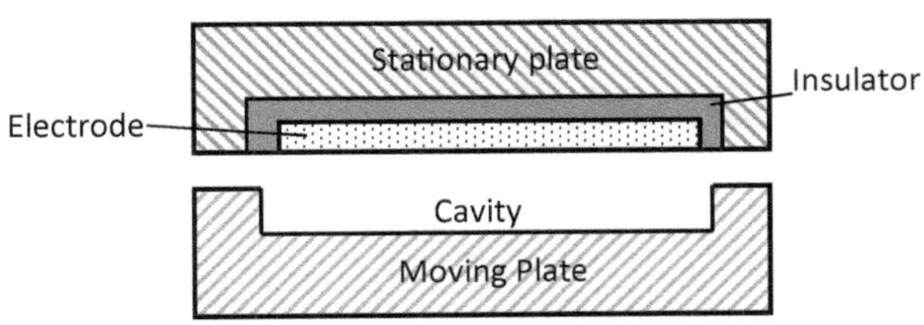


**Figure 8.12** Schematic illustration of mold design

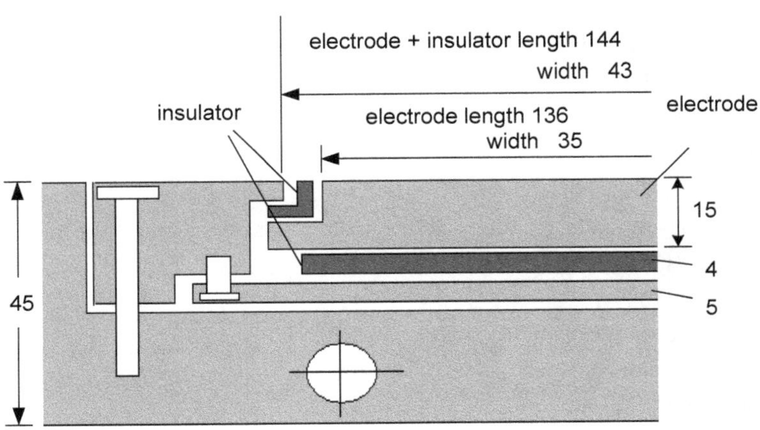


**Figure 8.13** Cross section graph of stationary mold side

In addition, a wire is welded at the back of the stainless steel electrode and extended out of the mold through a channel to connect with a measuring circuit. Another wire is connected with the grounded mold base.

The measured capacitance is a sum of two parts, as shown in Fig. 8.14. One is contributed by the parallel plate capacitor, $C_f$, whose medium is a composite of polymer and air in the filling stage, and another part, $C_i$, is caused by the insulating material.

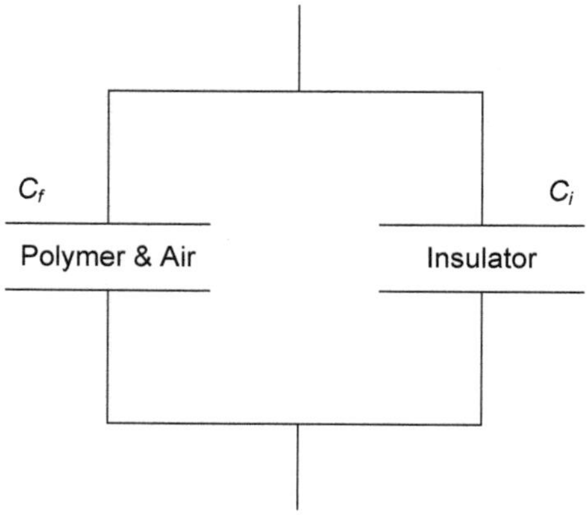


**Figure 8.14** Capacitance contribution

**Table 8.2** Physical Properties of the Selected Ceramic

| Physical Properties and Specifications | |
|---|---|
| Composition | 92% $Al_2O_3$ |
| Color | White |
| Density [g/cm$^3$] | 3.6 |
| Water absorption rate | 0 |
| Yield strength [MPa] (RT) | 390 |
| Compressive strength [MPa] | 2403 |
| Fracture toughness [Mpa m$^{1/2}$] | 3 |
| Elasticity ($\times 10^5$) [MPa] | 2.5 |
| Hardness [HV] [GPa] | 14 |
| Hardness [HRA] | 88 |
| Thermal expansion (RT–800°C) [$10^{-6}$/°C] | 7.6 |
| Thermal conductivity [W/(m·°C)] | 16.74 |
| Thermal shock resistance [°C] | 200 |
| Specific heat [J/(kg·°C)] | 790 |
| Maximum application temperature [°C] | 1500 |
| Relative dielectric coefficient | 10.2 |
| Volume resistivity [$10^{16}$ Ω cm] | $10^{16}$ |
| Dielectric strength [kv/mm] | > 5.9 |

To reduce the insulator contribution to the capacitance measurement, the dielectric constant and the temperature sensitivity of the insulating material should be small. In addition, as the pressure in the molding can be high, the insulator should have good mechanical properties, such as high hardness and impact resistance, to prevent early breakdown of the electrode system. Furthermore, the heat conductivity of the insulator should not be very different from that of the mold metal to prevent part warpage due to uneven heat conduction. A comprehensive search results in the selection of ceramic aluminum oxide as the insulator material for its steady dielectric performance, high hardness, and good heat conductivity. Table 8.2 lists the physical properties of the selected material provided by Mutugoro Corporation.

## 8.4.2 Mold Insert Design

A set of mold inserts was designed for the tests of the capacitive transducer. To study the characteristics of the melt development, several basic shapes along the melt flow path were covered in this set of molds. The insert components are shown in Fig. 8.15, which includes a constant cross-sectional area and abruptly or gradually varying area components. For the gradually varying area, both the linear and nonlinear profiles are included. Though the component shapes look simple, they may form a complex mold through their combinations. Furthermore, any complex mold may be decomposed into a combination set of simple shapes.

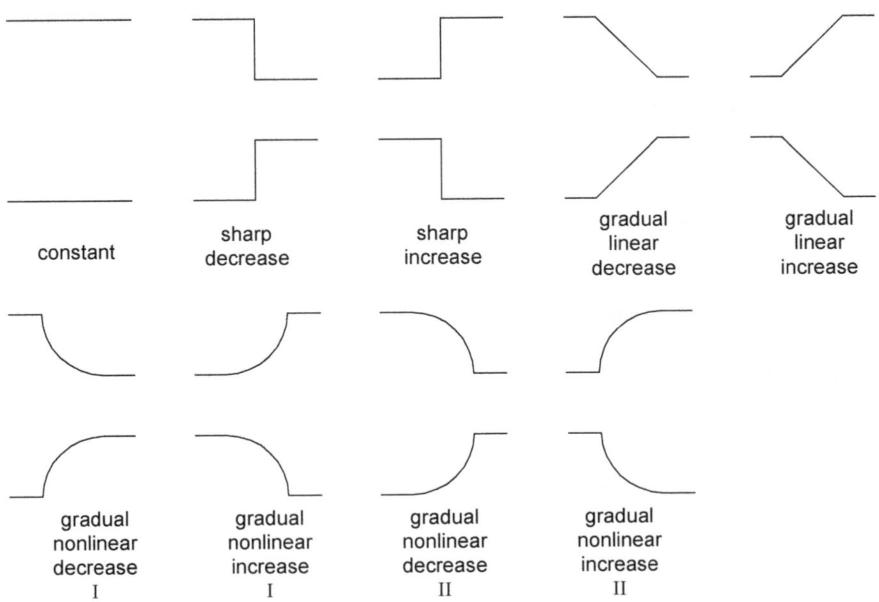


**Figure 8.15** Basic shape components

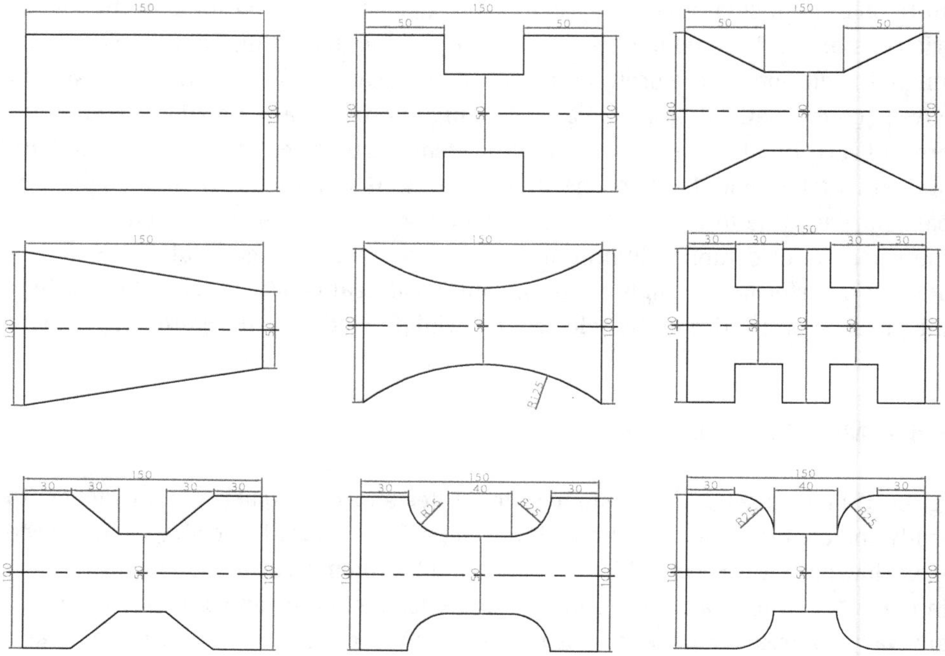

**Figure 8.16** Geometry of the mold inserts

Based on these shape components, a set of mold inserts was designed as shown in Fig. 8.16. It contains nine different inserts. The basic shape is a 150 mm × 100 mm rectangle, and the other eight are all modified plate inserts. With a same thickness, 2 mm, the inserts are referred as Mold 1 to Mold 9 in the following descriptions. Such a multiple insert design focuses on the melt flow study in basic shape molds. How the melt flow develops along the constant area, gradually increasing or decreasing area, and abruptly increasing or decreasing area can be observed by the capacitive transducer.

### 8.4.3 Capacitance Measurement

The electrodes are connected to a properly built measuring circuit that is designed based on the charge-and-discharge principle [156–159] to convert the capacitance signal to a voltage signal. The charge-and-discharge frequency is set at 800 Hz. As shown in Fig. 8.17, the voltage signal is amplified and filtered before it is connected to the computer data-acquisition system for further processing.

To calibrate the measuring system, four fixed-value capacitors with capacitance of 220 pf, 273 pf, 333 pf, and 395 pf are used to find the voltage-capacitance relation. The calibration data can be found in Table 8.3.

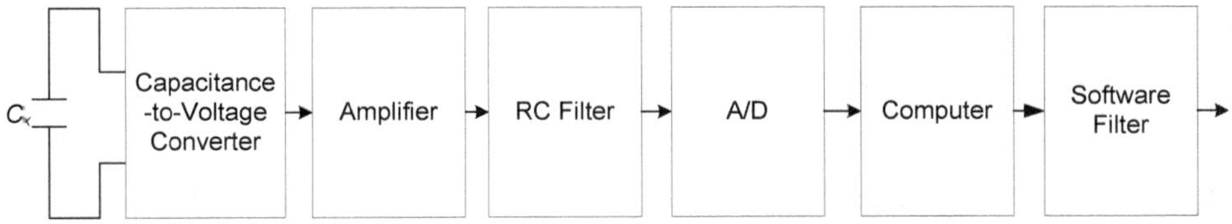


**Figure 8.17** Capacitance measurement scheme

**Table 8.3** Calibration Data of Capacitance Measuring Circuit

| Capacitance (pf) | Mean voltage (v) |
|---|---|
| 220 | 0.4486 |
| 273 | 2.8473 |
| 333 | 5.5444 |
| 395 | 8.3140 |

A linear least-squares data fitting shows that the measuring system has high linearity between the capacitance and voltage output, as shown in Fig. 8.18. The following formula can be used to correlate capacitance and voltage:

$$V = -9.430 + 0.0449 \times C \tag{8.12}$$

or

$$C = 210.02 + 22.27 \times V \tag{8.13}$$

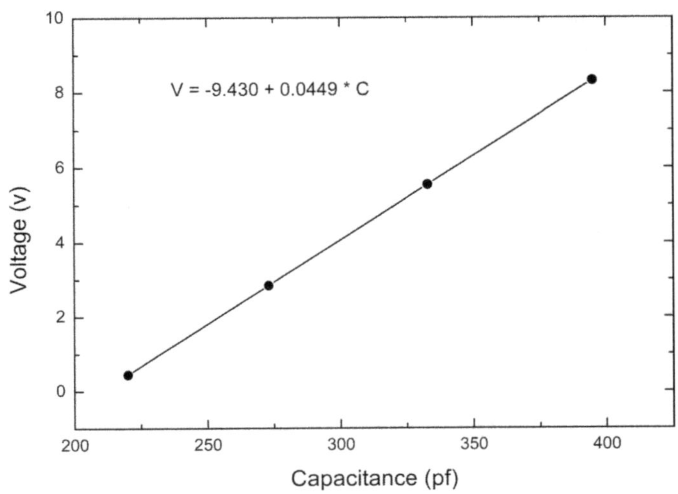


**Figure 8.18** Calibration of the capacitance measuring circuit

## 8.5 Applications in Melt Flow Detection during Filling Stage

Though it was designed for the average-flow-length measurement in the filling phase, the developed capacitive transducer has wide applications in injection molding [226]. To test the transducer performance, online experiments were conducted on the Chen Hsong reciprocating-screw injection molding machine. Both semicrystalline and amorphous materials, including high-density polyethylene (HDPE) (Marlex, HMN6060), polypropylene (PP) (HMC, PROFAX6331), and polystyrene (PS) (DOW, 685D), were used in the experiments under different injection velocity profiles with different mold inserts. The experimental application results, including detections of start of filling, V/P transfer, and melt flow in the cavity, are presented and analyzed as follows.

### 8.5.1 Detection of Filling Start

Injected by the screw forward movement, the polymer melt passes through nozzle, sprue, runner, and gate before finally entering the mold cavity. The time when the melt first enters the cavity is important for monitoring and controlling the melt flow. The developed capacitive transducer (CT) is capable of detecting this starting point. As described in the previous section, the CT output is proportional to the polymer melt length along the electrode installed in one side of the mold cavity. The CT output does not change until the polymer flows into the gap formed by the electrode, which corresponds to the time when the melt enters the mold cavity. This is clearly verified by the CT output during filling in Mold 1, as shown in Fig. 8.19.

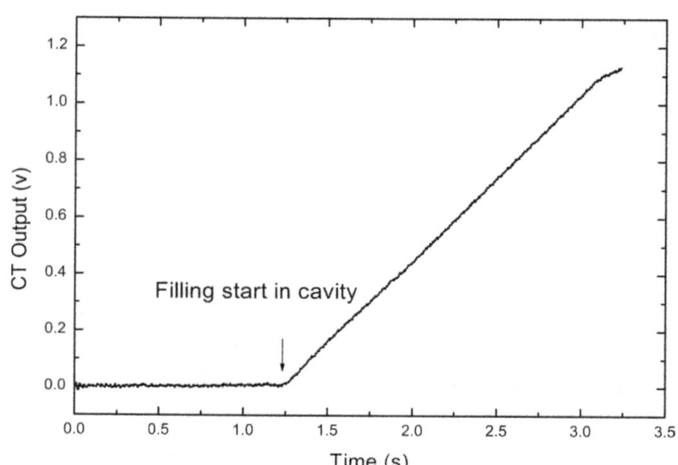


**Figure 8.19** Detection of filling start

In this case, the injection velocity is controlled at a constant speed of 15 mm/s during the entire filling phase.

The figure shows that there is a period of about 1.25 s at the beginning of filling during which the CT output curve remains zero. This period corresponds to the polymer filling the runner and sprue before it reaches the cavity. After the melt enters the cavity, the CT output increases with time. The time of the polymer melt entering the cavity can be clearly seen as marked in the figure.

### 8.5.2 Detection of V/P Transfer

After the mold cavity is filled, the process switches from filling to the packing-holding stage. Correspondingly, the control of velocity in the filling stage should be transferred to that of pressure at this critical time, commonly referred to as the V/P transfer point. Early transfer results in a short shot, whereas late transfer causes flash. Accurate detection of the V/P transfer is of importance for the proper operation of a molding process. By utilizing the process characteristic that the nozzle pressure significantly increases at the end of filling, a fuzzy system technique [225] has been successfully developed to control the V/P transfer in a timely way.

The capacitive transducer is found to also have such a potential for the V/P transfer detection. The measured capacitance increases with the melt development, and eventually it becomes a constant after the melt reaches the end of the electrode, if neglecting the temperature influence. When the electrode is designed to cover the last filled point of the mold cavity, proper V/P transfer can then be easily detected by observing the CT output.

A number of experiments were conducted by Chen et al. [140] with different injection stroke settings. The results using HDPE with Mold 1 are presented in Figs. 8.20–8.22, in which both the nozzle pressure and CT output are plotted for comparison. In the deliberate short-shot case shown in Fig. 8.20, both the pressure and capacitive signal increase with time, and no clear V/P transfer point can be observed. In Fig. 8.21, a case when the mold is nearly filled, an inflection in the capacitive signal can be observed, as marked by a gray circle, indicating that the melt has reached the end of the electrode, and the capacitance increment reaches saturation. The pressure signal in this case does not show any indication of filling end. In Fig. 8.22, the case in which the mold is just filled, inflections appear in both signals. The early appearance of the inflection point in the capacitive signal is owing to the fact that the electrode in our case does not cover the last filled point of the whole mold cavity. This is because the initial design function does not focus on the V/P transfer detection. But the results have indicated that the CT output has the potential to detect the V/P transfer, once it is purposely designed to cover the last filled point of the cavity.

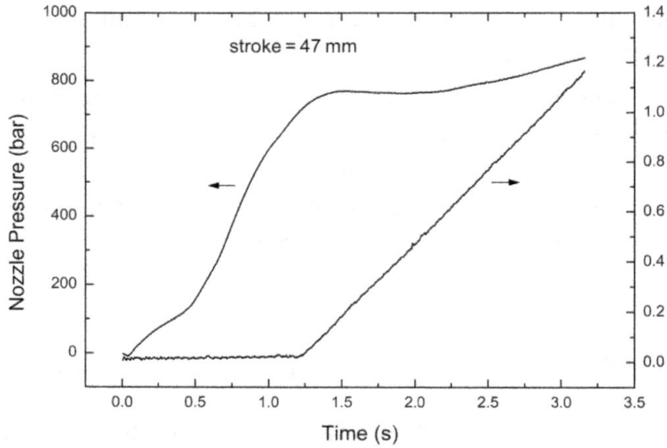


**Figure 8.20** Test of V/P transfer (stroke = 47 mm)

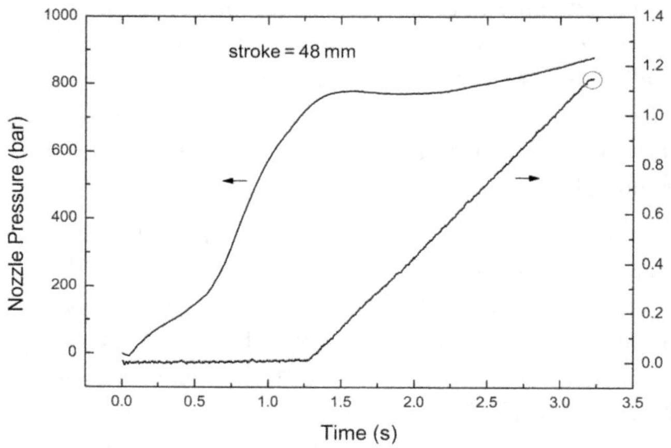


**Figure 8.21** Test of V/P transfer (stroke = 48 mm)

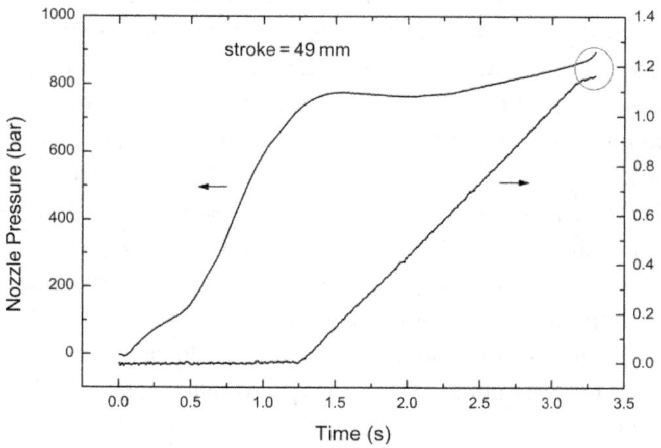


**Figure 8.22** Test of V/P transfer (stroke = 49 mm)

### 8.5.3 Detection of melt flow during filling

Equation 8.14 describes the relationship among injection velocity (*IV*), average-flow-length (*afl*), and melt-front area ($A_{mf}$) during filling:

$$IV \cdot A_b = \frac{dafl}{dt} \cdot A_{mf} \tag{8.14}$$

where $A_b$ is the constant barrel area. For Mold 1, the melt-front area is nearly constant throughout most of the filling, with minor exceptions in the beginning and end of the cavity. The average-flow-length (*afl*) during filling with a constant injection velocity (*IV*) should increase linearly with time according to Eq. 8.14. This has been verified by the results shown in the previous figures for the detections of filling start and V/P transfer, in which the CT outputs, representing the average-flow-length, increase in a straight line during filling. To further demonstrate that the capacitive transducer can quantitatively measure the average-flow-length change, a series of experiments are conducted using different velocity profiles, different mold inserts, and different materials in the following sections.

#### Test of Different Velocity Profiles

Experiments were conducted using HDPE and Mold 1 with constant injection velocities of 10 mm/s, 15 mm/s, 20 mm/s, 25 mm/s, and 30 mm/s, respectively. The CT outputs of these cases are shown in Fig. 8.23. As shown in the figure, different injection velocities result in differences in the time of the melt flow entering the cavity. After that, the CT outputs of all cases appear as straight lines for filling at the constant injection velocity of the constant cross-sectional area of the mold cavity.

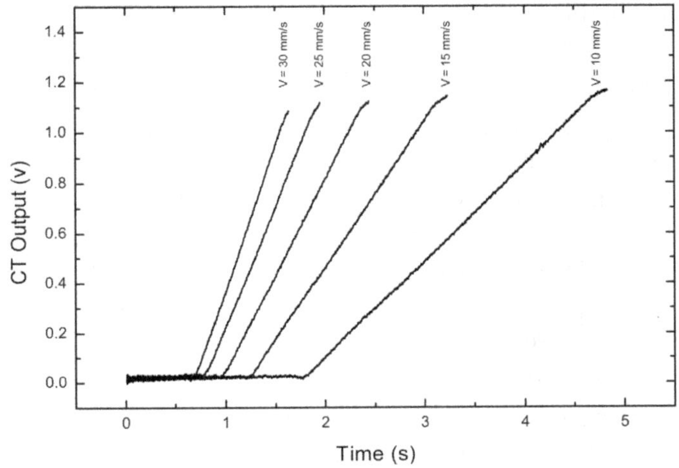


**Figure 8.23** CT outputs of different constant injection velocity settings

**Table 8.4** Comparison of Different Constant Velocity Profiles

| Injection Velocity (mm/s) | *R* Value | Slope | Slope/Velocity |
|---|---|---|---|
| 10 | 0.9999 | 0.3836 | 0.0384 |
| 15 | 0.9998 | 0.5780 | 0.0385 |
| 20 | 0.9998 | 0.7679 | 0.0384 |
| 25 | 0.9998 | 0.9643 | 0.0386 |
| 30 | 0.9998 | 1.1531 | 0.0384 |

A linear least squares regression is used to analyze the experimental data for the relation between the CT output and injection velocity. The CT outputs during filling of the cavity are well approximated by linear increasing ramps with the results shown in Table 8.4. The correlation coefficients ($R$ value), a measure of how well the data are fitted by the regressing function, are very close to one, which means a perfect correlation. This indicates a high linearity of the CT output. In addition, the regressed slope of each ramp function is listed in the third column of the table, together with the ratio of slope over injection velocity presented in the last column. Though the CT outputs are with different slopes at different injection velocities, the ratios of slope over injection velocity are nearly the same for all of the injection velocity settings. The largest discrepancy of these ratios is less than 1%, agreeing with Eq. 8.14 very well.

Experiments with step-change profiles were next conducted for further testing of the capacitive transducer. Four different step-change profiles, including both the step-up and step-down types, are used. The first one has a step change from 10 mm/s to 30 mm/s, the second from 15 mm/s to 25 mm/s, the third from 25 mm/s to 15 mm/s, and the fourth from 30 mm/s to 10 mm/s. All of the step changes are designed to happen around the time when the mold cavity is nearly half filled. The recorded CT outputs are shown in Fig. 8.24.

In contrast to the measurements with constant injection velocities, obvious inflections can be found in all of the CT output curves due to the step changes. Furthermore, the increasing and decreasing step changes in the injection velocity result in opposing inflection shapes: a concave shape for the step-up, and a convex shape for the step-down. The inflection magnitude is clearly associated with the size of the step change. In Fig. 8.24, the CT outputs of the 10 to 30 mm/s step-up and 30 to 10 mm/s step-down profiles show a more pronounced inflection than those of the other two step-change profiles with a smaller step size, the 15 to 25 mm/s step-up and 25 to 15 mm/s step-down.

A quantitative analysis is also conducted on the data from the step-change experiments. The CT outputs, together with the injection velocity, are shown in Fig. 8.25 for a step-up and Fig. 8.26 for a step-down, respectively, with the same step size of 20 mm/s.

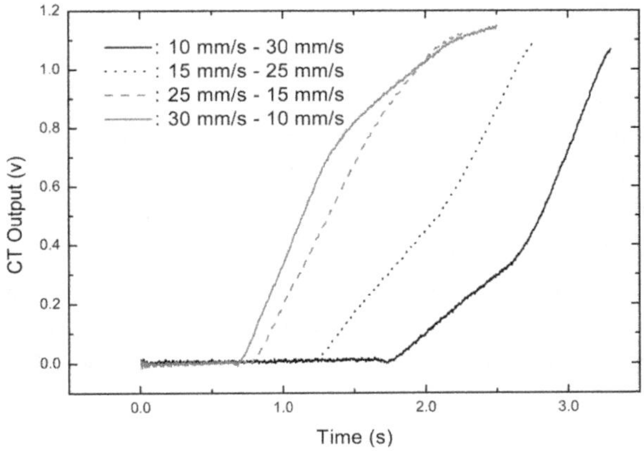


**Figure 8.24** CT outputs of different step-change profiles

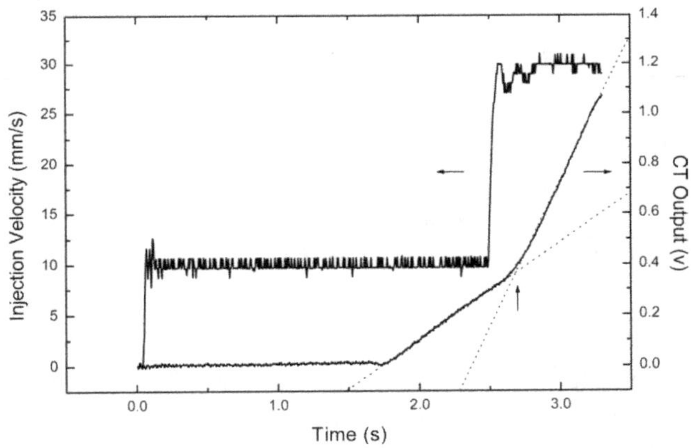


**Figure 8.25** Test of step-up profile (10 to 30 mm/s)

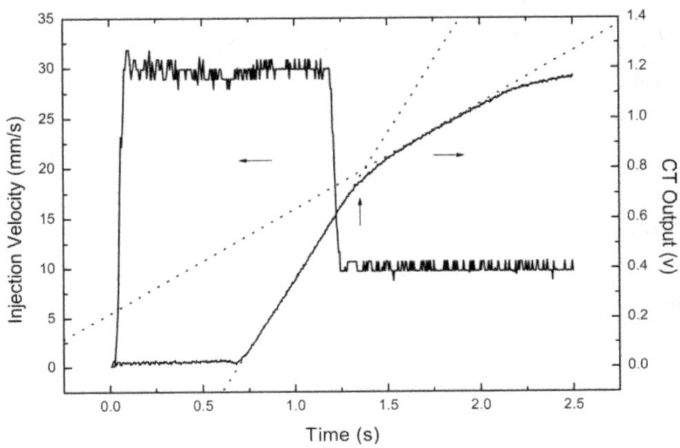


**Figure 8.26** Test of step-down profile (30 to 10 mm/s)

It can be clearly seen that the CT output inflection follows the step change in injection velocity. The inflected curve can be divided into two separate parts by the inflection point, as marked in the figures. Each part can be well approximated by a ramp function, as represented by the dotted lines in the figures. Tables 8.5 and 8.6 show the data regression results of the step-up and the step-down profiles, respectively. The regressed slope of each part is listed in the third column of the table. The real injection velocity is also calculated, as listed in the second column. The slope over velocity ratio is listed in the last column of the tables, which are nearly the same before and after the inflection (or the step change). This indicates that the step change in injection velocity can be quantitatively measured by the capacitive transducer.

**Table 8.5** Data Analysis for Step-Up Profile

| | Averaged Injection Velocity (mm/s) | Regressed Slope | Slope/Velocity |
|---|---|---|---|
| Before Inflection | 10.01 | 0.395 | 0.0395 |
| After Inflection | 29.69 | 1.164 | 0.0392 |

**Table 8.6** Data Analysis for Step-Down Profile

| | Averaged Injection Velocity (mm/s) | Regressed Slope | Slope/Velocity |
|---|---|---|---|
| Before Inflection | 28.46 | 1.157 | 0.041 |
| After Inflection | 9.96 | 0.416 | 0.042 |

In addition to the constant and step-change profiles, the ramp profiles were also tested. According to Eq. 8.14, for a mold with a constant cross-sectional area, a constant injection velocity results in a ramp CT output. A ramp velocity profile should result in a parabolic CT output. To verify the capacitive transducer performance, a ramp-up profile and a ramp-down profile were experimentally tested, with results shown in Figs. 8.27 and 8.28, respectively. In the experiments, the ramp settings of the injection velocity start only after the melt enters the mold cavity; before that, the injections are set at constant rates. An inspection of the figures indicates that the CT outputs are parabolic; they can indeed be well fitted by parabolic curves. The ramp-up and ramp-down result in opposing curvature directions of the two CT output curves. The CT output of the ramp-up injection profile can be approximated by a concave parabolic function, while that of the ramp-down profile looks like a convex one. The fitting results are illustrated in Fig. 8.29 and also presented in Table 8.7.

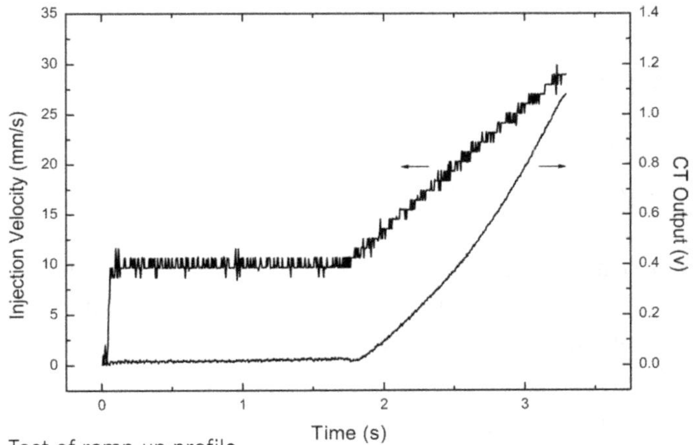


**Figure 8.27** Test of ramp-up profile

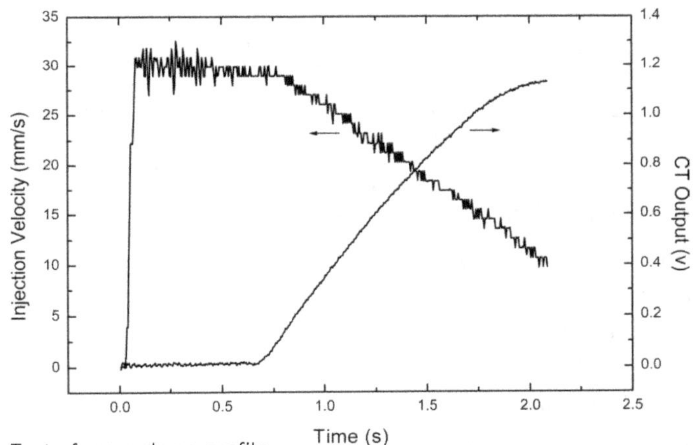


**Figure 8.28** Test of ramp-down profile

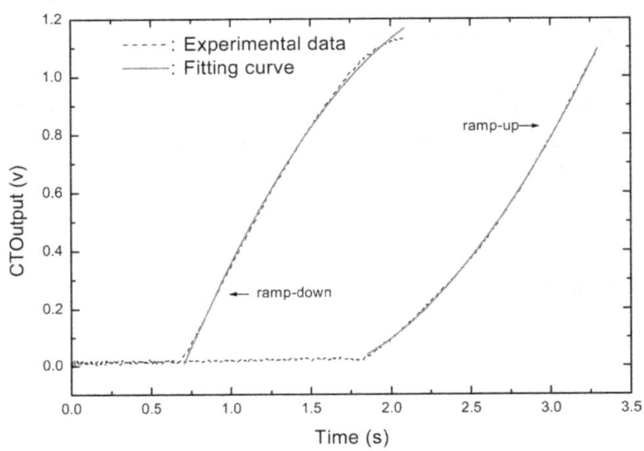


**Figure 8.29** Curve fitting of CT outputs under ramp injection profiles

**Table 8.7** Curve Fitting Results of CT Outputs under Ramp Profiles

| Injection velocity profile | $\frac{dIV}{dt}$ | CT output | $\frac{dIV/dt}{d^2C/dt^2}$ |
|---|---|---|---|
| $IV(t) = 10.85 + 12.41t$ | 12.41 | $C(t) = 0.253t^2 - 0.623t + 0.327$ | 49.15 |
| $IV(t) = 28.96 - 14.52t$ | -14.52 | $C(t) = -0.285t^2 + 1.538 - 0.938$ | 50.97 |

The injection velocity is approximated by a ramp function, as shown in the first column of Table 8.7. Differentiating both sides of Eq. 8.14, the following is obtained by assuming a constant melt-front area with Mold 1:

$$\frac{dIV}{dt} \cdot A_b = \frac{d^2 afl}{dt^2} \cdot A_{mf} \tag{8.15}$$

The injection velocity gradient, $\frac{dIV}{dt}$, therefore, is proportional to $\frac{d^2 afl}{dt^2}$. As the CT output, $C(t)$, represents the average-flow-length, $\frac{dIV}{dt}$ should also be proportional to $\frac{d^2C(t)}{dt^2}$, which is represented by the coefficient of the second-order term of the parabolic equation. The ratios, $\frac{dV_b}{dt} \Big/ \frac{d^2C(t)}{dt^2}$, of the two cases are listed in the last column of the table, which are very close to each other.

The above results of different types of injection profiles prove that the capacitive transducer can quantitatively measure the changes in average-flow-length caused by the changes in injection velocity.

#### Test of Different Mold Inserts

The previously-mentioned results were obtained with a constant cross-sectional area mold. The measurement performance of the capacitive transducer was also evaluated for melt flow position in molds with varying cross-sectional areas. Nine different mold inserts as shown in Fig. 8.16 were tested. Experiments were first conducted using HDPE with a constant 20 mm/s injection velocity for each insert, and the CT outputs are shown in Fig. 8.30. All nine inserts share the same sprue, runner, and gate system, resulting in the same start of filling at the same injection velocity, as shown in the figure. The melt flow development in different molds, however, is dependent on the shape of the insert. This is clearly reflected by the CT outputs. The fillings of different inserts not only end at different times, but also experience different melt flow rate histories.

The injection stroke of each mold insert, which represents the mold volume, is listed in Table 8.8 for the nine mold inserts. It is obvious that Molds 5 and 3 have the smallest volumes, as shown in Fig. 8.16; correspondingly, they have the shortest filling time, as shown in Fig. 8.30 at the same injection velocity.

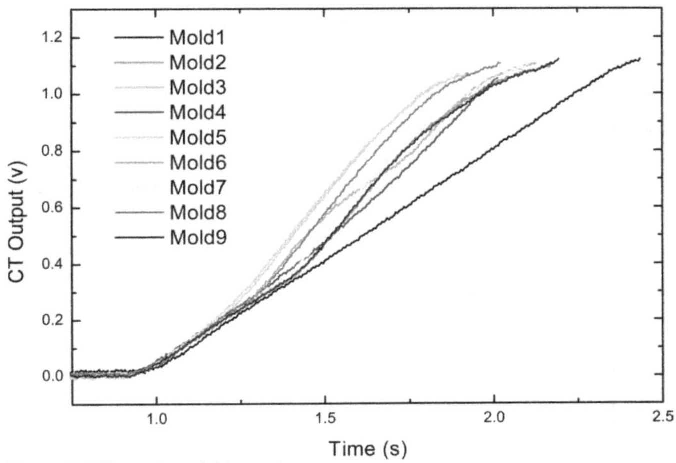


**Figure 8.30** Test of different mold inserts

**Table 8.8** Injection Stroke Settings for Different Mold Inserts

| **Mold No.** | 1 | 2 | 3 | 4 | 5 | 6 | 7 | 8 | 9 |
|---|---|---|---|---|---|---|---|---|---|
| **Stroke (mm)** | 48 | 42.5 | 37.5 | 39.5 | 37 | 41.5 | 41.5 | 39.5 | 43 |

Mold 1, on the contrary, with the largest volume, consequently results in the longest time to fill at the lowest CT output.

As represented by Eq. 8.14, at a constant injection velocity, the average-flow-length is directly related to the melt-front area that reflects the changes in mold insert shape. The CT output of Mold 1, the constant cross-sectional area mold, is linear with time, as shown in Fig. 8.30 at a constant injection velocity. The CT outputs for the other molds, however, are not linear with time at the constant injection velocity because those molds have varying cross-sectional areas along the melt flow paths. The shape changes in the CT output can reflect changes in the melt-front area caused by the mold geometry variation.

A clearer comparison of the CT outputs for Molds 1, 2, and 3 is given in Fig. 8.31. Different from the linear CT output of the constant cross-sectional area Mold 1, an obvious inflection point can be found in the CT output of Mold 2, corresponding to the change in melt-front area caused by a sudden constraint in the middle part of Mold 2. After this inflection point, the slope of the CT output, reflecting the melt-front velocity, obviously increases at the constant injection velocity. The slope of the CT output decreases again after the melt front passes the constraint of the mold, in correspondence with the increase in the cross-sectional area. The changes in the cross-sectional area of Mold 3 are more gradual than in Mold 2, resulting in more gradual changes in the slope of the CT output. As the middle constraint parts of Molds 2 and 3 are designed with a constant cross-sectional area, the middle part of the CT outputs of these two molds more or less appear as straight lines, as marked by a square in the figure. Furthermore, as the cross-sectional areas in the constraint part are the same in these two molds, the two straight lines in the middle part of the CT outputs are also nearly parallel.

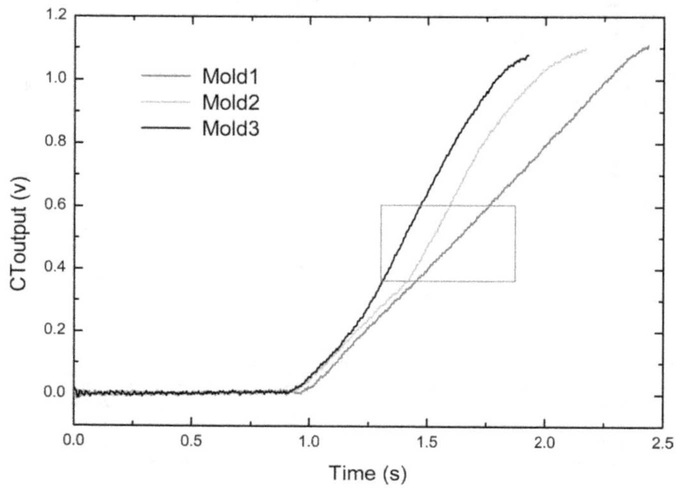


**Figure 8.31** Comparison of CT outputs for Molds 1, 2, and 3

Unique to Mold 4, its cross-sectional area decreases along the melt flow path, resulting in the acceleration in the CT output without any inflection point. The output gradient, reflecting the melt-front velocity, consistently increases, as shown in Fig. 8.32 for filling this continuously varying melt-front-area mold at the constant injection velocity.

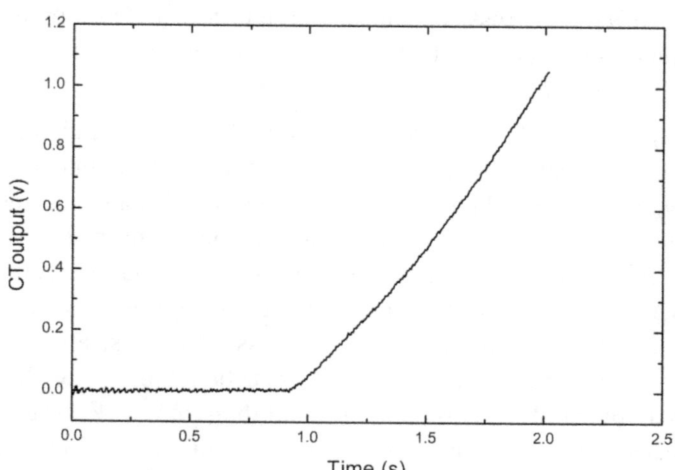


**Figure 8.32** CT output for Mold 4

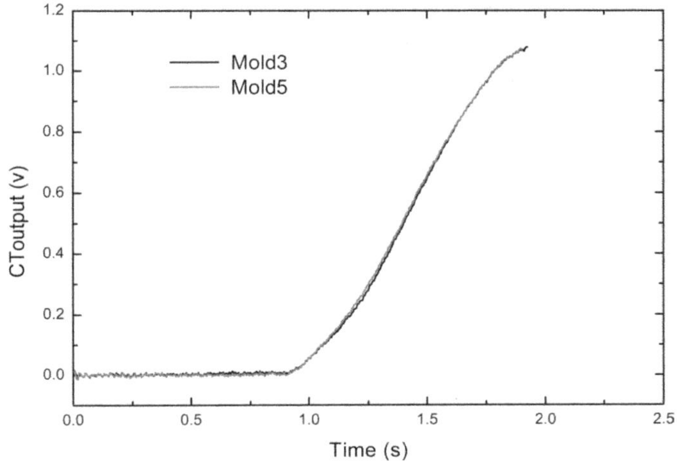


**Figure 8.33** Comparison of CT outputs for Molds 3 and 5

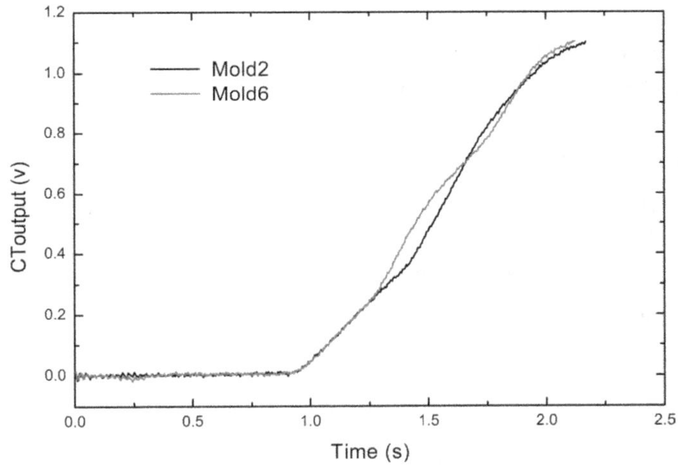


**Figure 8.34** Comparison of CT outputs for Molds 2 and 6

Molds 5 and 3 are similar geometrically, and so are their CT measurements, as shown in Fig. 8.33 with nearly overlapping outputs. In contrast to Mold 2, Mold 6 has more sudden changes in its cross-sectional area. Correspondingly, more inflections and deflections can be found in its CT output, as shown in Fig. 8.34.

Molds 7, 8, and 9 have the constraint size in the middle, but with different convergence and divergence rate; their CT outputs can be quite different, as shown in Fig. 8.35, due to their differences in those rates. The filling of the same beginning part results in the near overlapping of the CT outputs. After that, the CT outputs diverge with the different geometry change. The CT output of Mold 7 is between the outputs of Molds 8 and 9, corresponding to its geometrical relation with that of Molds 8 and 9.

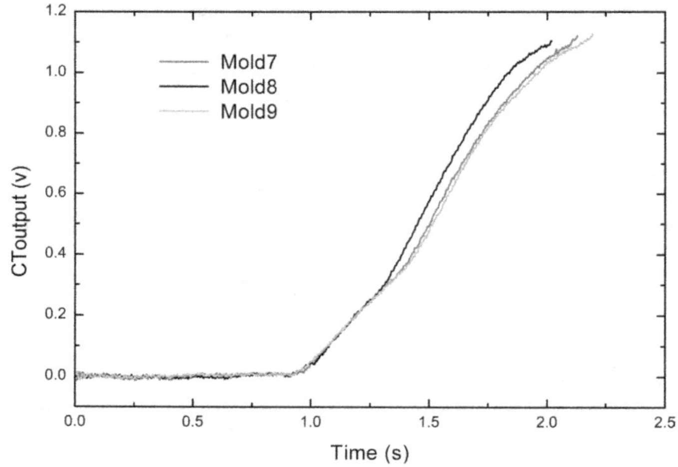


**Figure 8.35** Comparison of CT outputs for Molds 7, 8, and 9

The above results show that the melt development in molds with different geometry can be well reflected by the capacitive transducer.

### Test of Different Materials

In addition to HDPE used in the previous experiments, other materials such as PP and PS were also tested for the capacitive transducer developed. Figure 8.36 compares the experimental results using different materials with Mold 1 at a 20 mm/s injection velocity.

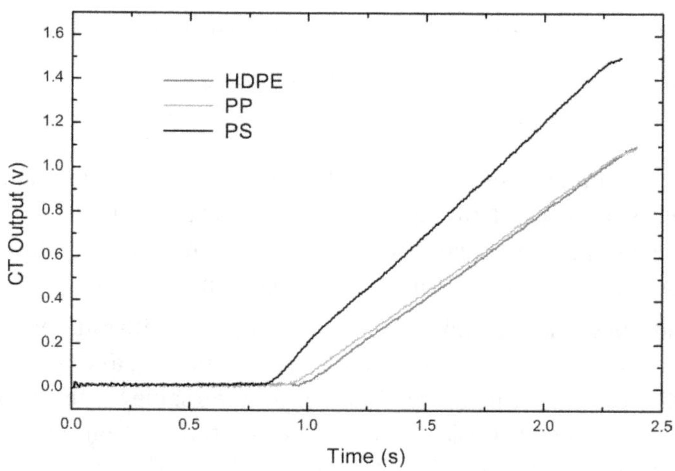


**Figure 8.36** CT outputs of different materials

The CT outputs of HDPE and PP are similar, lower in both the output and its increasing rate than those of PS. This result can be explained by the following equation derived from Eqs. 8.8 and 8.12:

$$\text{Slope} = \frac{dV}{dt} = 0.0449 \times \frac{dC}{dt} = 0.0449 \cdot (k-1) \cdot \frac{C_0}{L} \cdot \frac{dx}{dt} \tag{8.16}$$

where $C_0$ is the capacitance before the start of filling, $k$ the relative dielectric constant of the filled polymer, $x$ the average melt-flow length, and $L$ the total length of the plate. In these cases, $C_0$, $L$, and $\frac{dx}{dt}$ are the same, and the differences lie in $k$. This agrees with the fact that the dielectric constants of HDPE and PP are close and lower than that of PS as listed in Table 8.1.

## 8.6 Applications for the Packing and Cooling Stages

This section explores the application of the capacitive transducer beyond the filling stage. The results show that the capacitive transducer can be used for monitoring the molding status during the packing-holding stage and cooling stage as well.

### 8.6.1 Guide to Packing Pressure Setting

During the packing stage, extra material is packed into the mold to compensate for part shrinkage due to cooling. Proper setting of the packing pressure is important for producing quality parts. One of the difficulties lies with the lack of a convenient means of monitoring the material status in the mold during this stage. The capacitive transducer developed provides such a means to study packing effects on the in-mold polymer melt status [140].

To show how the CT output is affected by packing pressure and packing time, four different constant packing pressures, 300 bar, 400 bar, 500 bar, and 600 bar, were set for the experiments using HDPE. For each pressure level, four different packing durations, including 3 s, 5 s, 6 s, and 7 s, were tested. The comparisons of the CT outputs for different packing pressure settings are given in Figs. 8.37–8.40. In all cases, the filling of the cavity was conducted under the same conditions, resulting in overlapping CT outputs before packing. The CT output after filling, however, was pressure dependent, which can be clearly observed from the figures. The CT output after packing increased with the packing pressure level, as a denser part was molded under a higher packing pressure. The CT output at 600 bar packing pressure was significantly lower than the other CT outputs at the early packing. This may be explained by the overpacking associated with the 600 bar packing case. Overpacking at this pressure resulted in a slight opening of the mold, and consequently a larger electrode separation distance and a smaller capacitance. This can be confirmed by the observation of part flash at the 600 bar packing pressure. Parts molded at packing pressures of 300 bar, 400 bar, or 500 bar had no flash.

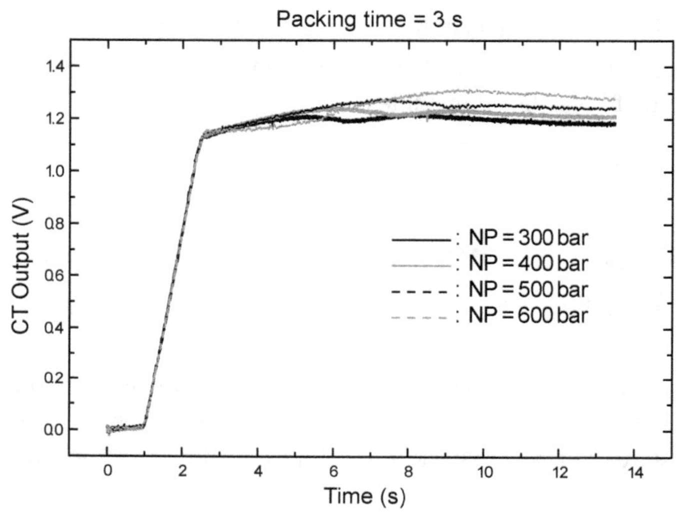


**Figure 8.37** CT outputs with 3 s packing time of HDPE

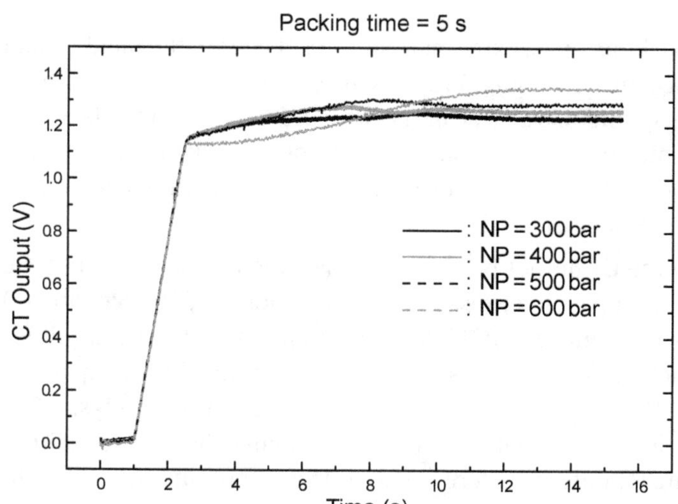


**Figure 8.38** CT outputs with 5 s packing time of HDPE

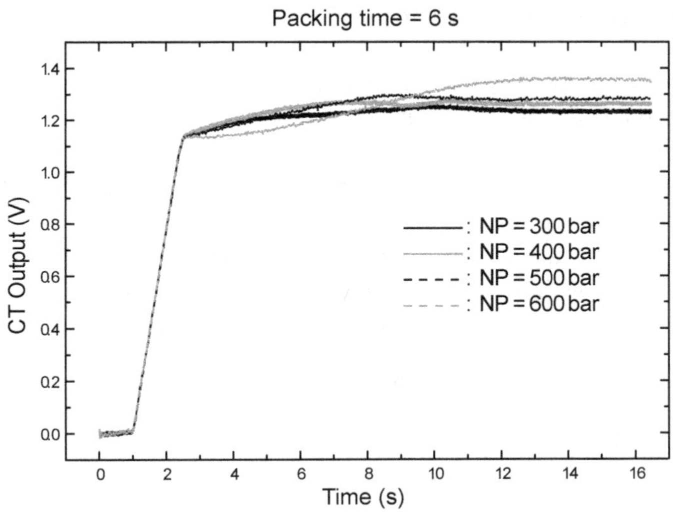


**Figure 8.39** CT outputs with 6 s packing time of HDPE

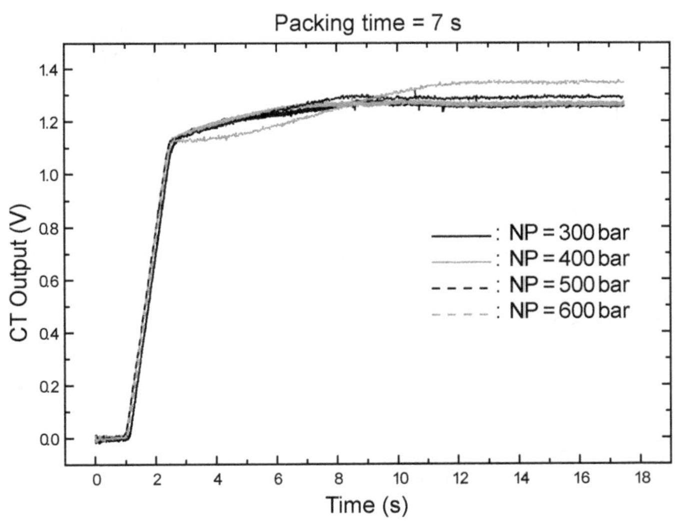


**Figure 8.40** CT outputs with 7 s packing time of HDPE

The CT signals can have at least the following two applications in the packing-holding state: (1) the end of packing CT signal for part weight monitoring, and (2) the early packing CT signal for monitoring of possible mold separation, that is, mold flash. In addition to the above, it can also be used for gate freezing detection, as described in the following section.

### 8.6.2 Detection of Gate Freezing-Off Time

After the gate is frozen, material can no longer be packed into the mold cavity. The packing should cease after this point. The gate freezing-off point can be detected by the capacitive transducer developed [140]. Figure 8.41 shows the CT outputs of HDPE at different packing pressures with the same 7 s packing time. The packing starts at about 2.6 s, and the CT output curves level off after about 8.6 s, indicating that the CT outputs reach a peak after 6 s of packing; after that the gate may have frozen. This is in agreement with the offline detection by comparing the part weight of different packing times, as shown in Fig. 8.42, which shows that little additional

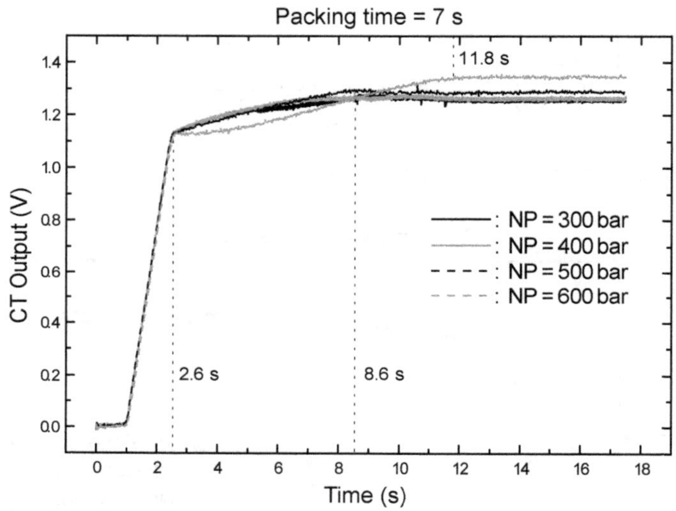


**Figure 8.41** CT outputs for gate freezing detection with HDPE

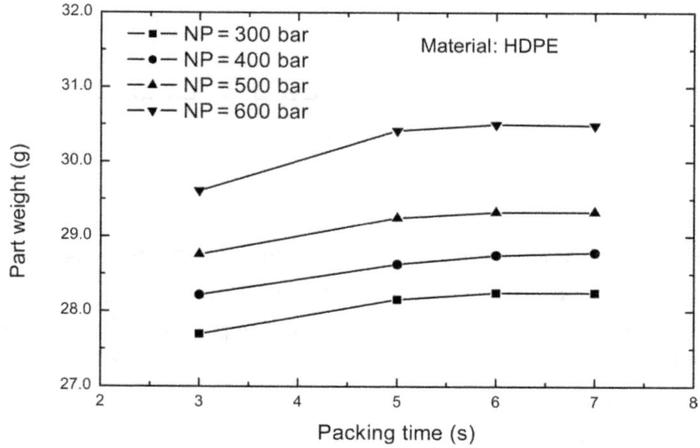


**Figure 8.42** Offline test of gate freezing-off point with HDPE

material can be packed into the mold after 6 s of packing. The exception is with the 600 bar packing case, which resulted in flash due to mold opening, as stated previously. Another experiment is conducted with PP at 400 bar packing pressure, with the CT output shown in Fig. 8.43. With this material, the CT output reaches the peak after about 2.1 s of packing, though the packing-holding stage keeps for 5 s. The offline test result shown in Fig. 8.44 demonstrates that after 2 s of packing, the part weight hardly increases and the gate is frozen. This series of experiments suggests that the online gate freezing can be detected by the CT measurement. The point when the CT output becomes flat or decreases may be considered as the point when the gate has frozen off, and the packing can cease after this point.

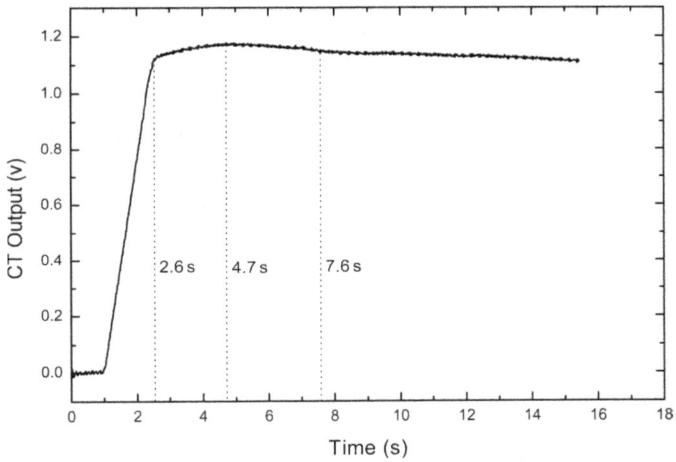


**Figure 8.43** CT outputs for gate freezing detection with PP

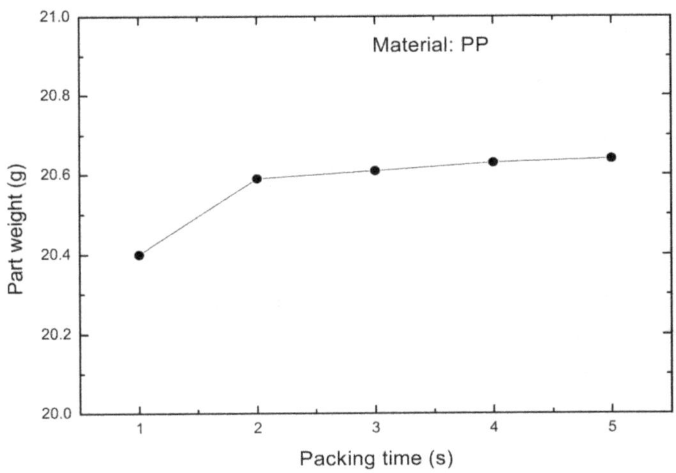


**Figure 8.44** Offline test of gate freezing-off point with PP

### 8.6.3 Solidification Rate Monitoring

Cooling time, which takes up most of the injection cycle time, has critical effects on the quality and productivity of parts. Insufficient cooling time leads to inconsistent moldings, while an excessive one increases the cost of the injection molding process [160]. Solidification monitoring to reduce unnecessary energy and time consumption for cooling is therefore essential. The capacitive transducer has been validated to be capable of monitoring the solidification rate [141].

With the capacitive transducer, the signal captured by injecting HDPE showed an interesting pattern during the cooling stage (Fig. 8.45). It can be categorized into three regions: (1) the CT signal first increases slightly after packing ends, (2) followed by a gentle drop, and (3) it then comes to a plateau. The increase of CT was caused by the solidification of HDPE, while the decrease of CT was due to the shrinkage and formation of an air gap, coupled with the factor of polymer solidification. The final CT output plateau for times after 18 s indicated the end of solidification. A similar pattern of CT output is expected for other semicrystalline and amorphous polymers, providing a means for solidification rate monitoring of molded parts.

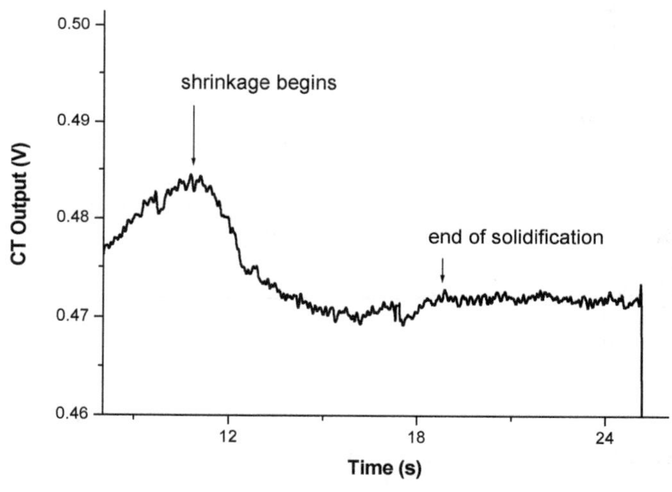


**Figure 8.45** Zoom-in figure of CT output from end of packing to mold opening

The parallel capacitor principle suggests that the dielectric constant of the material inside the cavity is contributed to by different layers of polymer melt and solid, in the thickness direction. A governing equation to determine the layer growth of solidification can therefore be derived:

$$\frac{y_s}{D} = \left(\frac{1}{k} - \frac{1}{k_m}\right) \Big/ \left(\frac{1}{k_s} - \frac{1}{k_m}\right) \tag{8.17}$$

where $y_s$ is the growth of a solidified layer over the thickness of molded part $D$, $k$ is the experimental dielectric constant of the material inside the cavity, and $k_m$ and $k_s$ are the dielectric constant of the polymer melt at its processing state and the dielectric constant of the final solidified layer, respectively.

During cooling, the molded part was supposed to shrink, considered in the thickness direction, as the CT signal experienced a gradual drop (Fig. 8.45). This was further proven by the data of cavity pressure, measured by the transducers installed in the mold. At around the CT signal peak, the cavity pressure reached a very small value and became zero shortly, indicating that the molded part no longer exerted a force onto the mold walls, and thus shrinkage began to start.

Taking the air layer into account, the governing equation has to be modified to

$$\frac{y_s}{D}=\left(\frac{1}{k}-\frac{1}{k_m}\right)\Bigg/\left(\frac{1}{k_s}-\frac{1}{k_m}\right)-\left[\left(\frac{1}{k_{air}}-\frac{1}{k_m}\right)\Bigg/\left(\frac{1}{k_s}-\frac{1}{k_m}\right)\right]\frac{y_{air}}{D} \quad (8.18)$$

where $k_{air}$ is the dielectric constant for air. Since the continuous growth of the solidified layer and air layer are indeed competing with each other, Eq. 8.18 is difficult to solve explicitly. Provided that the degree of solidification at the CT peak is known from the proposed hypothesis, let us denote this time by $t_0$, and the solidification is represented by $y_{s0}$. The growth of the air gap ($y_{a1}$) at the next time step $t_1$ is approximated from $y_{s0}$ and the current experimental value $k_1$ from Eq. 8.18. Meanwhile, the amount of polymer melt, which solidifies and leads to that certain volume contraction and cannot be compensated for by packing, is estimated from the conservation of mass. The additional growth of the solidified layer from $t_0$ to $t_1$ can then be computed and hence the value of solidification at $t_1$ ($y_{s1}$). The procedures are repeated until the degree of solidification and air gap development attain the steady state.

Experiments were conducted using PP (polypropylene) to demonstrate the above hypothesis. The mold temperature was controlled constantly at 25°C throughout the entire injection cycle. The plot of normalized solidified layer growth, in the part thickness direction, was shown in Fig. 8.46, in which the packing phase ended at 7 s. The growth of the solidified layer, which was proportional to the CT output according to Eq. 8.17 increased with a similar trend as the CT output. However, at around the 12th second, the rate of solidification was slowed down by the part shrinkage. The sudden increase in thermal resistance reduced the efficiency of heat conduction from the part to the cold mold wall and thus led to a slower rate of solidification. Simultaneously, the normalized growth of the air layer was estimated and presented in Fig. 8.47. Part solidification has completed, and there is no further shrinkage by solidification from the 18th second onward.

Appling the layer-by-layer solidification model to estimate the growth of the solidified layer in a mold offered a possibility to monitor the solidification rate and part shrinkage. It can also be embedded to further control a scheme to optimize the cooling time in the future.

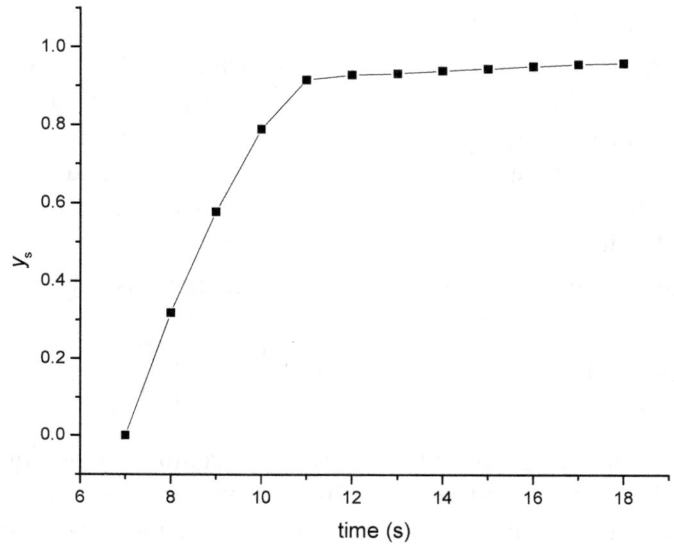


**Figure 8.46** Plot of the growth of the solidified layer ($y_s$) from packing to end of solidification

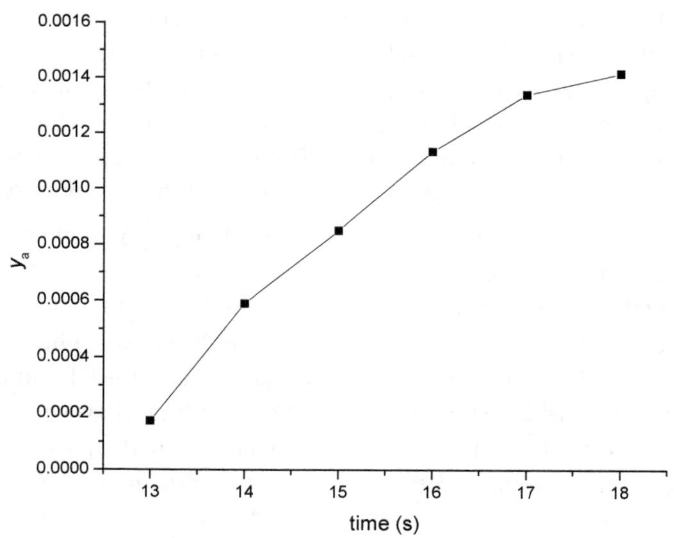


**Figure 8.47** Plot of growth of the air layer ($y_a$) from beginning of shrinkage to end of solidification

## 8.7 Online Part Weight Prediction Using the Capacitive Transducer

Besides the monitoring of the process status, the developed CT can also be applied for online quality control with the help of a data-based regression technique [142]. A typical CT output trajectory of a normal injection cycle, for instance, with a constant packing pressure of 450 bar, is shown in Fig. 8.48. At a constant injection speed, the CT output increases linearly, corresponding to the filling of the mold of a constant cross-sectional area. After filling, the CT output increases at a modulated speed during the packing stage before the gate is frozen. During packing, the capacitance continues to increase, but at a much slower rate, because the rate of material entering the mold is smaller than that in the injection phase. The CT output and the amount of materials in the mold cavity are expected to be at their maximum at the time when the gate is frozen. The signal of CT output at this time, as indicated by the circled region in Fig. 8.49, would well represent the product weight. This, in turn, can be applied for part weight prediction and monitoring. The delta CT output, the signal change between the maximum and the start of the CT output, is then used for online weight prediction. Two molds with different shapes are used for demonstration. One is a rectangular mold, as shown in the first subplot of Fig. 8.16; the other has a decreasing cross-sectional area along the flow path, as shown in the fourth subplot in Fig. 8.16. They are referred to as Mold 1 and Mold 4, respectively, in the following description.

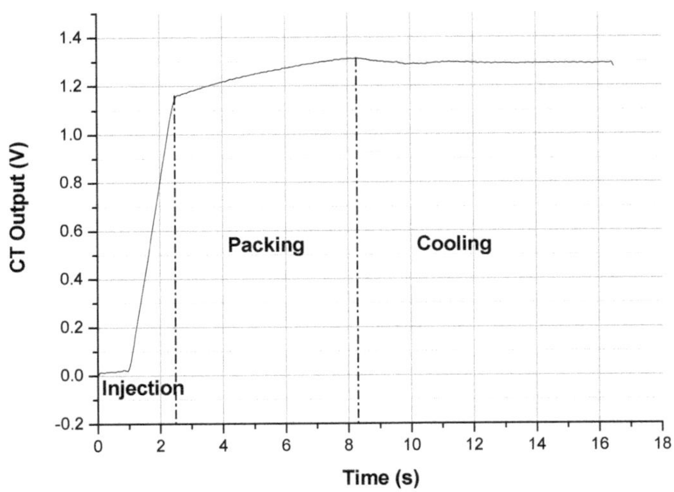


**Figure 8.48** A typical plot of changes of CT output in an injection cycle

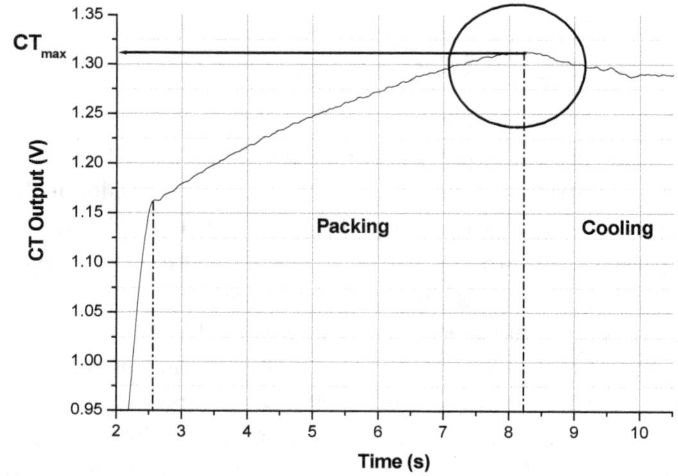


**Figure 8.49** Typical CT output trajectory of an injection cycle, with the packing stage zoomed-in

With the aim of investigating the relation between part weight and CT output, experiments with varying packing profiles were first conducted on Mold 1, and a linear relation between the part weight and delta CT output was thus established, as shown in Fig. 8.50. The packing profiles included different constant levels, and ramp-up and ramp-down profiles as well. It is clear that the data are distributed evenly along the linear fit for all profiles. The linear relation is also confirmed by a high $R$ value of 90.41%. The final predicted weight ($Wt_1$) is represented by

$$Wt_1 = 16.96 \times \Delta CT + 6.85 \tag{8.19}$$

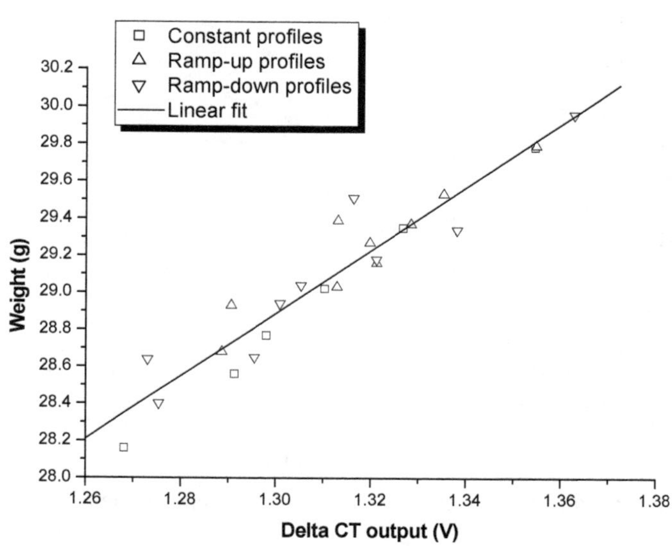


**Figure 8.50** Part weight against delta CT output under ramp packing profiles for Mold 1

Data from experiments not used for model fitting were further tested to validate the accuracy of the part weight prediction. Parts with different packing pressure conditions were checked, and the test results are shown in Fig. 8.51. The squares connected by the solid line represent the actual weight measurements, while the circles connected by the dashed line represent the weight prediction using Eq. 8.19, taking the CT output as an input. In general, the trend of the part weight variation can be well predicted by the CT sensor. The deviations of the predicted value from the actual weight, ranging from -0.29 g to +0.27 g, which correspond to a relative error of -0.96% to +0.93%, respectively, are small.

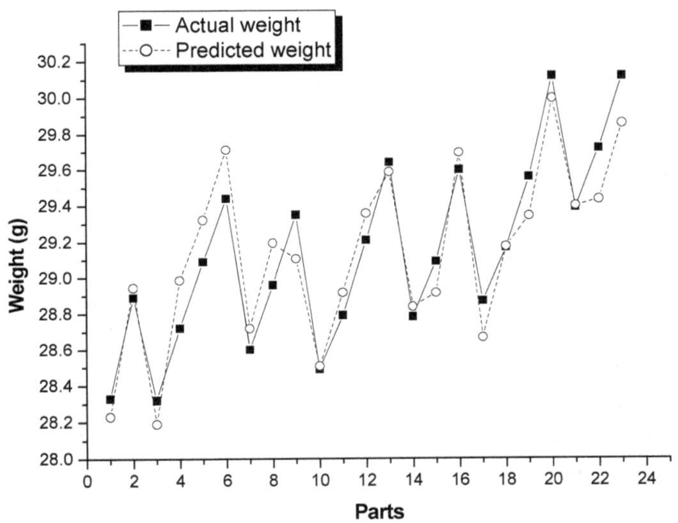


**Figure 8.51** Comparison between actual and predicted part weight for Mold 1

In addition to the rectangular Mold 1, Mold 4 with decreasing cross-sectional area along the flow path was also utilized. The experiments were conducted with 36 varying packing pressure profiles with start and end pressures chosen from the pressure set of {300, 350, 400, 450, 500, 550} bar. Profiles with 600 bar were not included as the parts are found to be flashed. Half of the data set was used to develop the correlation between part weight and CT output, and the other half was used for validation. The part weight prediction results are consistent as before. That is, a linear relationship ($R^2$ = 91.76%) between part weight and delta CT output is established and is shown in Fig. 8.52, with the predicted weight ($Wt_2$) represented by the following equation:

$$Wt_2 = 11.76 \times \Delta CT + 7.04 \tag{8.20}$$

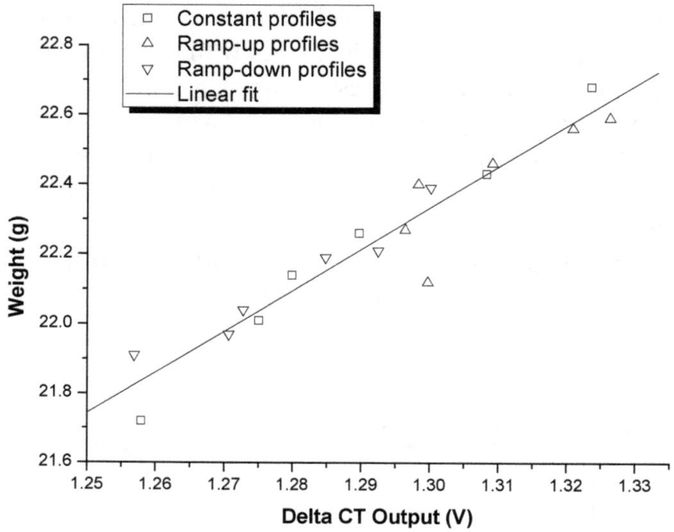


**Figure 8.52** Part weight against delta CT output under ramp packing profiles for Mold 4

The slope of the fitted line in Fig. 8.52, which represents changes in part weight with respect to delta CT output, is 11.76 with the nonconstant mold geometry, while the slope from the rectangular mold is 16.96. The smaller slope computed in the latter case can be explained by the smaller part weight from the nonconstant mold geometry.

For the part weight prediction using the mold with decreasing cross-sectional area, the actual weights and their trends can also be well predicted by the established linear relation, as represented by Fig. 8.53. The deviation of the predicted value from the actual weight ranges from -0.14 g to +0.13 g, which corresponds to a relative error of -0.63% to +0.57%, respectively. This shows that the proposed part weight prediction method is applicable to various mold geometries. The results for the two molds are shown in Table 8.9. The high prediction accuracy for both molds shows that the in-mold weight prediction by the capacitive transducer is mold-independent.

**Table 8.9** Part Weight Prediction Results Comparison between Mold 1 and Mold 4

| | Mold 1 | Mold 4 |
|---|---|---|
| Fitted weight equation (g) | $=16.96(\Delta CT)+6.85$ | $=11.76(\Delta CT)+7.04$ |
| $R^2$ for linear fit (%) | 90.41 | 91.76 |
| MSE for checking (%) | 3.01 | 0.53 |
| Range of actual weight variation (g) | 1.80 | 0.96 |
| Range of actual CT variation (mV) | 94.71 | 69.40 |
| Deviation of predicted to actual weight (g) | -0.29 to +0.27 | -0.14 to +0.13 |
| Relative prediction error (%) | -0.96 to +0.93 | -0.63 to +0.57 |

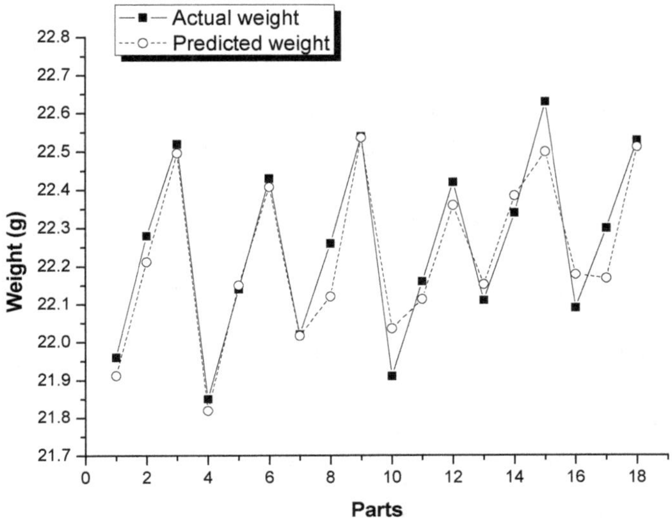


**Figure 8.53** Comparison between actual and predicted part weight for Mold 4

Deviation or scattering of the actual part weights from the CT predictions may be attributed to the measurement or process factors. Sources of measurement errors include the following. First, the weight was measured with a balance of a limited resolution of 0.01 g. Second, the part was weighed after the sprue and the runner were cut off. Efforts have been made to ensure good repeatability of the cutting process, but some measurement errors of weight may still be introduced here. Third, the range of CT output changes is relatively small, and this affects the output resolution. Finally, the signal processing and the identification of the maximum CT output value may also be imprecise.

In terms of the process, there is a small degree of solidification taking place during the filling and packing phases. A packing profile may have certain small effects on the solidification pattern. The solidified and melt polymers exhibit slight differences in their dielectric properties, and this may also introduce prediction errors. Having said the above, the CT sensor can still predict the weight very accurately, giving a linear relation with errors well within acceptable ranges. Measurement improvements may further enhance the accuracy of online prediction.

## References


[130] F. Gao, W.I. Patterson, and M.R. Kamal, *Polym. Eng. Sci.*, **36**, 1272, 1996

[131] F. Gao, W.I. Patterson, and M.R. Kamal, *Intern. Polym. Process., Ind. Eng. Chem. Res.*, 147, 1993

[132] H. Yokoi, T. Hayashi, K. Tado, and N. Morikita, SPE *Annu. Tech. Conf. Papers*, **34**, 329, 1988

[133] H. Yokoi, S. Kamata, and T. Kanematsu, SPE *Annu. Tech. Conf. Papers, SPE*, **37**, 358, 1991

[134] H. Yokoi, S. Kamata, and T. Kanematsu, SPE *Annu. Tech. Conf. Papers, SPE*, **37**, 367, 1991

[135] H. Yokoi and Y. Inagaki, SPE *Annu. Tech. Conf. Papers, SPE*, **38**, 457, 1992

[136] T.J. Bress and D.R. Dowling, *J. Reinf. Plast. Compos.*, **17**, 1374, 1998

[137] H. Wang, B. Cao, C.K. Jen, K.T. Nguyen, and M. Viens, *Polym. Eng. Sci.*, **37**, 363, 1997

[138] C.L. Thomas, A.O. Adebo, and A.J. Bur, SPE *Annu. Tech. Conf. Papers, SPE*, **40**, 2236, 1994

[139] A.J. Bur and C.L. Thomas, *Polym. Eng. Sci.*, **37**, 1430, 1997

[140] X. Chen, G. Chen, and F. Gao, *Polym. Eng. Sci.*, **44**, 1571, 2004

[141] H.Y. Wong, K.T. Fung, and F. Gao. *Sens. Actuators, A*, **141**, 712, 2008

[142] K.T. Fung, F. Gao, and X. Chen, *Polym. Eng. Sci.*, **47**, 347, 2007

[143] R. Puers, *Sens. Actuators, A*, **37**, 93, 1993.

[144] W.C. Heerens, *J. Physics E: Sci. Instrum.*, **19**, 897, 1986

[145] H. Kobayashi, and Y. Hosokawa, *Sens. Actuators, A*, **24**, 27, 1990

[146] R.K. Pearson, *IEEE Trans. Instrum. Meas.*, **39**, 421, 1990

[147] D.D. Denton, C.N. Ho, and S. He, *IEEE Trans. Instrum. Meas.*, **39**, 502, 1990

[148] J. Fraden, *AIP Handbook of Modern Sensors, Physics, Designs & Applications*, American Institute of Physics, New York, 1993

[149] L.K. Baxter, *Capacitive Sensors Design & Applications*, IEEE Press, 1997

[150] A.R. Blythe, *Electrical Properties of Polymers*, Cambridge University Press, 1979

[151] T. Osswald and G. Menges, *Materials Science of Polymers for Engineers*, Hanser Publishers, 1996

[152] D.A. Seanor, (Ed.), *Electrical Properties of Polymers*, Academic Press, 1982

[153] D.W. Van Krevelen, *Properties of Polymers* (3rd ed.), Elsevier Science Publishers, 1990

[154] J.M.G. Cowie, *Polymers: Chemistry and Physics of Modern Materials* (2nd ed.), Blackie and Son Limited, 1991

[155] M. He, W. Chen and X. Dong, *Polymer Physics*, Fudan University Press, 1990

[156] D.L. Feucht, *Handbook of Analog Circuit Design*, Academic Press, 1990

[157] J. Davidse, *Analog Electronic Circuit Design*, Prentice Hall, 1991

[158] J. Sigdell, *IEEE Trans. Instrum. Meas.*, **21**, 60, 1972

[159] J.C. Lotters et al., *IEEE Trans. Instrum. Meas.*, **48**, 89, 1999

[160] C.L. Thomas, A.A. Tseng, A.J. Bur, and J.L. Rose, *Adv. Polym. Technol.*, **15**, 151, 1996

[225] Gao F, C. Zhao, *Method and apparatus for the control of injection molding*, US Patent 6,309,571[P], 2001

[226] X. Chen, Ph.D, Thesis, Hong Kong Univ. of Sci. and Technol., 2002

# 9 Profile Setting of Injection Velocity

The quality of an injection molded part depends on the characteristics of the material, the performance of the molding machine, the mold design, and processing conditions. For a given material, mold, and machine, the part quality is mainly determined by its processing conditions, among which injection velocity during the filling stage plays a particularly important role. Studies [161–170] have shown that injection velocity can affect the degree of orientation and residual stress, as well as surface properties of the molded parts. In view of its importance, advanced control techniques [172–174] have been developed for the control of injection velocity. With such advanced controllers, the injection velocity can accurately follow any reasonable profile. But for a given mold and material, how should the injection velocity be profiled to produce a good-quality part? A method for profiling the injection velocity based on a constant melt-front-velocity strategy will be presented in this chapter.

## 9.1 Constant Melt-Front-Velocity Strategy

Considering the melt flow in a mold cavity as schematically illustrated in Fig. 9.1, where $IV$ is the screw injection velocity, $V_m$ the average melt-front velocity in the mold, $A_b$ the cross-sectional area of the barrel, and $A_{mf}$ the corresponding melt-front area inside the mold, the melt-front development is greatly influenced by the mold geometry. With a given driving force, the melt-front velocity changes with the geometry. Physically, when the melt front reaches a restricted region of the mold, the melt front develops quickly, whereas when it reaches a less restricted region, the flow slows down. To minimize part defects caused by nonuniform melt development in a mold cavity, a constant melt-front velocity, $V_m$, through the filling of the entire mold cavity is desirable [166–170]. This requires that the screw injection velocity, $IV$, should be profiled in a way that harmonizes the melt-front velocity with the melt-front area that is affected by the molding geometry. A generally accepted and practical implementation method for such velocity profiling, however, is still lacking due to the difficulties in the online measurement of the melt-front velocity or the melt-front area.

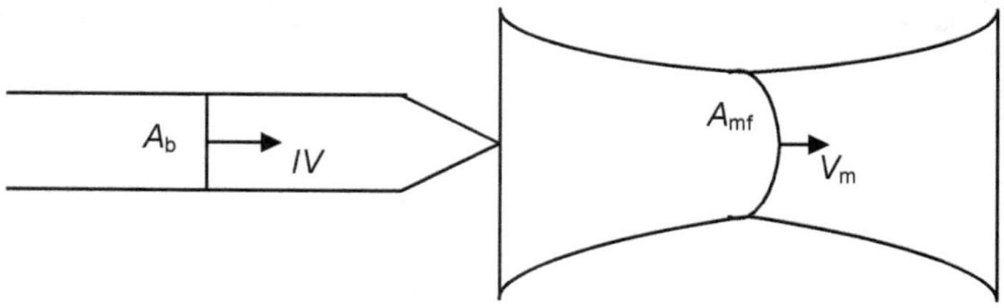


**Figure 9.1** Schematic of mold filling

The melt-front areas vary during the filling because of the geometry change. The following mass balance holds, assuming that the polymer melt is incompressible and the leakage through the check valve is negligible:

$$IV \cdot A_b = V_m \cdot A_{mf} \tag{9.1}$$

The cross-sectional area of the barrel, $A_b$, is constant; the melt-front velocity, $V_m$, can be kept constant during filling if the screw injection velocity, $IV$, is profiled and controlled proportional to the melt-front area, $A_{mf}$. If the latter can be measured online, the profile setting of the injection velocity can be implemented according to the block diagram in Fig. 9.2, where $\overline{IV}$ is the screw injection velocity profile.

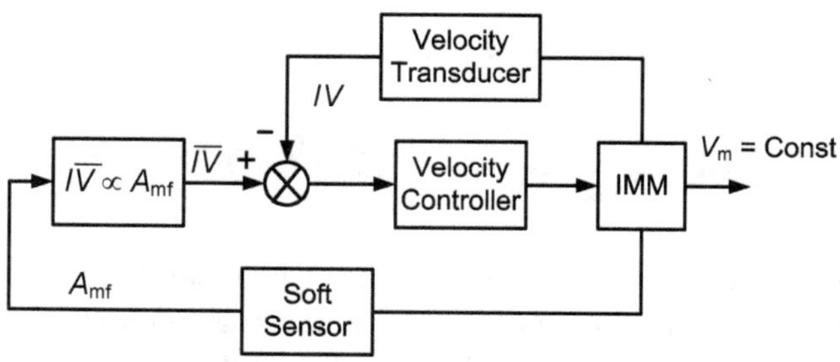


**Figure 9.2** Velocity profile setting based on the melt-front area

In addition to the regular velocity feedback controller, there exists a profile-setting module based on the measurement of the melt-front area. This module provides a velocity profile, $\overline{IV}$, to the injection velocity controller to keep the melt in the mold cavity developing at a constant rate. The key to the system, therefore, is the online measurement of the melt-front area, which is represented by the soft-sensor module block in Fig. 9.2.

A soft-sensor scheme using online-measurable variables to predict the melt-front area was proposed by Chen and Gao [175]. To construct such a proper soft sensor for the melt-front area measurement, the relationship between the melt-front area and its correlation variables is first analyzed. The screw velocity and average melt-front velocity may be expressed, respectively, by the following differential equations:

$$\begin{cases} IV = \dfrac{\mathrm{d}SD}{\mathrm{d}t} \\ V_m = \dfrac{\mathrm{d}afl}{\mathrm{d}t} \end{cases} \tag{9.2}$$

where *SD* is the screw displacement, and *afl* is the average-flow-length of the melt inside the cavity. Substituting Eq. 9.2 into Eq. 9.1, the melt-front area can be represented by

$$A_{\mathrm{mf}} = \frac{\mathrm{d}SD}{\mathrm{d}afl} \cdot A_{\mathrm{b}} \tag{9.3}$$

For a given machine, $A_{\mathrm{b}}$ is a constant. The screw displacement, *SD*, can be measured online by a displacement transducer. The melt-front area, $A_{\mathrm{mf}}$, is thus predictable if the average-flow-length, *afl*, is known. The above equation can also be expressed in a discrete form:

$$A_{\mathrm{mf}_n} = \frac{SD_n - SD_{n-1}}{afl_n - afl_{n-1}} \cdot A_{\mathrm{b}} \tag{9.4}$$

where the subscript *n* represents the *n*th sampling instant during a filling cycle.

Equation 9.4 suggests that at any time instant the melt-front area can be directly calculated, providing the average-flow-length is known. The problem is thus transformed into the determination of average-flow-length.

Physically, when the melt front reaches a restricted region of the mold, the melt-front velocity will increase according to the mass balance when the injection velocity is kept constant. At the same time, the restricted area will provide more resistance to the melt flow, which consequently results in a higher demand on injection pressure. In contrast, when the melt flow reaches a wider region, the melt-front velocity decreases in conjunction with a relatively smaller pressure increase. The melt-front velocity, therefore, can be reflected by the online-measurable variables such as nozzle pressure, injection velocity, and screw displacement. Average-flow-length, a time integral of the melt-front velocity, can be correlated to those variables.

For simplicity, the filling is assumed to have constant melt and mold temperatures because the filling time is relatively short. With a given mold and material, the average-flow-length at the *n*th sampling instant, $afl_n$, may be mathematically expressed in a formula as the following:

$$afl_n = f\left(afl_{n-1}, NP_n, \Delta NP_n, SD_n, \Delta SD_n, IV_n\right) \tag{9.5}$$

where $afl_{n-1}$ is the average-flow-length of the previous sample, $NP_n$ is the nozzle pressure representing the driving force for the mold filling, $\Delta NP_n$ represents the additional force required for the flow-length increment, $SD_n$ represents the total amount of melt in the mold, $\Delta SD_n$ denotes the amount of melt entering the mold during the short time increment, and $IV_n$ is the injection velocity, which also obviously influences the flow length. The function $f$ in Eq. 9.5 is complicated, representing the combination of the mass, momentum, and constitutive relations. This relation, being difficult to express in an analytic form, can be treated as a black box input-output relation represented by neural networks.

Based on the above-mentioned consideration, Chen & Gao [175] proposed a combined neural network model, as shown in Fig. 9.3. It consists of four major components: a dynamic network to predict the average-flow-length from the inputs directly measurable online, a static network to improve the prediction of the average-flow-length using the dynamic network output as an input, a deterministic model to calculate the melt-front area from the predicted average-flow-length of the network output, and a filter using the average of the past four points to smooth the calculated melt-front area. The average-flow-length, a variable monotonously increasing with time, was used in this soft-sensor model as an intermediate variable. This is because it is easier for neural networks to represent a monotonous function because the transfer functions used in the networks are typically monotonous [176]. When the average-flow-length is converted to the melt-front area according to Eq. 9.4, the prediction error is largely brought to the model due to the difference operation. The filter is, therefore, necessary to smooth the prediction. By using such a soft sensor, the melt-front area during a filling cycle can be predicted online by using the measurable variables as listed in the inputs.

Since the melt-front area data are difficult to obtain from experiments, data generated from simulations using C-Mold [177], a well-accepted commercial software package for simulating the injection molding filling process, were tested. Nine molds with different simple geometrical features were designed to generate C-Mold data for the training of the neural network model. In addition, nine different injection velocity profiles covering a large range of velocity distributions were used for the simulations of each mold. Polystyrene was selected as the testing material. In the simulation, the nozzle pressure is substituted for by gate pressure because the former depends on the machine as well as the mold. During each C-Mold simulation, up to 50 data pairs are obtained for the injection velocity, screw displacement, gate pressure, average-flow-length, and melt-front area. Good agreements were achieved between the C-Mold and the neural network predictions for all of the nine training molds. To further validate the developed model, C-mold simulations were conducted on two additional molds, as illustrated in Fig. 9.4, with geometries different from the training molds.

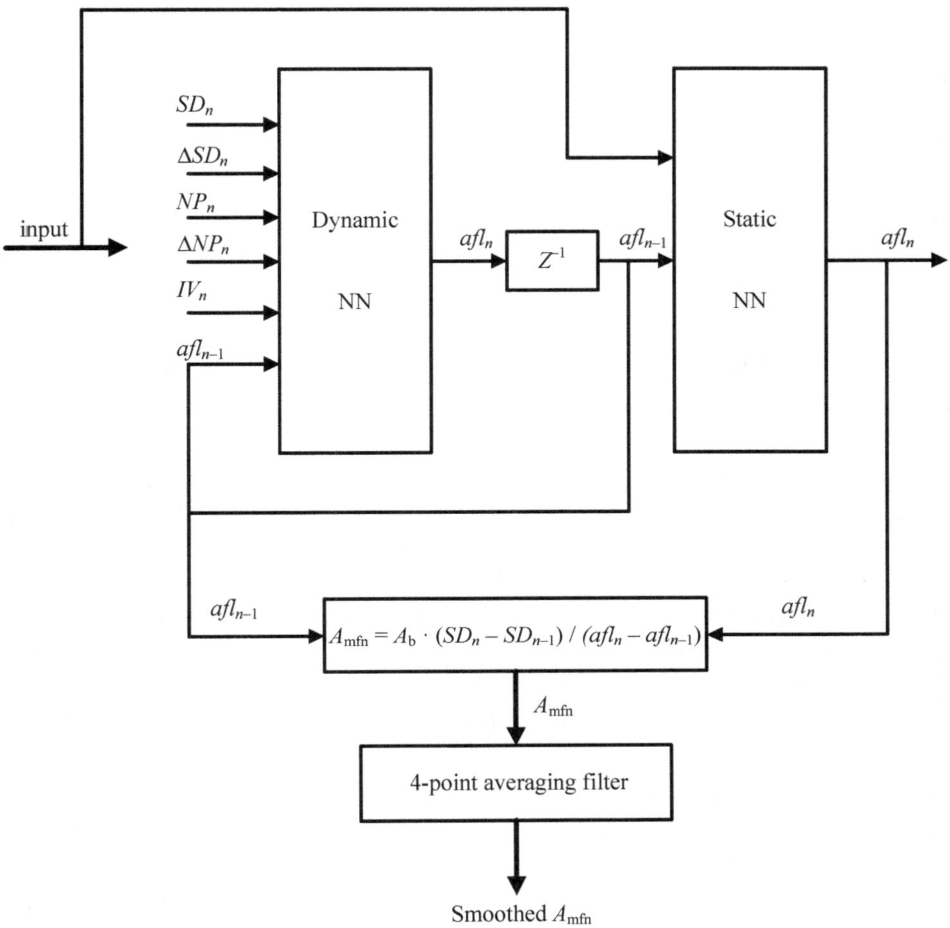


**Figure 9.3** Structure of the soft-sensor model

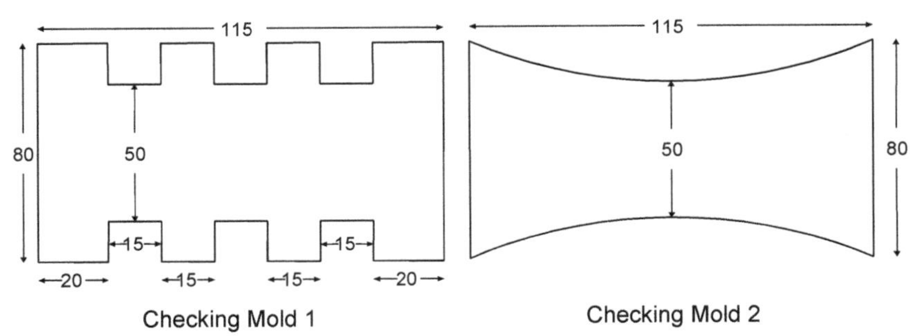


**Figure 9.4** Geometry of the checking molds

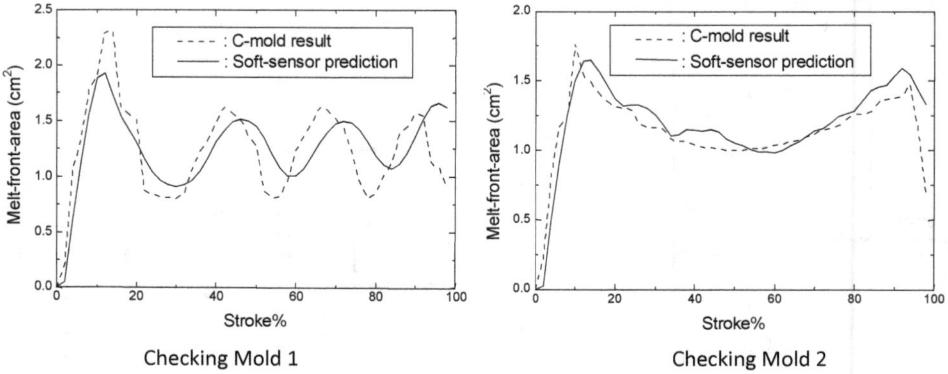


**Figure 9.5** Results of checking molds

The prediction results of the melt-front areas of the checking molds are presented in Fig. 9.5. The *x* axis represents the relative injection stroke during the mold filling as a percentage. The dashed lines are the melt-front area obtained from the C-Mold simulation results; the solid lines present the predictions of the developed soft-sensor model. It is shown that the soft-sensor predictions are in good agreement with the C-Mold simulations, despite the fact that the checking molds were not presented in the network training. This suggests that the properly trained network can be implemented online as a neural-network-based soft sensor for real-time prediction of the melt-front area.

The above has shown that the strategy is feasible for online predictions of the melt-front area, and profiling the injection velocity based on the melt-front area is also feasible. But the training and verification of this soft-sensor model is based on CAD simulation data. The accuracy of the soft-sensor model is thus dependent on the accuracy of the CAD software predictions. On the other hand, because the melt-front area is difficult to measure online, experimental verification is also difficult. Alternatives to the above are presented in the following for possible and easy implementation.

# 9.2 Scheme Based on Average-flow-length

According to Eq. 9.2, the melt-front velocity is a derivative of the flow length, the distance that the melt front has traveled in the mold from the gate. As the melt front is characterized by many points, the averaged value is used to describe this distance. Considering the relation between the melt-front velocity and the average-flow-length described in Eq. 9.2, the former can be kept constant if the latter can follow a ramp relation during filling. Therefore, if the average-flow-length can be measured online, the constant melt-front velocity strategy can be implemented by controlling the average-flow-length following a ramp profile, as illustrated in Fig. 9.6.

The system consists of two control loops: an inner traditional injection velocity control loop, and an outer average-flow-length (AFL) control loop. The output of the outer AFL controller provides an injection velocity profile for the inner velocity controller. A simple ramp profile for the AFL controller will keep the melt-front velocity constant. In view of the cyclic nature of the process, an iterative learning control method [178, 179] can be used for design of the AFL controller. In such a learning procedure, information from the last cycle can be used to improve the control of the current cycle. After a number of cycles, the AFL controller will produce a stable output, representing the proper velocity profile. The AFL controller can then be removed, and the constant melt-front velocity can be ensured by running the inner velocity control loop with the learned profile setting. The profile-seeking problem can be implemented by a controller learning procedure, in which the online measurement of the average-flow-length is the key.

The capacitive transducer is such a key to provide the essential measurement. As described in the previous chapter, the transducer is designed by considering the dielectric properties of polymers. It provides a direct measurement of the average-flow-length in a mold cavity, together with some other potentially useful applications for the injection molding process. With the online measurement from the developed capacitive transducer, Yang [180] successfully implemented such a controller to realize a uniform mold filling based on the strategy proposed in Fig. 9.6.

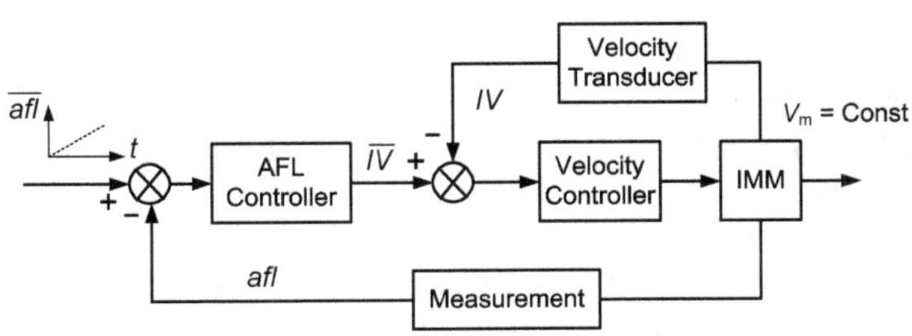


**Figure 9.6** Block diagram of the average-flow-length controller system

## 9.3 Neural Network Model of Average-flow-length

It is not economical and practical for all molds to be fitted with a capacitive transducer as developed in the preceding section. This section discusses the model of the average-flow-length for injection velocity profiling. A first-principles model for the filling process has been under active research over the last three decades. It has been developed from simple one-dimensional models [181–197] to two-dimensional models [198–206] and more complete and complicated three-dimensional models [207–210]. The complexity of those first-principles models precludes any analytical solution. Numerical methods such as finite element methods have to be employed in solving those models. In those models, mold geometry has to be carefully modeled. This type of first-principles model is ill-suited for online prediction of melt flow behaviors. In modeling nonlinear and complex processes, a neural network approach has become a popular tool due to its fast computing and learning-by-example features. To profile the injection velocity, a soft-sensor scheme using neural networks was introduced to correlate the melt-front area (or average-flow-length) during filling with other measurable variables. With the developed capacitive transducer, a soft sensor based on the actual experimental data can be built.

### 9.3.1 Inputs and Output of the Neural Network Model

Different from the model based on simulation data as described in Eq. 9.5, nozzle pressure was used to replace the gate pressure as one input for the relatively convenient instrumentation at the nozzle. Furthermore, nozzle temperature, which was ignored in the simulation, was also included as an additional input; it could affect the melt viscosity and consequently the melt flow. Those changes resulted in the following relation for modeling the average-flow-length:

$$afl_n = f\left(afl_{n-1}, NP_n, \Delta NP_n, SD_n, \Delta SD_n, IV, \Delta IV, NT_n, \Delta NT_n\right) \tag{9.6}$$

where subscript $n$ represents the $n$th sampling instant during filling, and symbol $\Delta$ stands for the variable increment during the past sampling instant. In the equation, $NP_n$ is the nozzle pressure representing the driving force for mold filling; $\Delta NP_n$ represents the additional force required for the average-flow-length increment; $SD_n$ is the screw displacement, which represents the total amount of melt in the mold; $\Delta SD_n$ denotes the amount of melt entering the mold during the last time increment; $IV_n$ is the screw injection velocity, which will directly affect the melt-front velocity and consequently influence the average-flow-length; $\Delta IV_n$ is the injection velocity increment; and $NT_n$ and $\Delta NT_n$ are nozzle temperature and its increment, respectively. Injection time is not a separate input as it can be inferred from the screw displacement and injection velocity. Treated as a black box input-output relation, this model was represented by a neural network.

The neural network model has nine inputs, as listed in the right side of Eq. 9.6, and its output is the average-flow-length of the current sample. To derive a more general model and also to facilitate the neural network learning, all of the input and output variables have been normalized to be in a range of 0–1. The neural network output becomes the normalized average-flow-length, which represents the relative flow length in percentage. Similar is the screw displacement input, $SD_n$. Divided by the injection stroke required for the mold filling, it is also normalized in a percentage form. The other inputs, such as the nozzle pressure, nozzle temperature, and injection velocity, are divided by their maximums provided by the injection molding machine during filling. After normalization, all of the input and output variables become dimensionless.

### 9.3.2 Architecture of the Neural Network Model

Architecturally, neural networks can be categorized into two types [211]: feed-forward and recurrent.

The feed-forward neural network, shown in Fig. 9.7(a), is widely used, typically consisting of three types of layers: input layer, hidden layer, and output layer. The neurons in each layer of a feed-forward network have only the output signals of the preceding layer as their inputs. The nodes in the input layer supply the input signals to the nodes in the second layer (the hidden layer). The outputs of the second layer act as inputs to the third layer (the output layer).

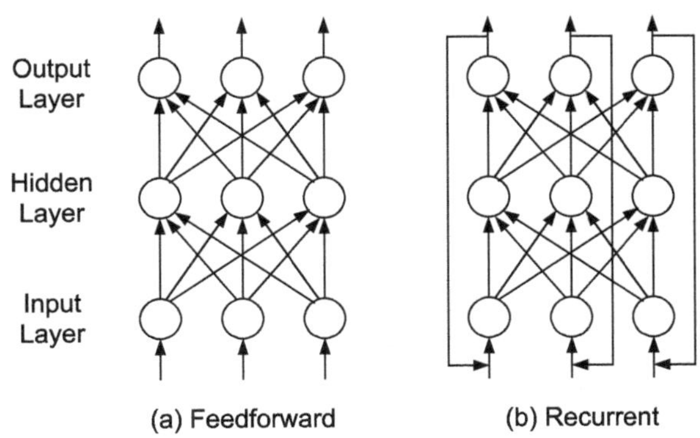


**Figure 9.7** Neural network architectures

A recurrent neural network, on the other hand, distinguishes itself from a feed-forward network in that it has at least one feedback loop. For example, a recurrent neural network consists of a layer of neurons, some of which feed their outputs back as the inputs to neurons in the previous layer, as illustrated in Fig. 9.7(b). Owing to its particularly recurrent structure, the recurrent network can store information for future reference and thus is able to learn temporal as well as spatial patterns. This makes it useful in signal processing and prediction, where time plays a dominant role.

In Eq. 9.6, the average-flow-length of the $n$th sample, $afl_n$, depends on not only the information of the same sample, but also the last measurement ($afl_{n-1}$). The recurrent neural network, depicted in Fig. 9.8, is a suitable structure to model such a dynamic relation. Training of such a recurrent neural network, however, is difficult, as most existing training algorithms are derivative-based. For a feed-forward network, it is easy to obtain the error derivative with respect to each weight, but for a recurrent network, the training becomes complicated due to the interaction among the parameters of different times; as a result, the training tends to be trapped at local minimums. A feed-forward neural network structure is, therefore, recommended for the training of recurrent networks in many cases [212–216]. A feed-forward neural network shown in Fig. 9.9, consisting of two hidden layers with 15 neurons in the first layer and 20 in the second, is used in the training stage. It has been shown that a two-hidden-layer neural network can approximate any continuous nonlinear relations [213]. The number of neurons used is obtained based on a combination of experience and trial-and-error tests.

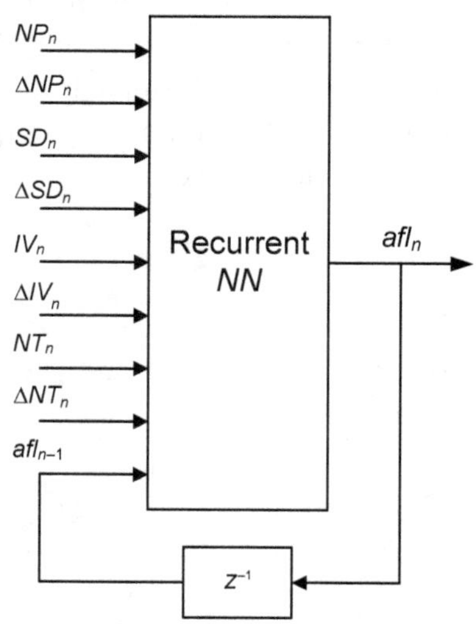


**Figure 9.8** Structure of recurrent neural network model

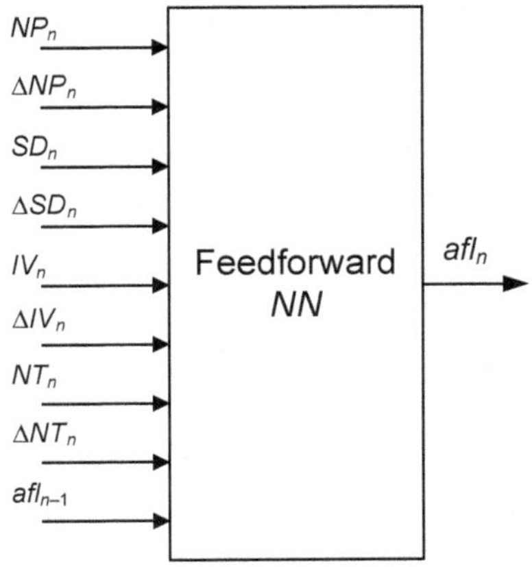


**Figure 9.9** Structure of feed-forward neural network model

### 9.3.3 Training Algorithm

During the training stage, the weights and biases of the network were iteratively adjusted to minimize a defined network objective function, typically the mean squared error (MSE): the average squared error between the network outputs and the target outputs. A prescribed set of well-defined rules for the solution of the training problem is called a training algorithm. As one can expect, there are many algorithms that can be used for the network training, differing from each other in the way in which the weights are adjusted. The Levenberg-Marquardt (LM) algorithm, a training algorithm well known for its fast convergence speed and small residual training error [215–218], was employed in this work. The updating rule of the LM algorithm is represented as

$$\Delta W = \left(J^T J + \mu I\right)^{-1} J^T e \tag{9.7}$$

where $\Delta W$ is the weight increment, $J$ the Jacobian matrix of derivatives of each error with respect to each weight, $I$ the identity matrix, $\mu$ a scalar, and $e$ the calculated MSE. When $\mu$ is zero, the LM becomes a Gauss-Newton method using an approximate Hessian matrix; when $\mu$ is large, the LM becomes the gradient descent method with a small step size. As the Gauss-Newton method is faster and more accurate in approaching the error minimum, $\mu$ is adjusted in such a way to shift the LM toward the Gauss-Newton method as quickly as possible. The value of $\mu$ is decreased after each successful step (reduction in MSE) and is increased only when a tentative step increases the error. In this way, it ensures a rapid reduction of the objective function. This method has a faster converging rate and a smaller residual training error compared with the commonly used back-propagation (BP) algorithm in network training [217].

### 9.3.4 Data Collection of Training and Validation Samples

The training and validation data were collected from experiments conducted on the reciprocating-screw injection molding machines using mold inserts with the capacitive transducer. The material used in the experiments was high-density polyethylene (HDPE) (Marlex, HMN6060).

Two sets of mold inserts were designed for the experiments. The first set, containing nine different inserts, are shown in Fig. 9.10. This set of inserts covers the basic mold shapes that the melt flow typically encounters, including constant area, gradually increasing/decreasing area, and abruptly increasing/decreasing area. A complex mold may be considered as a combination of those basic shapes. This set of mold inserts is referred to as Molds 1 to 9; they were used to generate the training data samples for the network.

Another set of molds with the same thickness but more complex geometry was also designed, as illustrated in Fig. 9.11, and referred to as Molds 10 and 11 for validation data generation. These molds combine several basic shapes of the first set. For example, the cross section of Mold 10 first decreases before it becomes a constant, similar to the shape of the first part of Mold 3; the middle section of Mold 10 is similar to Mold 9 before it is restored to the shape of the last part of Mold 3. Similarly, Mold 11 is geometrically similar to the combination of Molds 6 and 8. The design objective of this second set of molds is to demonstrate that the neural network model trained with data from the first simple set of molds can be used to predict the flow behavior for complex molds like Molds 10 and 11.

The injection velocity range for the machine was limited, due to its limited hydraulic power. Eleven different injection velocity profiles were designed to cover as wide as possible molding conditions for each mold, as illustrated in Fig. 9.12. They could be classified into three types: constant, step-change, and ramp profiles. There were five constant profiles, in which the injection velocity was set at 10 mm/s, 15 mm/s, 20 mm/s, 25 mm/s, and 30 mm/s, respectively. The step-change group consisted of two step-up and two step-down profiles, including 10 to 30 mm/s, 15 to 25 mm/s, 25 to 15 mm/s, and 30 to 10 mm/s. The step change was introduced at different times for different velocities and molds, but all around the time when the mold was half filled. For two ramp profiles, 10 to 30 mm/s ramp-up and 30 to 10 mm/s ramp-down, the ramp started at the point when the melt just entered the mold cavity; prior to that, the velocity was set at a constant rate for filling of sprue and runner. During filling, the sampling rate was set at 5 ms, which could guarantee that more than 150 samples could be collected, even for the fastest filling case. To avoid the local-minimum problem, different initial values were given in the network training; the results with the small errors were employed for prediction.

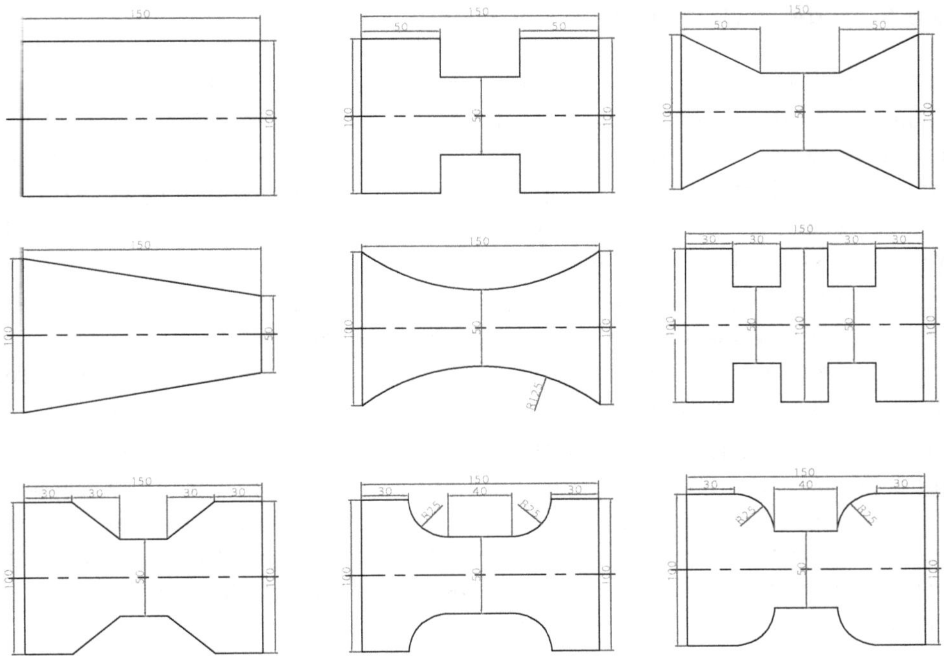

Figure 9.10 Geometry of the training mold inserts

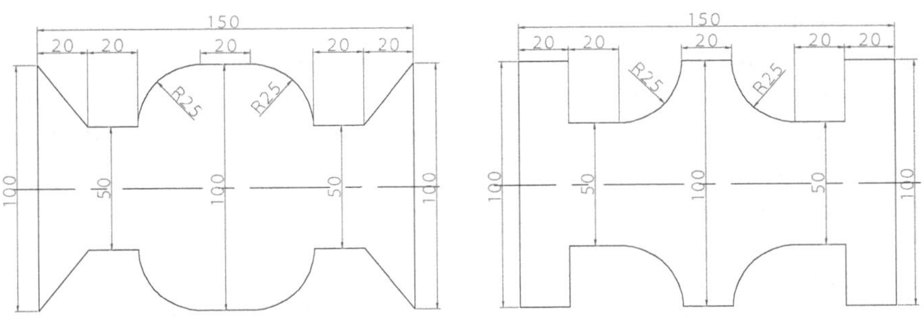


Figure 9.11 Geometry of validation molds

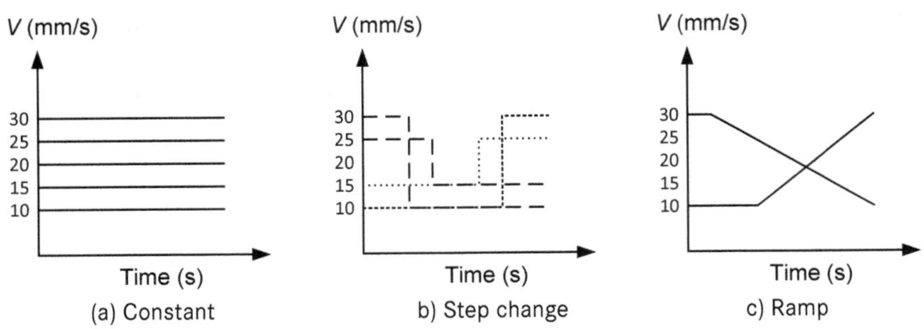


**Figure 9.12** Injection velocity profiles

### 9.3.5 Model Performance

Different from the training stage, the recurrent neural network model structure should be used for the soft-sensor prediction because the preceding average-flow-length is not available. This results in an output error different from that of the training, even for the training samples. Because of this, small training errors may not necessarily guarantee a good approximation. Errors for both the training and validation samples are given in the following.

Figure 9.13 compares the predicted average-flow-length and the actual measurement of the capacitive transducer for Molds 1 to 9 with constant 20 mm/s injection velocity. With different mold geometry, the curve shapes of the average-flow-length are also different; they can be well predicted with a small error for all nine training molds.

The average-flow-length curve is influenced by not only the mold geometry, but also the injection velocity profile. Figure 9.14 presents the results with a step-change profile, in which the injection velocity changes from 30 mm/s to 10 mm/s during filling. Though the shapes are very different from those in Fig. 9.13, good agreement between network predictions and actual data can be obtained. This suggests that the neural network model has been well trained to predict the velocity change.

Experimental data from Molds 10 and 11 with 11 different injection velocity profiles are shown in Figs. 9.15 and 9.16, respectively. Although a slightly larger discrepancy exists between the predicted results and the CT measurements in comparison with those of the training samples, we still can see that the network outputs are in agreement with the true measurements in nearly all cases. The results of Mold 11 are better than those of Mold 10, possibly because Mold 11 is geometrically more similar to the training molds. If we define a geometry set to include the basic geometry elements appearing in the molds, we can see that both Mold 10 and Mold 11 are composed of seven geometry elements along the melt flow path. All seven geometry elements of Mold 11 have appeared in the training molds. Mold 10, however, is different from the training geometry set in the early converging and the last diverging sections. Though similar geometry elements are also present in Mold 3, the converge and diverge rates of the cross section in Mold 10 are larger. This suggests that the performance of the developed neural network model can be improved for prediction of the average-flow-length if the geometry set for training can be further enriched.

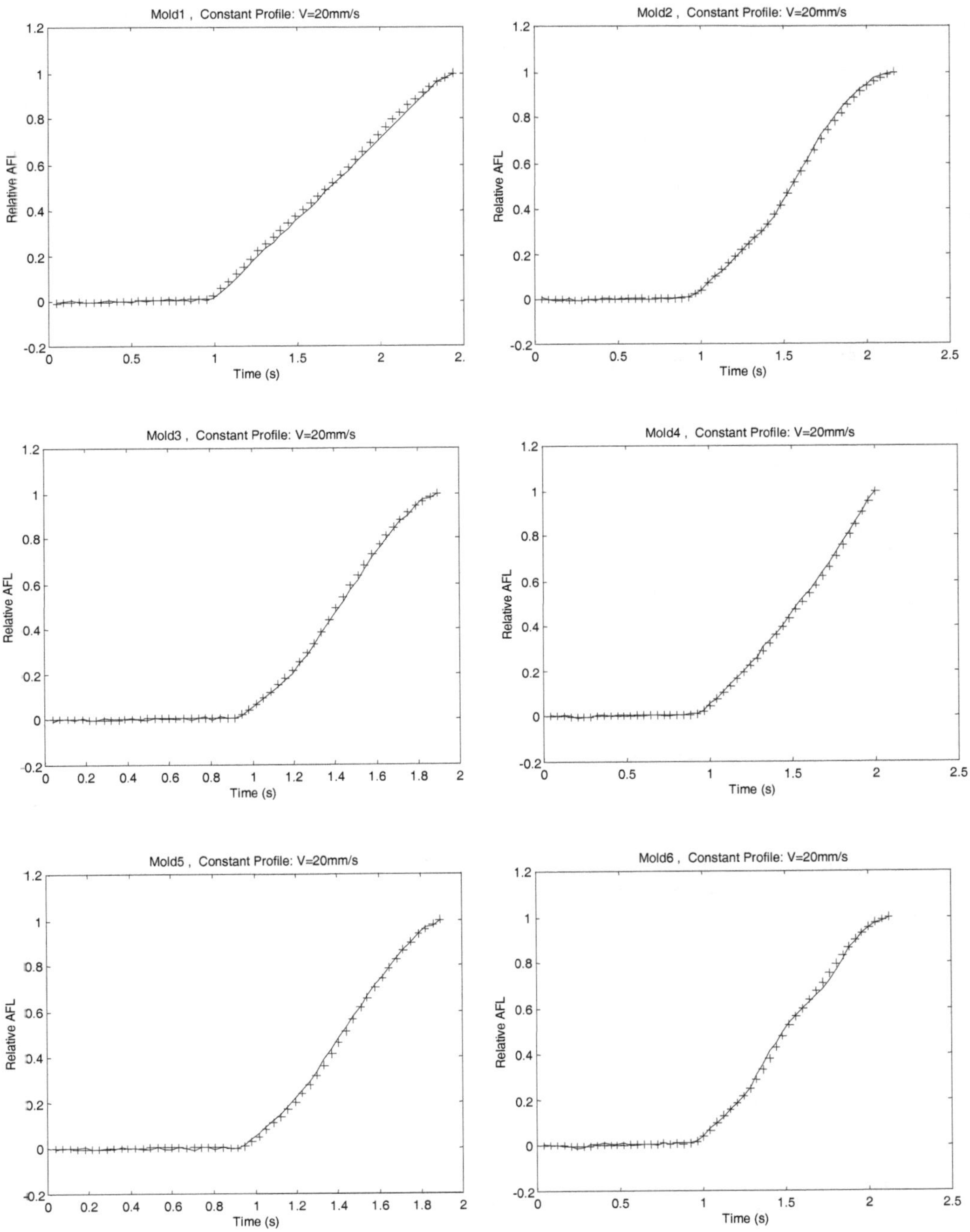


**Figure 9.13** Results of different molds under a constant 20 mm/s profile
(+: network output; –: CT measurement)

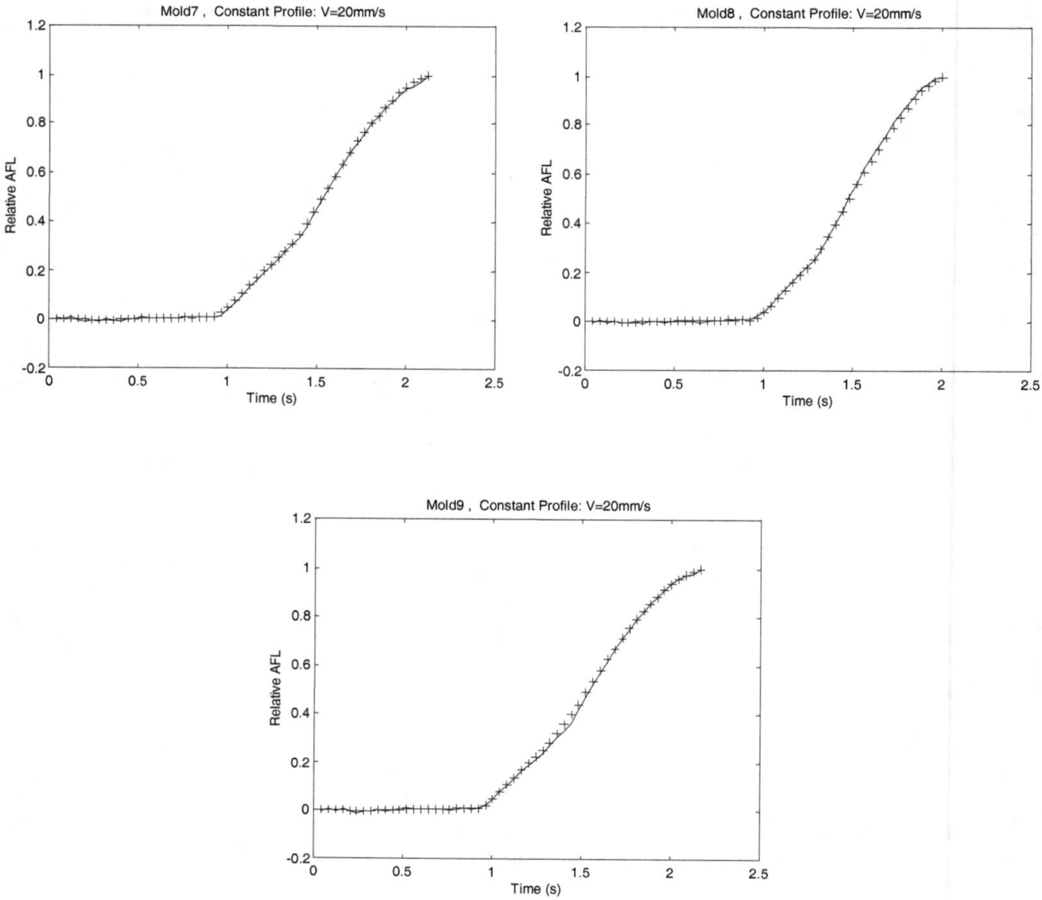


**Figure 9.13** (*continued*) Results of different molds under a constant 20 mm/s profile (+: network output; -: CT measurement)

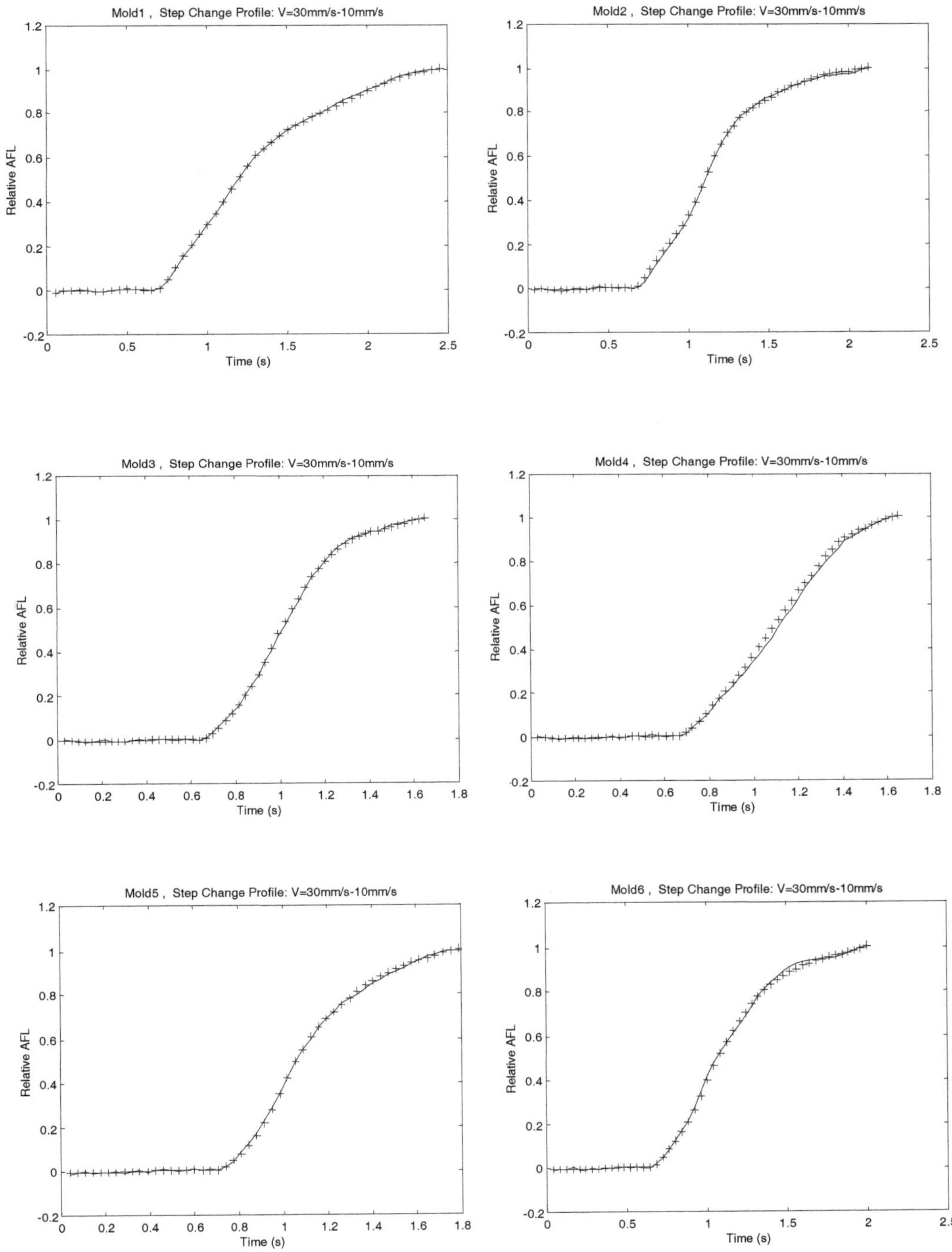


**Figure 9.14** Results of different molds under a step-change profile (+: network output; –: CT measurement)

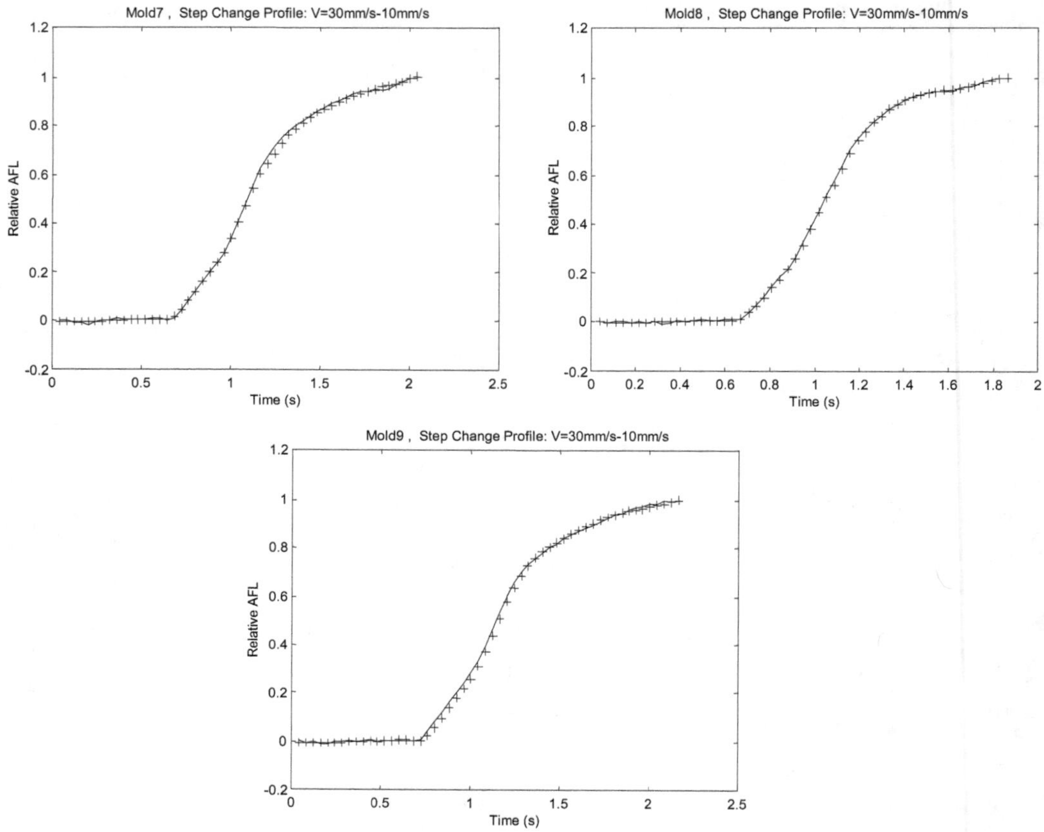


**Figure 9.14** (*continued*) Results of different molds under a step-change profile (+: network output; –: CT measurement)

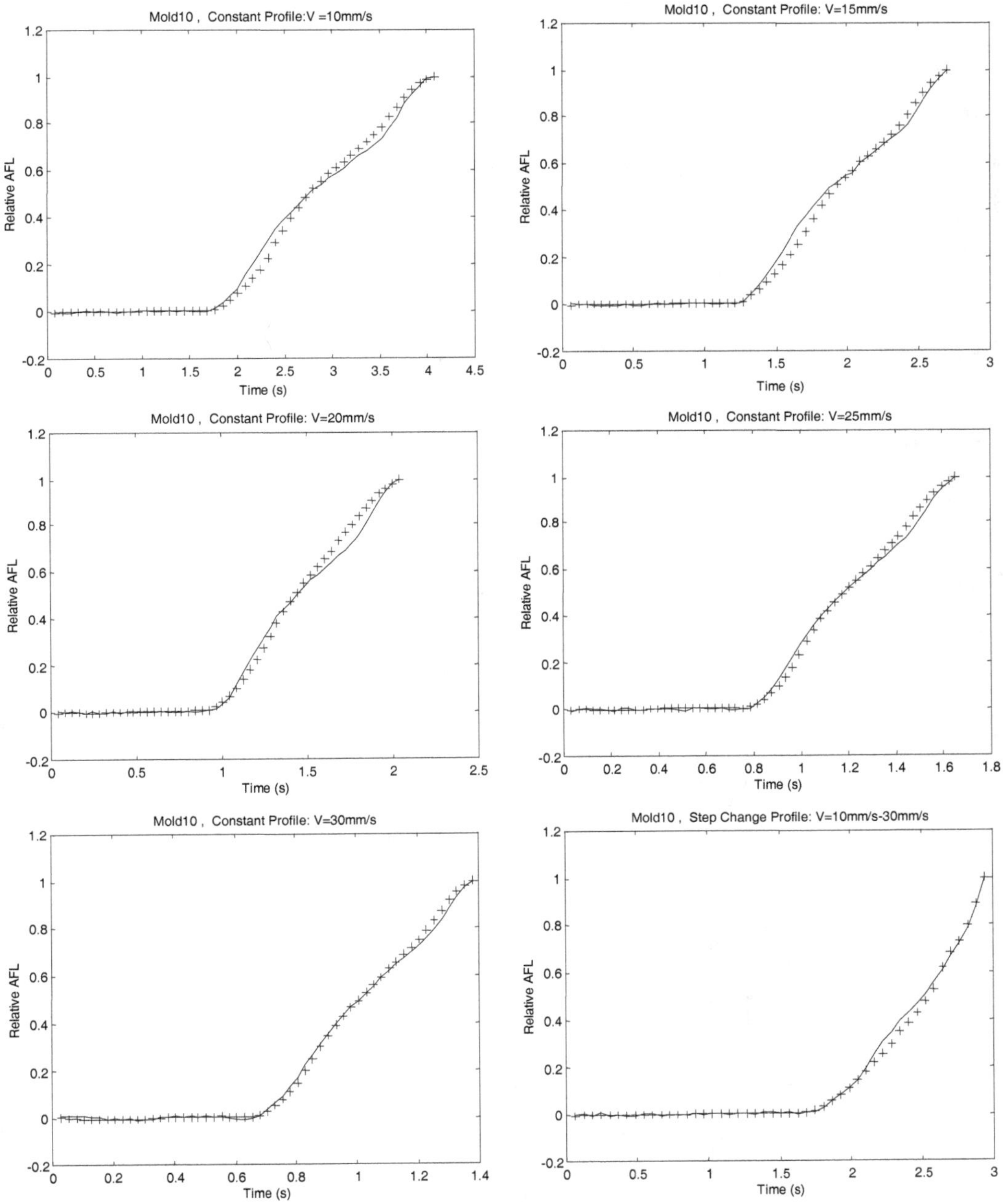


**Figure 9.15** Validation results of Mold 10 (+: network output; –: CT measurement)

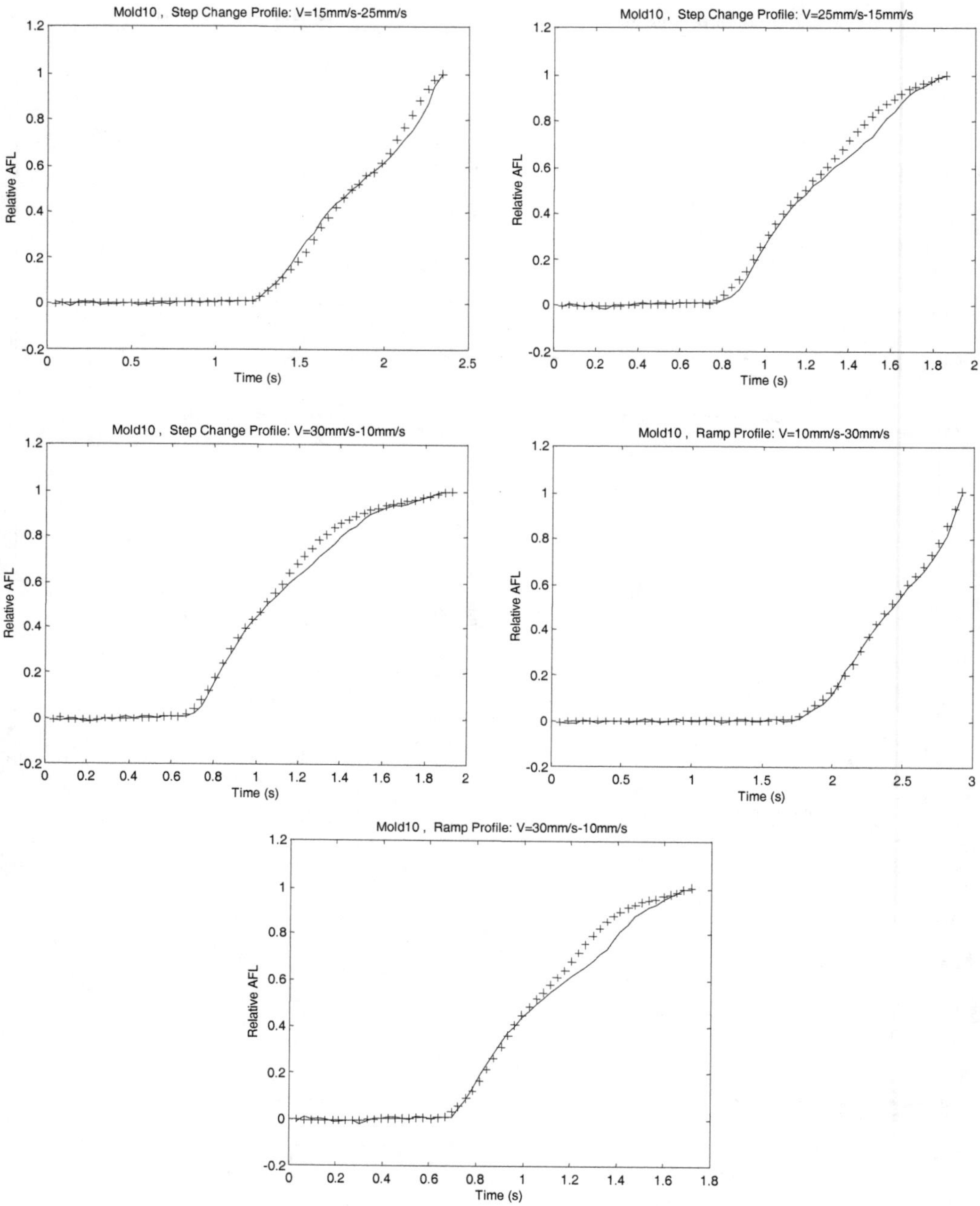


**Figure 9.15** (*continued*) Validation results of Mold 10 (+: network output; –: CT measurement)

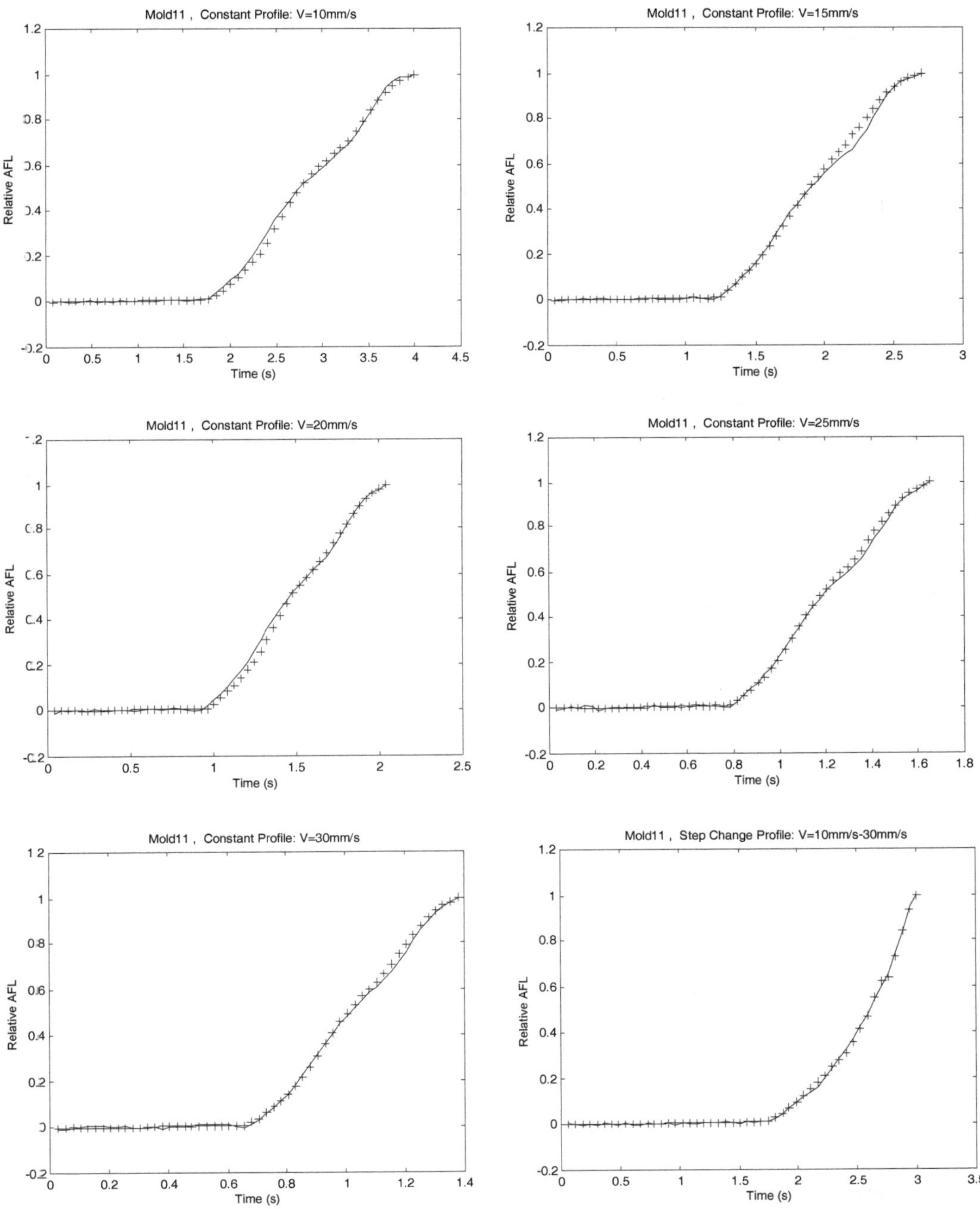


**Figure 9.16** Validation results of Mold 11 (+: network output; –: CT measurement)

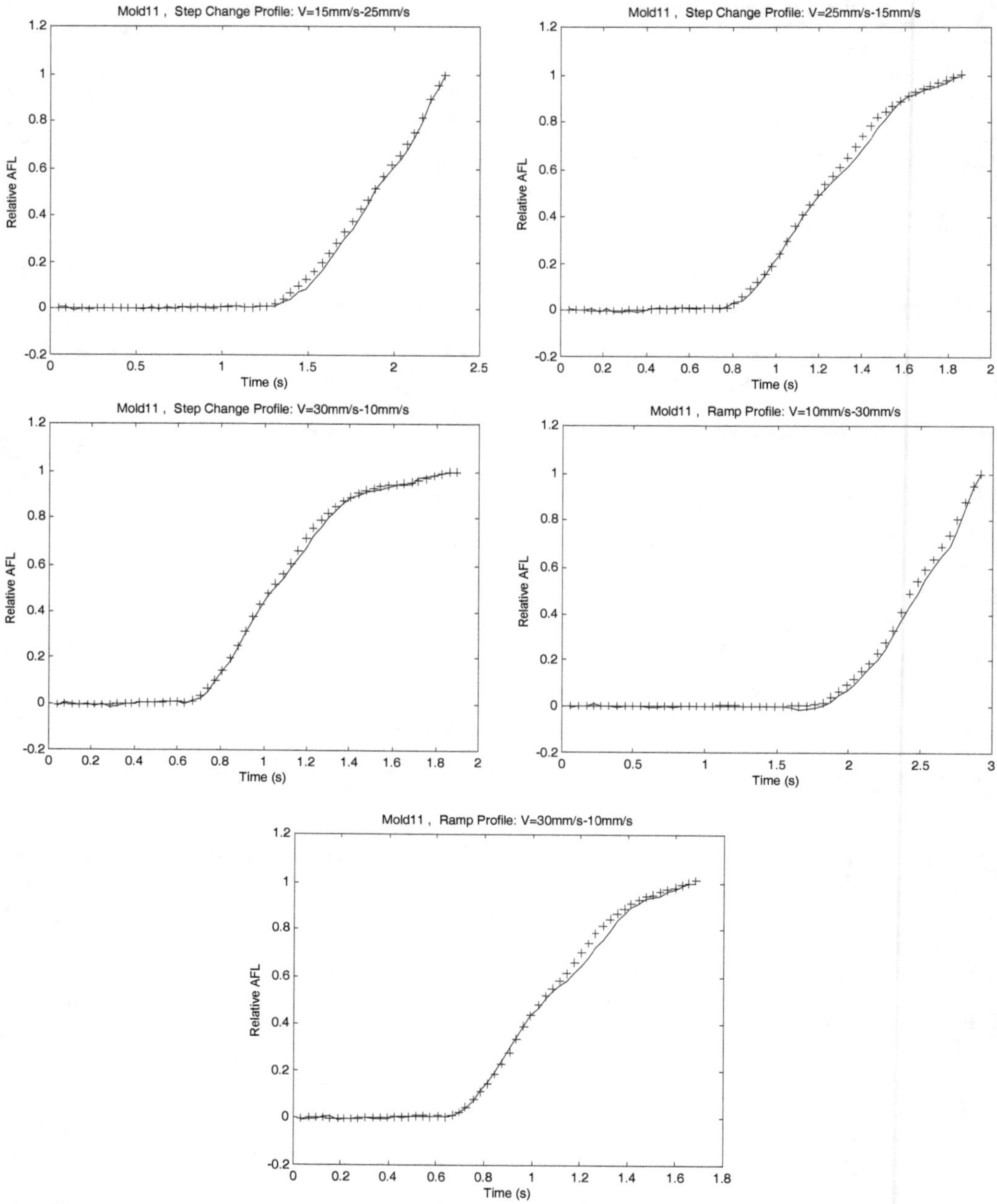


**Figure 9.16** (*continued*) Validation results of Mold 11 (+: network output; –: CT measurement)

## 9.4 Profiling Strategy via Optimization

Development of the capacitive transducer allows a direct measurement of the average-flow-length, which makes the direct control of the melt-front velocity possible, provided that this sensor can be installed in all molds. As illustrated in Fig. 9.6, profiling of the injection velocity can thus be converted to a controller design problem with the online feedback measurement of the average-flow-length. In practice, however, such an online tuning of the injection velocity to output a straight average-flow-length curve is difficult for at least the following reasons. First, not all molds are equipped with such a capacitive sensor; second, there exists a significant delay from the injection velocity to the average-flow-length; third, direct manipulation of the melt-front velocity may result in excessive changes in the injection velocity within a short time due to measurement noises. An offline profiling method based on the neural network model developed is presented in this section, together with the experimental verifications.

By assuming the injection velocity follows a relation with filling time in a parametric form as

$$IV(t) = f(K,t) \tag{9.8}$$

where $f$ is such a relation with time $t$, the injection velocity can then be determined by $K$, the parameter vector in relation $f$. During the offline profiling, the screw displacement starts from zero and ends when reaching the stroke setting of this mold; in between, it can be obtained by integrating the velocity profile as shown below:

$$SD(t) = \int_0^t IV(t)\,\mathrm{d}t \tag{9.9}$$

The nozzle pressure and nozzle temperature cannot be measured during the offline procedure, but for a specific mold they may be correlated with the screw displacement and injection velocity data:

$$NP(t) = f_{\mathrm{P}}(IV, SD) \tag{9.10}$$

$$NT(t) = f_{\mathrm{T}}(IV, SD) \tag{9.11}$$

As the average-flow-length can be predicted by the neural network model using information on injection velocity, screw displacement, nozzle pressure, and nozzle temperature

$$afl(t) = NN(IV, SD, NP, NT) \tag{9.12}$$

by substituting Eqs. 9.8 to 9.11 into 9.12, the average-flow-length can thus be represented by a function of the parameter vector $K$ as shown below:

$$afl(t) = g(K,t) \tag{9.13}$$

The profile setting problem can now be converted to find the optimal values of vector $K$ to result in a straight average-flow-length curve corresponding to a constant melt-front velocity in filling, as illustrated in Fig. 9.17. Described in a mathematical form, it is

$$\begin{aligned} &\min_K \int_0^{t_{\text{end}}} abs\left(g(K,t) - \alpha \cdot (t - t_0)\right) \mathrm{d}t \\ &\text{s.t.:} \\ &K_{\min} \le K \le K_{\max} \\ &V_{\min} \le IV(t) \le V_{\max} \end{aligned} \tag{9.14}$$

where $\alpha$ is a predetermined slope of the average-flow-length ramp, corresponding to the constant melt-front velocity rate in the cavity. Since the injection velocity cannot be arbitrarily set, constraints are made. Here, $V_{\min}$ and $V_{\max}$ are the lower and upper bounds of the injection velocity, and $K_{\min}$ and $K_{\max}$ are the bounds of the vector $K$. A Matlab program based on the sequential quadratic programming (SQP) method [219–223] can be employed to solve the constrained nonlinear optimization problem.

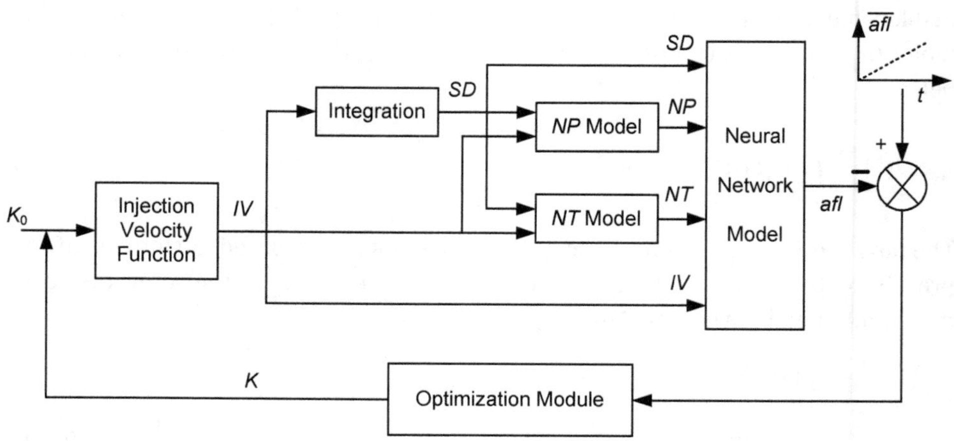


**Figure 9.17** Schematic of offline profiling based on optimization

## 9.5 Parabolic Velocity Profile

The profiling method can be conducted with the parabolic profile for its simple expression. If the injection velocity is kept constant at 25 mm/s before the melt enters the cavity, and later it changes following a parabolic profile, the injection profile during mold filling can thus be described as

$$IV(t) = a(t - t_0)^2 + b(t - t_0) + 25 \tag{9.15}$$

where $t_0$ is the start time when the melt reaches the cavity. Two parameters, $a$ and $b$, can then determine the injection velocity profile.

The offline profiling method can be demonstrated by taking Mold 4 of Fig. 9.10 as an example. Mold 4 has a decreasing cross-sectional area along the melt flow path. In correspondence, parameter $a$ is set greater than zero, and parameter $b$, in contrast, lower than zero, to ensure a decreasing velocity profile. Due to the machine limitations, the bounds of velocity, $V_{min}$ and $V_{max}$, are set at 10 mm/s and 30 mm/s, respectively.

With the initial values of $a$ and $b$ set as 20 and -20, respectively, the initial injection velocity profile takes the shape as shown in Fig. 9.18, resulting in an average-flow-length profile in Fig. 9.19 for the first iteration. By using the established procedure, parameters $a$ and $b$ are optimized.

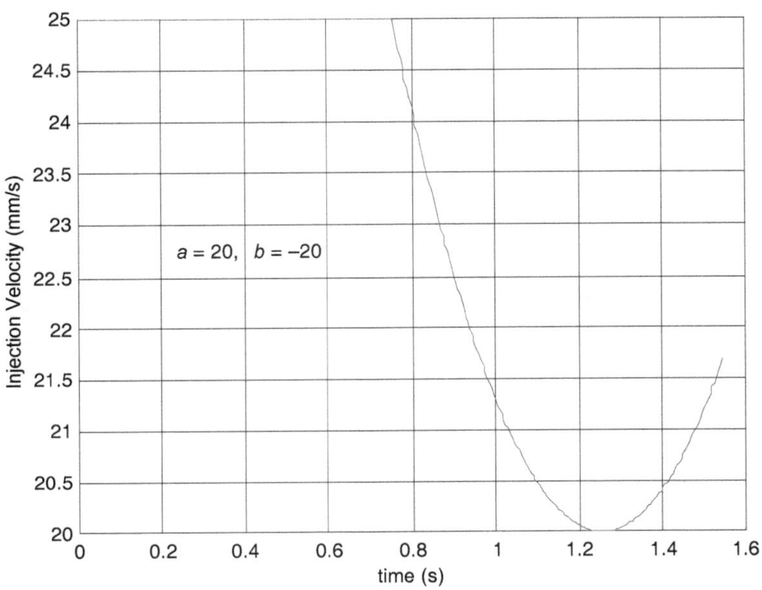


**Figure 9.18** Initial parabolic velocity profile

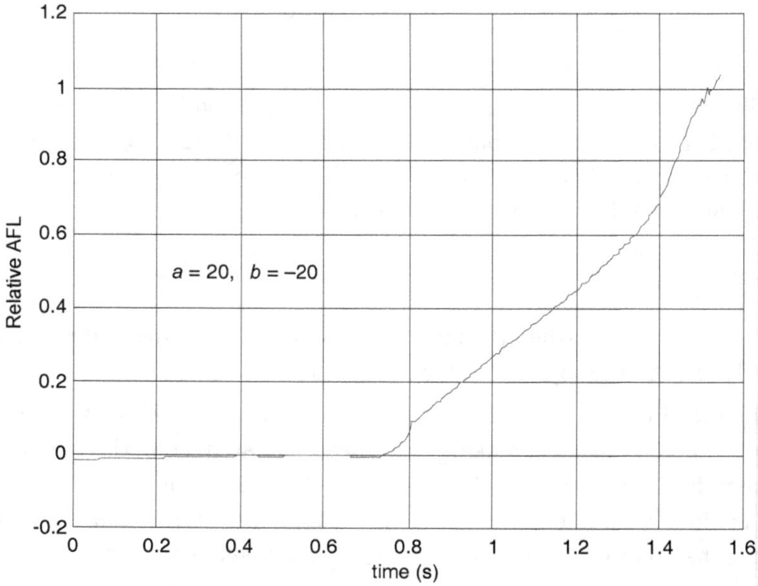


**Figure 9.19** Average-flow-length with initial parabolic velocity profile

The intermediate and final results of the optimization are shown in Figs. 9.20 to 9.23, respectively, for the injection velocity profile and the corresponding average-flow-length.

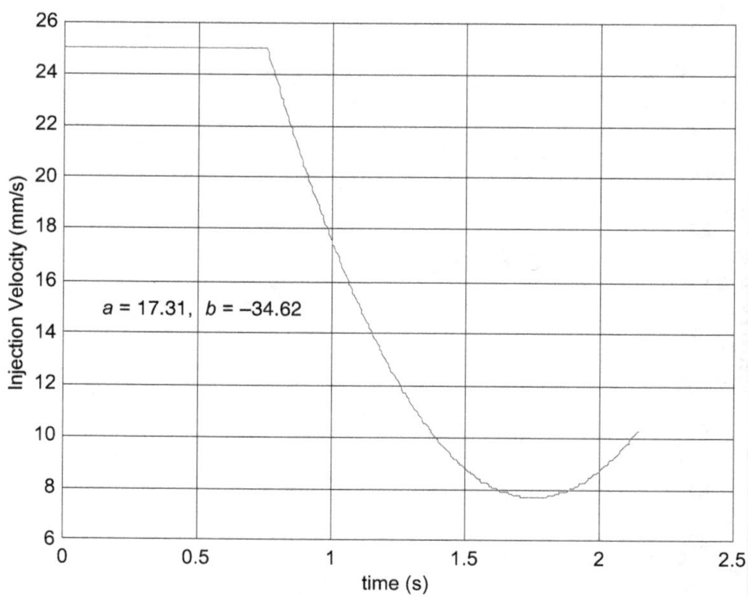


**Figure 9.20** Intermediate parabolic velocity profile

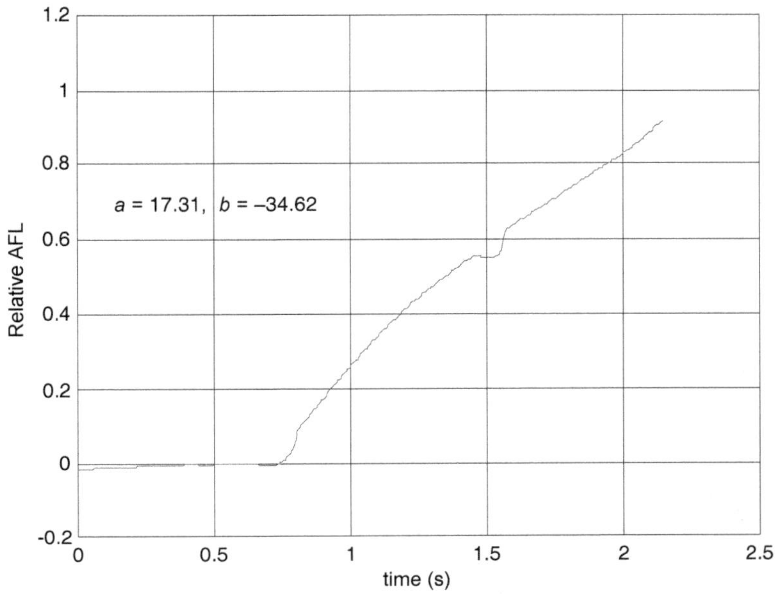


**Figure 9.21** Average-flow-length with intermediate parabolic profile

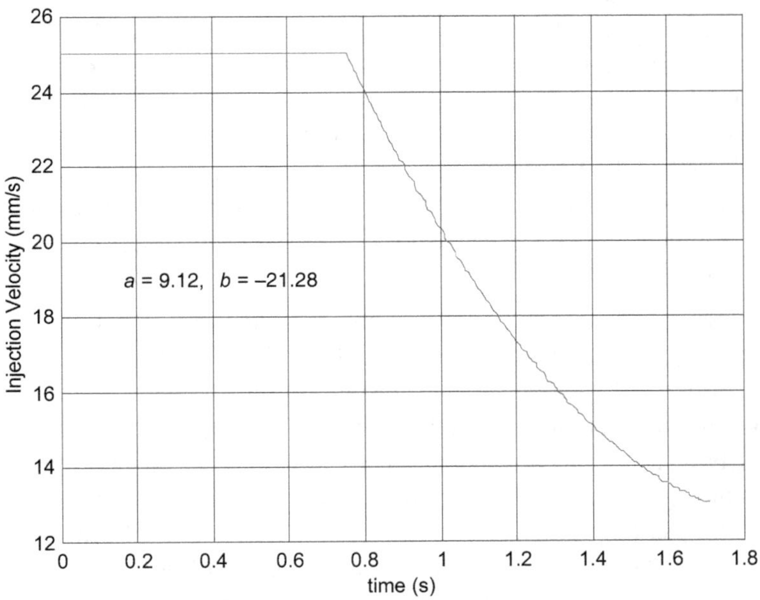


**Figure 9.22** Final parabolic velocity profile

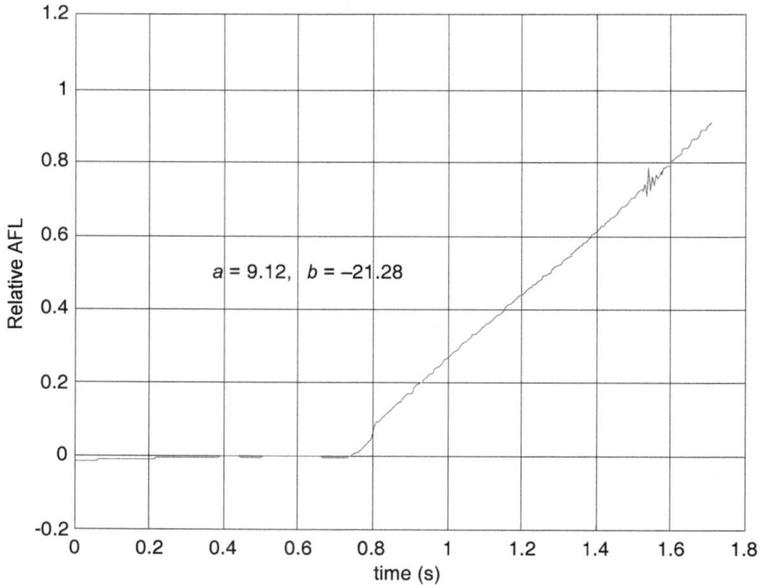


**Figure 9.23** Average-flow-length with final parabolic profile

The optimization process gradually tunes parameters *a* and *b* until the average-flow-length becomes a straight line. The spikes in the average-flow-length of Fig. 9.23 are believed to be errors caused by the average-flow-length model, as the corresponding injection velocity does not have any oscillations. This will not happen in molding. The injection velocity profiles, as a consequence of the optimization procedure, change gradually from the initial decreasing-and-increasing parabolic curve to a consistently decreasing one at the end. This final injection velocity profile is harmonious with the shape of Mold 4, which has consistently decreasing cross-sectional areas along the flow path.

Experiments are also conducted on an injection molding machine to test the offline-derived profiles. Both the initial profile of Fig. 9.18 and the optimal profile of Fig. 9.22 are tested, together with a constant profile of 25 mm/s for comparison. The CT measurements, representing the average-flow-length, are shown in Fig. 9.24. It can clearly be seen that the CT output with the optimal injection profile is the best in producing a straight average-flow-length (i.e., a constant melt-front velocity); the constant injection profile is the worst; and the initial parabolic injection profile is in between as the mold shape has been considered in the profile initialization. The results prove that a constant injection velocity cannot produce a consistent melt-front velocity for a mold with a varying cross-sectional area; on the other hand, profiling of the injection velocity based on the neural network model can bring about a uniform melt-front velocity throughout filling of the mold cavity with varying cross-sectional area.

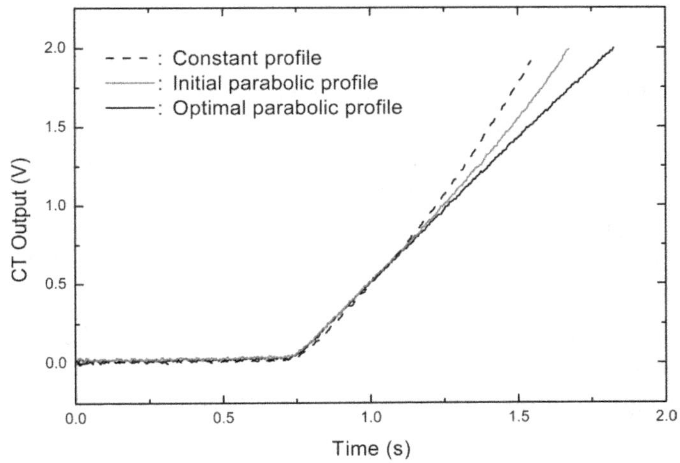


**Figure 9.24** Comparison of parabolic injection profiles

## 9.6 Piece-Wise Ramp Velocity Profile

The parabolic profile can be followed by certain injection molding machines that have been fitted with sophisticated controls. In the industry, however, the injection velocity is often profiled in a series of steps or ramps. Piece-wise ramp profiles, as illustrated in Fig. 9.25, are more practical. The injection stroke rather than injection time was used as the $x$ axis. Like before, for simplicity, the injection velocity was set at a constant before the melt enters cavity. The filling of the cavity was then evenly divided into several periods in terms of the injection stroke. Within each period, the injection velocity was set as a ramp starting with the ending velocity of the previous period. The ramp slope, however, is adjustable.

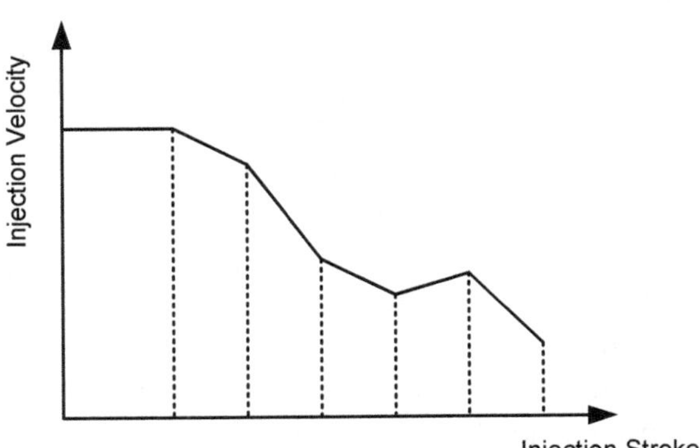


**Figure 9.25** Schematic of a piece-wise ramp profile

Taking again Mold 4 for illustration, a five-piece ramp function was used to seek the "optimal" injection profile. The injection velocity was set at a constant of 25 mm/s before the melt enters the mold cavity. The optimization procedure was similar to the parabolic case, with the exception that the optimization parameters were changed to the five slopes. Considering the nature of the decreasing cross-sectional area of Mold 4, the slope vector was initialized as $X_0 = [-20, -18, -15, -12, -10]$, corresponding to a decreasing velocity profile. After a number of iterations, it converged to an "optimal," with slopes $X^* = [-14.45, -12.82, -10.63, -9.47, -6.7]$. The CT output for the experiment conducted with this "optimized" injection profile was shown in Fig. 9.26, in comparison with the CT output with a constant 25 mm/s injection profile. It can be clearly seen that the CT output with the "optimized" injection profile is very close to a ramp, suggesting that the cavity is filled at a near-constant melt-front velocity.

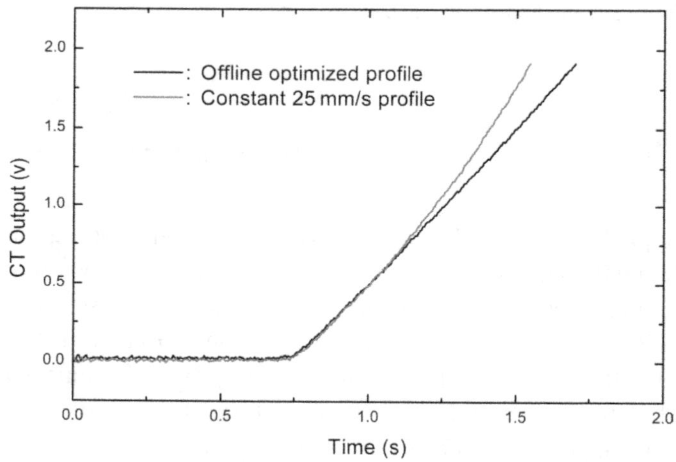


**Figure 9.26** Comparison of injection profiles with Mold 4

The proposed profiling method was also tested with another mold, Mold 5, using the piece-wise ramp function. The cross-sectional area of this mold decreases in the first half and increases in the second half, as shown in Fig. 9.10. Considering this mold geometry feature, the slope vector was initialized as $X_0 = [-30, -10, 10, 20, 30]$. After optimization, the slope vector eventually converged to a new "optimal" with the slope values of each period as $X^* = [-50.0, -7.80, 22.36, 26.67, 24.57]$, a result agreeing with the geometry features. The velocity is profiled to first decrease before it increases. It should be interesting to note that the optimized injection velocity profile was not symmetrical to the center point, even though the mold geometry is. Due to the melt-front development along the flow path, the melt flow does not take a plug flow shape during mold filling. The experiment showed that the "optimized" profile could result in a ramp CT output, as shown in Fig. 9.27, indicating that the cavity was filled at a constant melt-front velocity with this injection velocity profile. For comparison purposes, the CT output is also plotted with a constant screw injection in Fig. 9.27. Obviously, the injection velocity needs to be profiled to result in a constant melt-front velocity for such a mold.

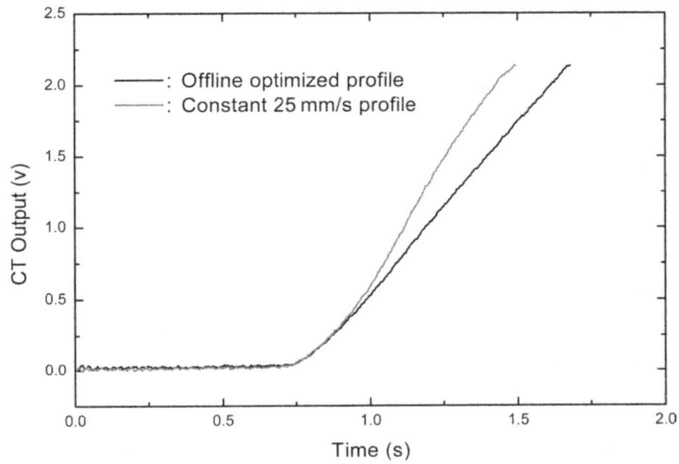


**Figure 9.27** Comparison of injection profiles with Mold 5

The above results demonstrate that the proposed profiling method can ensure a uniform melt-front velocity for the filling of molds with varying geometry.

Regarding the stage division for the optimal profile, Chen et al. [224] proposed an intelligent method. The nonlinear response of melt development in a mold cavity is first acquired with a constant screw injection velocity. Based on this response, a stage division method is proposed to determine an appropriate stage number of a piece-wise linear profile and the locations of the switch points. An optimization based on a sequential simplex search is then conducted to search for the optimal profile of injection velocity. The details can be found in the literature.

## 9.7 Conclusions

A profiling method for injection velocity has been developed and demonstrated in this chapter, to ensure the cavity is filled at a constant melt-front velocity. Relationships among injection velocity, melt-front velocity, melt-front area, and average-flow-length are analyzed. The profiling of injection velocity based on melt-front area has been discussed. The feasibility of using a neural network for modeling the melt-front area is first demonstrated by using simulation data. This has later been improved to velocity profiling based on average-flow-length. A capacitive transducer has been developed to measure online the average-flow-length during filling, among its many potential applications in molding monitoring. A soft sensor to replace the capacitive sensor installation is also developed via a neural network model. Based on this soft-sensor model, a profiling optimization method is proposed and experimentally demonstrated to be effective in ensuring uniform mold filling.

## References

[161] A. Boldizar, J. Kubat, and M. Rigdahl, *J. Appl. Polym. Sci.*, **39**, 63, 1990

[162] F. Johnnaber, *Injection Molding Machines* (2nd ed.), Hanser Publishers, 1985

[163] C.-P. Chiu, Shih L.-C., and J.-H. Wei, *Polym. Eng. Sci.*, **31**, 1417, 1991

[164] H.W. Cox and C.C. Mentzer, *Polym. Eng. Sci.*, **26**, 488, 1986

[165] A. Haddout and G. Villoutreix, *Composites*, **2**, 147, 1994

[166] D.B. Hunkar, SPE *Annu. Tech. Conf. Papers, SPE*, **21**, 161, 1975

[167] L.W. Fritch, SPE *Annu. Tech. Conf. Papers, SPE*, **21**, 15, 1979

[168] L.R. Schmidt and J.L. Maxam, SPE *Annu. Tech. Conf. Papers, SPE*, **39**, 1070, 1993

[169] L.S. Turng, H.H. Chiang, and J.F. Stevenson, SPE *Annu. Tech. Conf. Papers, SPE*, **41**, 668, 1995

[170] J.C. Rowland, F. Gao, On-line Quality Monitoring Coupling Computer Analysis and Actual Measurement, *Japanese Society of Polymer Processing*, Tokyo, June, 1994

[171] R.G. Speight, A.J. Monro, and A. Khassapov, SPE *Annu. Tech. Conf. Papers, SPE*, **45**, 520, 1998

[172] Y. Yang and F. Gao, *Int. Polym. Process.*, **XIV**, 196, 1999

[173] Y. Yang and F. Gao, *Control Eng. Pract.*, **8**, 1285, 2000

[174] H.-P. Tsoi and F. Gao, *Polym. Eng. Sci.*, **39**, 3, 1999

[175] X. Chen and F. Gao, SPE *Annu. Tech. Conf. Papers, SPE*, **45**, 633, 1999

[176] T. Hryces, *Neural Control, Towards an Industrial Control Methodology*, John Wiley & Sons, 1997

[177] *C-MOLD Reference Manual*, Advanced CAE Technology Inc., New York, 1995

[178] K.L. Moore, *Iterative Learning Control for Deterministic Systems*, Adances in Industrial Control, Springer-Verlag, Berlin, 1993

[179] K.L. Moore and J-X. Xu, (Eds.), *Iterative Learning Control*, Taylor and Francis, London, 2000

[180] Yi Yang, Ph.D. Thesis, Hong Kong Univ. of Sci. and Technol., 2004

[181] R.S. Spencer and G.D. Gilmore, *J. Colloid Sci.*, **6**, 118, 1951

[182] R.L. Ballman, T. Shusman, and H.L. Toor, *Ind. Eng. Chem.*, **51**, 847, 1959

[183] R.L. Ballman, T. Shusman, and H.L. Toor, *Mod. Plast.*, **37**, 105, 1959

[184] R.L. Ballman, T. Shusman, and H.L. Toor, *Mod. Plast.*, **37**, 115, 1959

[185] R.L. Ballman and T. Shusman, *Mod. Plast.*, **37**, 126, 1959

[186] R.L. Ballman and H.L. Toor, *Mod. Plast.*, **38**, 113, 1960

[187] G.B. Jackson and R.L. Ballman, *Soc. Plast. Eng. J.*, **16**, 1147, 1960

[188] H.L. Toor, R.L. Ballman, and L. Cooper, *Mod. Plast.*, **38**, 117, 1960

[189] J. Pearson, *Mechanical Principles of Polymer Melt Processing*, Pergamon, Oxford, 1966

[190] M.R. Kamal and S. Kenig, *Polym. Eng. Sci.*, **12**, 294, 1972

[191] M.R. Kamal and S. Kenig, *Polym. Eng. Sci.*, **12**, 302, 1972

[192] J.L. Berger and C.G. Gogos, *Polym. Eng. Sci.*, **13**, 102, 1973

[193] G. Williams and H.A. Lord, *Polym. Eng. Sci.*, **15**, 553, 1975

[194] H.A. Lord and G. Williams, *Polym. Eng. Sci.*, **15**, 569, 1975

[195] D.H. Harry and R.C. Parrott, *Polym. Eng. Sci.*, **10**, 209, 1970

[196] P. Thienel and G. Menges, *Polym. Eng. Sci.*, **18**, 314, 1978

[197] A.I. Isayev and C.A. Hieber, *Rheol. Acta*, **19**, 168, 1980

[198] S. Richardson, *J. Fluid Mech.*, **56**, 609, 1972

[199] C.A Hieber and S.F. Shen, *Israel J. Technol.*, **16**, 248, 1978

[200] J.L. White, *Polym. Eng. Sci.*, **15**, 44, 1975

[201] P.G. Lafleur and M.R. Kamal, *Polym. Eng. Sci.*, **26**, 92, 1986

[202] R.A. Behrens et al., *AIChE J.*, **33**, 1178, 1987

[203] D. Fauchon, H.H. Dannelongue, and P.A. Tanguy, *Int. Polym. Process.*, **VI**, 13, 1991

[204] S.F. Shen, *Int. J. Numer. Methods Fluids*, **4**, 171, 1984

[205] P. Kennedy, *Flow Analysis of Injection Molds*, Hanser, 1995

[206] *C-MOLD Reference Manual*, Advanced CAE Technology Inc., New York, 1995

[207] M. Vincent, in J.A. Covas et al., (Ed.), *Rheological Fundamentals of Polymer Processing*, NATO ASI Series, Kluwer Academic Publishers, Dordrecht, 1995

[208] B. Friedrichs and S.I. Güceri, *J. Non-Newtonian Fluid Mech.*, **49**, 141, 1993

[209] K. Sagae, M. Koizumi, and M. Yamakawa, *JSME Int. J.*, **37**, 531, 1994

[210] J.-F. Hétu et al., *Polym. Eng. Sci.*, **38**, 223 1998

[211] K. Warwick, G.W. Irwin, and K.J. Hunt, *Neural Networks for Control and Systems*, Peter Peregrinus Ltd., 1992

[212] S. Haykin, *Neural Networks: A Comprehensive Foundation*, IEEE Computer Society Press, 1994

[213] B.A. Pearlmutter, *IEEE Trans. Neural Networks*, **6**, 1212, 1995

[214] G. Cybenko, *Math. Contr. Signal Syst.*, **2**, 303, 1989

[215] K. Levenberg, *Q. Appl. Math.*, **2**, 164, 1944

[216] D. Marquardt, *J. Soc. Ind. Appl. Math.*, **11**, 431, 1963

[217] M.T. Hagan and M.B. Menhaj, *IEEE Trans. Neural Networks*, **5**, 989, 1994

[218] MathWorks Inc., *Neural Network Toolbox User's Guide*, 1998

[219] K. Schittowski, *Ann. Oper. Res.*, **5**, 485, 1985

[220] M.C. Biggs, in L.C.W. Dixon and G.P. Szergo, (Eds.), *Towards Global Optimization*, North-Holland, 1975

[221] S.P. Han, *J. Optim. Theory Appl.*, **22**, 297, 1977

[222] M.J.D. Powell, in A. Bachem, M. Grotschel, and B. Korte, (Eds.), *Mathematical Programming: The State of the Art*, Springer Verlag, 1983

[223] MathWorks Inc., *Optimization Toolbox User's Guide*, 1998

[224] X. Chen et al., *Polym. Eng. Sci.*, **50**(7), 1358, 2010

# 10 Profile Setting of Packing Pressure

After a mold cavity is filled, the injection molding process switches from the filling to the packing-holding stage. Proper profile setting of packing pressure is essential to part quality. A complete packing profile is determined by three key elements: start point of the packing, end point of the packing, and packing pressure levels in between. Several methods have been proposed and developed for detection of the start point of the packing, which is also the end point of the filling stage. For instance, the fuzzy V/P transfer technique [225], which is a technique based on the online detection of the abrupt pressure change at the end of filling through the fuzzy method. The capacitive transducer, as described in Chapter 8, is also capable of detecting this transfer point. In this chapter, the packing profiling work focuses on the other two elements.

## 10.1 Online Autodetection of Gate Freezing-Off Point

Ideally, the packing should end at the time when the gate is just completely frozen, immediately following the gate freezing-off point. With insufficient packing time, the melt in the mold may flow back, resulting in sink marks. An excessively long packing time, on the other hand, is also unnecessary, as the melt in the mold can no longer be influenced after the gate freezes off. Extending packing beyond this point will waste energy as well as time. Accurate prediction of the gate freezing-off point is, therefore, necessary for a proper setting of a packing profile.

Two existing approaches are commonly used to determine the gate freezing-off point, both of which are based on trial and error. The first approach records and compares the packing pressure and cavity pressure with different packing times. The freezing-off point is identified as the time when the cavity pressure can no longer be affected by the packing pressure. This method requires pressure transducers to be fitted both in the cavity and at the nozzle (or in the hydraulic cylinder).

The second approach does not have this requirement. In this approach, a series of experiments with different packing times is first conducted. The part weights are measured and plotted against the packing time. The gate freezing-off point is then identified as the time when the part weight first levels off. Both methods mentioned above are time-consuming, requires repetitive experimentation, and the result is determined offline. Changes in mold or molding conditions require a re-identification of the packing time. Alternatively, as presented in Chapter 8, the capacitive transducer can also be used to detect the gate freezing-off point. Experiments are conducted to find the time after which the capacitive transducer output curve levels off or starts to fall. This method is online and efficient, but it requires the sensor to be instrumented for such a mold. Another robust online system, without mounting any sensor in the mold cavity to detect the gate freezing-off point [226, 227], is presented in detail as follows.

### 10.1.1 Gate Freezing-Off Detection

The injection molding system during the packing-holding stage may be described as in Fig. 10.1, a dynamic system with time-varying characteristics due to continuous gate freezing. The input to this system is the valve opening (*VO*) that provides the driving force for packing. The system outputs are process variables such as nozzle pressure (*NP*), screw displacement (*SD*), and screw velocity (*SV*).

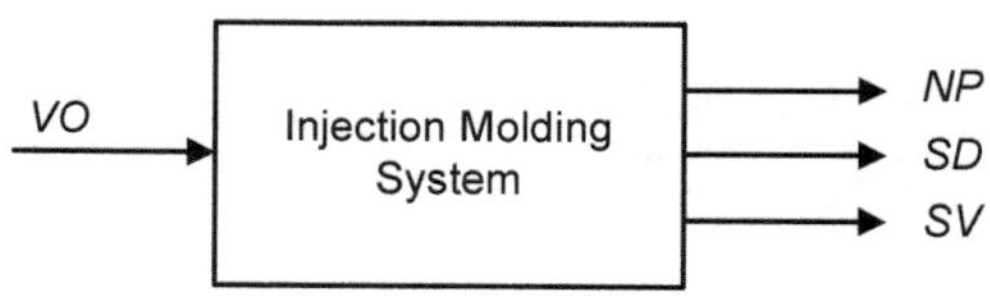


**Figure 10.1** Schematic of gate freezing process

During the packing-holding stage, the system gain of the melt flow packed into the mold cavity over the packing pressure should gradually decrease with time as a result of the gate freezing. The screw movement, which is manipulated by the valve opening, is characterized by the screw velocity and screw displacement. Before the gate is frozen, the screw keeps moving during packing, but the movement is not as obvious as that during filling. After the gate freezing-off point, the screw should be "stationary" with a constant packing pressure, if the leakage flow is not significant. The variation of the screw movement around the gate freezing-off point is also dependent on the nozzle pressure, which may be closed-loop controlled. Therefore, it enables detection of the gate freezing-off point by monitoring changes in these process variables. Among the available measurements, the data of the valve opening and screw velocity may be difficult to use directly for such detection, as they are varying in a very narrow range during the packing-holding. Typically, the valve opening changes in a range of 0–2%, and the velocity 0–1 mm/s. This results in a large noise-to-signal ratio. Furthermore, not all machines are equipped with the velocity transducer. Measurements of the screw displacement and nozzle pressure, on the other hand, can be readily available in most cases. These two process variables, therefore, are selected to develop the online gate freezing-off point detection system. The gate freezing characteristics and their influence on these two measurements are analyzed below.

Figure 10.2 presents a simplified schematic of the barrel and screw system, where the shadowed area represents the melt in front of the screw tip. The melt flow can be assumed as a plug flow. The dotted line shown in the figure represents the screw position at the start of packing, and the solid line the screw position during packing.

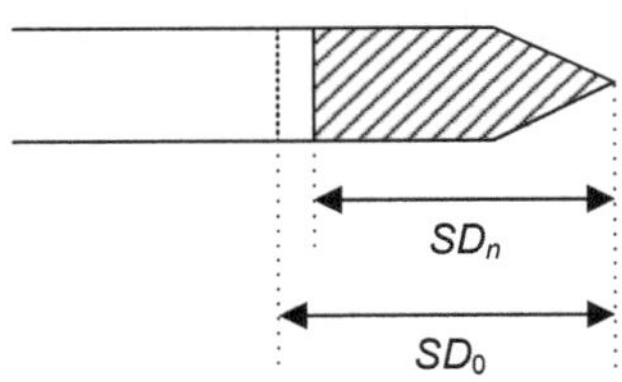


**Figure 10.2** Schematic of screw movement during packing-holding

For a constant packing pressure, the mass balance for material in front of the screw tip results in the following equation, ignoring the leakage through the check-ring valve:

$$m_0 - m_n = \rho \cdot A_b \cdot (SD_0 - SD_n) \tag{10.1}$$

where $m$ is the mass of the material in front of the screw tip, $\rho$ the density of the material, and $SD$ the screw displacement measured from the nozzle side. The subscript "0" represents the value at the beginning of packing, while the subscript $n$ represents the current sample during packing. The assumption of the constant packing pressure results in the density, $\rho$, being constant as the temperature variation is small. Defining the following new variables as

$$\Delta m_n = m_0 - m_n \tag{10.2}$$

$$\Delta SD_n = SD_0 - SD_n \tag{10.3}$$

and plugging the above into Eq. 10.1 results in

$$\Delta m_n = \rho \cdot A_b \cdot \Delta SD_n \tag{10.4}$$

where $\Delta m_n$ is the total mass change of the material in front of the screw tip, that is, the amount of material packed into the mold if the leakage can be ignored; $\Delta SD_n$, reflecting the screw movement during packing, can be measured via a screw displacement transducer. The sum of material packed into the mold, $\Delta m_n$, will increase with the packing time and finally reach a maximum and level off after the gate is completely frozen. The screw displacement change, $\Delta SD_n$, should have the same trend because it is proportional to $\Delta m_n$. The actual screw displacement after the gate freezing-off point may not be constant due to the leakage and measurement noises. But the screw movement characteristics before and after the gate is frozen will be definitely different. This feature can thus be used to predict the gate freezing-off point for a case of constant packing pressure.

When the packing pressure is not constant, the density variation caused by pressure change must be considered. The P-V-T relationship of polymers [228, 230] may be employed in this case. Unlike Eq. 10.1, the application of mass balance for the material in front of the screw tip results in the following equation:

$$\frac{m_0}{\rho_0} - \frac{m_n}{\rho_n} = A_b \cdot \Delta SD_n \tag{10.5}$$

where $\rho_n$ is the density of the current sample. Then the density can be approximated by a function of pressure in the following:

$$\rho_n = \rho_0 + \overline{\frac{\partial \rho}{\partial P}} \cdot \Delta P_n \tag{10.6}$$

where

$$\Delta P_n = P_n - P_0 \tag{10.7}$$

represents the pressure change since the start of packing; $\overline{\frac{\partial \rho}{\partial P}}$ is a material parameter reflecting the pressure compression coefficient during packing.

Equation 10.5 can thus be transformed as

$$\begin{aligned} A_b \cdot \Delta SD_n &= \frac{m_0 - m_n}{\rho_0} + \frac{m_n}{\rho_0} - \frac{m_n}{\rho_n} \\ &= \frac{\Delta m_n}{\rho_0} + \frac{m_n(\rho_n - \rho_0)}{\rho_0 \cdot \rho_n} \end{aligned} \tag{10.8}$$

Substituting Eq. 10.6 into 10.8 and also dividing by $A_b$ for both sides, the following equation can be derived:

$$\Delta SD_n = \frac{\Delta m_n}{\rho_0 A_b} + \frac{m_n}{\rho_0 \cdot A_b} \cdot \frac{1}{\rho_n} \cdot \overline{\frac{\partial \rho}{\partial P}} \cdot \Delta P_n \tag{10.9}$$

Equation 10.9 can be further transformed into the following dimensionless form:

$$\frac{\Delta SD_n}{SD_0} = \boxed{\frac{\Delta m_n}{\rho_0 \cdot A_b \cdot SD_0}} + \boxed{\frac{m_n}{\rho_0 \cdot A_b \cdot SD_0} \cdot \frac{P_0}{\rho_n} \cdot \overline{\frac{\partial \rho}{\partial P}}} \cdot \frac{\Delta P_n}{P_0} \tag{10.10}$$

Therefore, the relationship between $\frac{\Delta SD_n}{SD_0}$ and $\frac{\Delta P_n}{P_0}$ can be viewed as a pseudo-linear function with varying intercept and slope as boxed in Eq. 10.10:

$$\text{Intercept} = \frac{\Delta m_n}{\rho_0 \cdot A_b \cdot SD_0} \tag{10.11}$$

$$\text{Slope} = \frac{m_n}{\rho_0 \cdot A_b \cdot SD_0} \cdot \frac{P_0}{\rho_n} \cdot \overline{\frac{\partial \rho}{\partial P}} \tag{10.12}$$

where $P_0$, $\rho_0$ and $SD_0$ are the initial state variables, and $A_b$ is the barrel cross-sectional area. All of them are constant during the packing-holding stage. As the total amount of material packed into the mold, $\Delta m_n$, will increase during packing and reach a steady value after the gate is completely frozen, the intercept term, $\frac{\Delta m_n}{\rho_0 \cdot A_b \cdot SD_0}$, which is proportional to the mass of material packed into the mold, should have the same trend. The slope term, $\frac{m_n}{\rho_0 \cdot A_b \cdot SD_0} \cdot \frac{P_0}{\rho_n} \cdot \overline{\frac{\partial \rho}{\partial P}}$, however, depends on $\rho_n$ and $\overline{\frac{\partial \rho}{\partial P}}$, both of which relate to the pressure setting. But after the gate freezing-off point, since the intercept ideally becomes a constant, the slope then equals the ratio of relative volume change over relative pressure change, which relates to the material compression coefficient:

$$\text{Slope} = \frac{\left(\Delta SD_n - \Delta SD_{n-1}\right)/SD_0}{\left(\Delta P_n - \Delta P_{n-1}\right)/P_0} \tag{10.13}$$

If the intercept and slope terms can be derived during packing-holding, the gate freezing-off point in a varying packing profile case can thus be predicted by identifying their changes.

### 10.1.2 Development of Autodetection System

With a constant packing pressure profile, the detection principle is straightforward, as described in the previous section. In an ideal noise-free situation, the screw does not move after the gate is frozen. The gate freezing-off point can be detected by comparing the screw displacement signal. Considering the measurement noises in practice, the standard deviation of the screw displacement during a certain period of time, defined by Eq. 10.14 below, is proposed as a detecting signal for the gate freezing-off point:

$$std_n = \sqrt{\sum_{i=n-n_d+1}^{n} \left(SD_i - \overline{SD}_n\right)^2 \Big/ \left(n_d - 1\right)} \tag{10.14}$$

where $\overline{SD}_n$ is the average of the past $n_d$ samples, defined by

$$\overline{SD}_n = \sum_{i=n-n_d+1}^{n} SD_i / n_d \tag{10.15}$$

After the gate is frozen, the proposed standard deviation, $std_n$, will decrease in comparison with those before that point. This is true whether there is leakage or not. This can be used as a detection rule for the gate freezing-off point:

Rule 1: *If the ratio of current standard deviation over the average of the past is small, then the gate is frozen.*

By defining

$$\overline{std}_n = \sum_{i=1}^{n} std_i / n \tag{10.16}$$

and

$$NSD = std_n / \overline{std}_n \tag{10.17}$$

the standard deviation is normalized as represented by *NSD*. Rule 1 can then be mathematically represented in a form as follows:

Rule 2: *If NSD < SMALL, then the gate is frozen.*

where *SMALL* is a preset threshold that will be discussed later.

To prevent false prediction due to measurement noises, Rule 2 can be improved as

Rule 3: *If NSD < SMALL is holding for m consecutive samples, then the gate is frozen.*

where $m$ is a parameter preset by customers. Though a delay ($m$ times sampling period) is introduced to the detection, it prevents immediate termination of the packing after the gate is frozen, which ensures that the frozen gate is rigidly secure. To further reduce the noise effect, a second-order Butterworth filter [231, 232], as presented in Eq. 10.18, is used to smooth out the measurement data after analyzing the signal frequency:

$$\begin{aligned} y(n) = {} & 0.0009 \times x(n) + 0.0019 \times x(n-1) + 0.0009 \times x(n-2) \\ & + 1.9112 \times y(n-1) - 0.9150 \times y(n-2) \end{aligned} \tag{10.18}$$

where $x$ is the input and $y$ is the filtered output.

With a nonconstant packing pressure, the detection algorithm is more complicated. Both the screw displacement and nozzle pressure measurements are used for the prediction. Define

$$x_n = \frac{\Delta P_n}{P_0} \tag{10.19}$$

$$y_n = \frac{\Delta SD_n}{SD_0} \tag{10.20}$$

According to Eq. 10.10, the relationship between the dimensionless data pairs of past $n_d$ samples, $(x_i, y_i)_{i=n-n_d+1,\cdots,n}$, can be approximated by a pseudolinear function:

$$y_i = a_n \cdot x_i + b_n \tag{10.21}$$

The slope, $a_n$, and the intercept, $b_n$, can be derived from the data pairs via the least squares regression, as shown in the following:

$$a_n = SXY/SXX \tag{10.22}$$

$$b_n = \overline{y}_i - a_n \cdot \overline{x}_i \tag{10.23}$$

where $\overline{x}_i$ and $\overline{y}_i$ are the average of the $x_i$ and $y_i$ values, respectively; $SXX$ is the corrected sum of squares for the $x_i$ values defined by

$$SXX = \sum(x_i - \overline{x}_i)^2 \tag{10.24}$$

and $SXY$ is the corrected sum of cross products defined by

$$SXY = \sum(x_i - \overline{x}_i)(y_i - \overline{y}_i) \tag{10.25}$$

According to Eq. 10.11, the intercept is proportional to the amount of material packed into the mold during packing, which increases with time until the gate is frozen. Like the detection for the constant-pressure case, a similar technique is applied to identify the intercept change for the non-constant-pressure case, with the replacement of the standard deviation by a difference operation as

$$Diff_n = b_n - b_{n-n_d} \tag{10.26}$$

A rule similar to Rule 3 can be introduced:

Rule 4: *If* $Diff_n / \overline{Diff}_n < SMALL$ *is valid for 10 consecutive samples, then the gate is completely frozen.*

where $\overline{Diff}_n$ is the average of the past $Diff_i, i = 1,\cdots,n$. Notice that $Diff_n$ is used in the detection, so Rule 4 is valid for molding even with the leakage passing through the check-ring valve. To reduce the possibility of false prediction, two more rules are introduced:

Rule 5: *When the gate is completely frozen, the derived slope should be in the range* $(0, \varepsilon_a)$.

Rule 6: *When the gate is completely frozen, the derived intercept should be in the range* $(\varepsilon_b, \infty)$.

where $\varepsilon_a$ and $\varepsilon_b$ are two thresholds of the slope and intercept, respectively. Unlike Rules 1–4, these two rules are not in the form of IF-THEN. Rule 4 can be regarded as the sufficient condition of the frozen gate; these two rules, however, are the necessary conditions. According to Eq. 10.11, the intercept term, $\dfrac{\Delta m_n}{\rho_0 \cdot A_b \cdot SD_0}$, approximates the ratio of the polymer amount packed into the mold over the polymer amount in front of the screw tip at the beginning of packing. It relates to the cushion-setting and material-shrinkage feature. The slope term after the gate is frozen, as presented

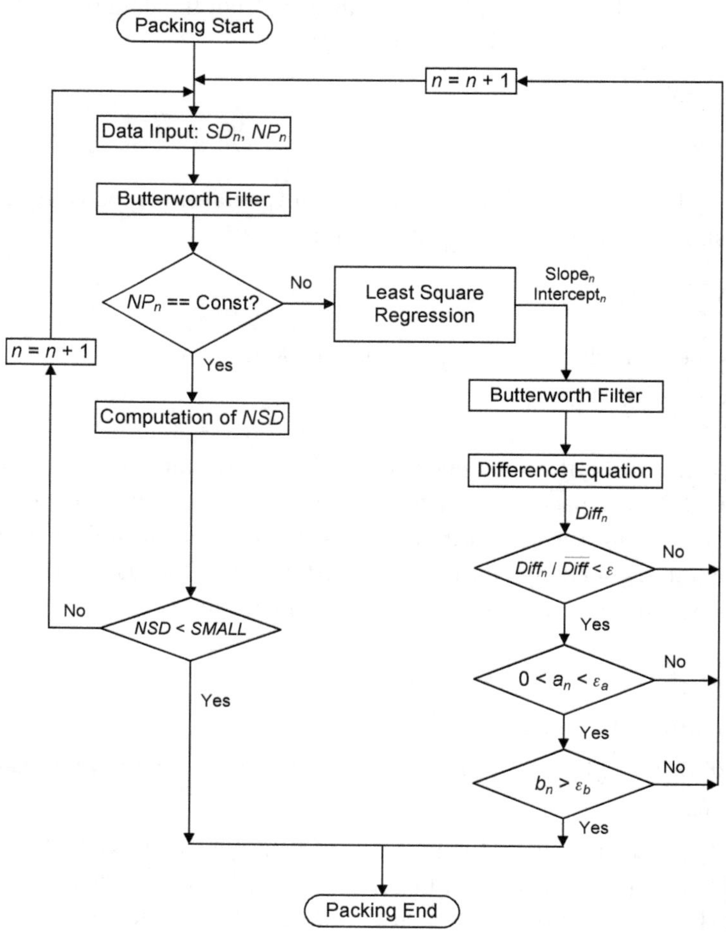


**Figure 10.3** Simplified flowchart of the online detection system

in Eq. 10.13, represents the ratio of relative volume change over relative pressure change. It relates to the material compression capability. Wide ranges are selected for these two coefficients to allow wide application conditions. In the tests, $\varepsilon_a$ is set as 30% and $\varepsilon_b$ 20%.

A simplified flowchart of the online detection system is illustrated in Fig. 10.3. Different detection algorithms and paths are taken for the different packing cases. The system first determines if the packing pressure is constant. This can be prior known information, or it can be determined by the detection system by comparing the pressure measurements. Packing ceases when the corresponding conditions are met.

### 10.1.3 Tests of Constant Packing Pressure Cases

Most existing machines have the packing pressure set at a constant during packing-holding. The detection of the gate freezing-off point with a constant packing profile is first conducted. A box mold with a spider gate as shown in Fig. 10.4 is used for the tests. The tests are conducted with different materials, including high-density polyethylene (HDPE), low-density polyethylene (LDPE), polypropylene (PP), and polystyrene (PS). In the following, unless otherwise specified, the results are obtained with the box mold, HDPE, and a 200 °C nozzle heater temperature setting. In all experiments, the screw rotation speed during plastication is set at 80 RPM, the injection velocity 30 mm/s, and the sampling period of packing 15 ms. The end of packing is determined online by the detection system, and the results are compared with the established offline measurements.

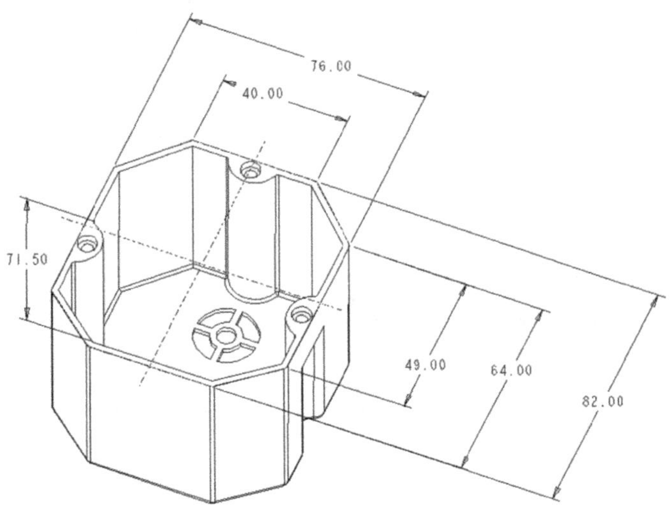


**Figure 10.4** Geometry of box mold

**Gate Freezing Detection for a Constant Packing Pressure**

The first test is conducted with a closed-loop packing-holding control at a 300 bar constant pressure. The threshold representing *SMALL* in Rule 3 is set at 10%. The gate freezing-off point is detected to be 10.7 s by the autodetection system, according to Rule 3. The screw displacement is shown in Fig. 10.5, in which both the raw and filtered data are illustrated. A small delay is introduced by the filter, but it does not affect the detection result significantly. In addition, to avoid the possible pressure variation around the V/P transfer, data for the first half-second of packing are not used in the detection. As previously analyzed, the figure indicates that the screw displacement levels off just before packing ends. The corresponding normalized standard deviation, defined in Eq. 10.17, is illustrated in Fig. 10.6, in which the values larger than one are truncated for better illustration. As suggested by the proposed Rule 3, a rapid decrease of the defined standard deviation can be observed just before the packing stops.

To evaluate the developed detection system, a number of moldings are made with the same pressure setting but different packing times. Their part weights are measured and compared with that of the autodetection system, as listed in Table 10.1. It can be seen that the part weight increases with the packing time, and the increase in packing time from 10 s to 14 s is small. The part weight of the autodetection system is 47.78 g, the same as that packed for 14 s. This suggests that the gate has already frozen when packing beyond the time suggested by the online detection system. On the other hand, a significant weight difference between that of the autodetection and those packed under 10 s indicates that gate has not frozen before 10 s of packing.

In the detection system, a threshold representing *SMALL* for the normalized standard deviation used in Rule 3 was set at 10%. Experiments have also been conducted to test the sensitivity of the detection system with different thresholds. The results show that the system is not very sensitive to the threshold, as the part weight variation is small. The best threshold can be determined by finding the minimal packing time but with a maximum weight. This system has also been successfully tested with different packing pressures and different molds. Detailed results can also be found in reference [226].

**Table 10.1** Part Weight Comparison of 300 bar Constant Packing Pressure

| | Results with fixed packing time | | | | | Results of detection system |
|---|---|---|---|---|---|---|
| Packing time (s) | 3 | 6 | 9 | 10 | 14 | 10.7 |
| Part weight (g) | 45.85 | 46.90 | 47.60 | 47.72 | 47.78 | 47.78 |

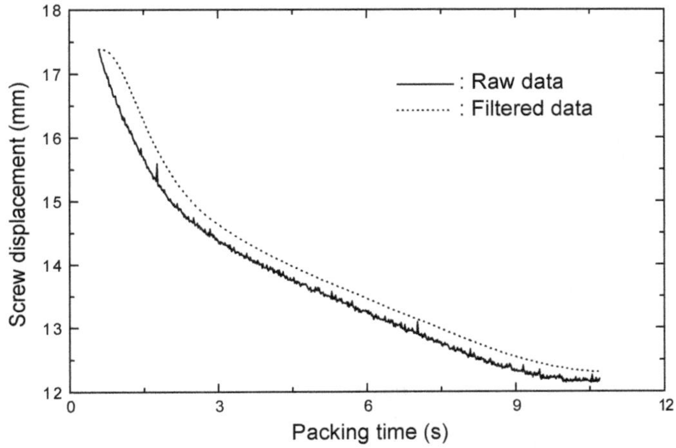


**Figure 10.5** Screw displacement with 300 bar constant packing profile

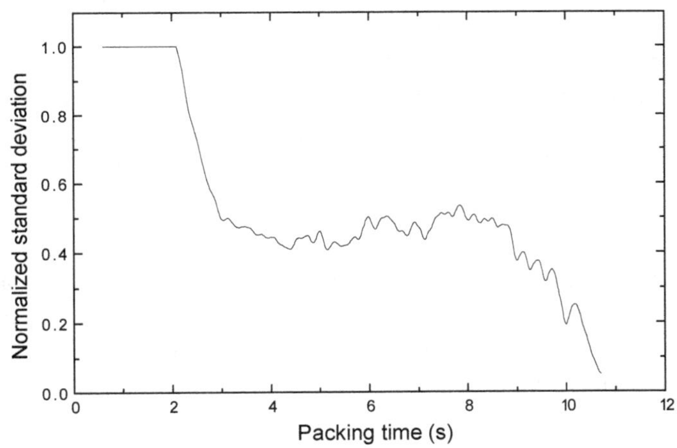


**Figure 10.6** Plot of normalized standard deviation

### Online Detection Results during Warm-Up Period

In practice, the machine takes a number of cycles to reach the steady state after it starts. This is referred to as the warm-up period. Typically, the part weight varies during this period if it is packed with a fixed length. The gate freezing time during this transient period may vary cycle to cycle. The autodetection system can compensate for the warm-up by automatically varying the packing time to keep a relatively constant part weight.

Experiments are conducted for a series of cycles immediately after the machine starts. The autodetected gate freezing-off times, together with the corresponding part weights, are shown in Fig. 10.7.

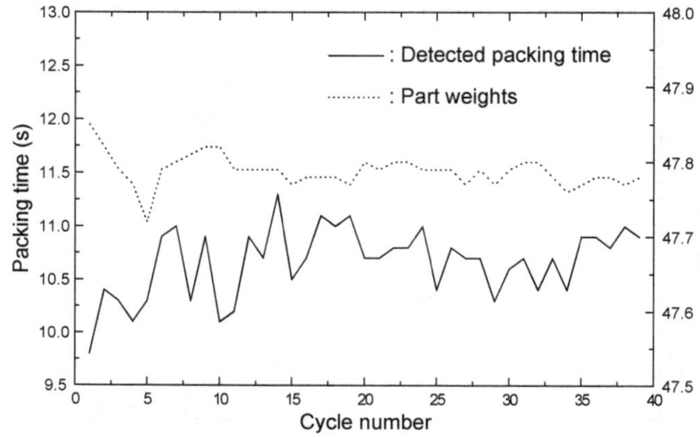


**Figure 10.7** Variations of packing time and part weights cycle to cycle

As the mold coolant temperature is controlled by a cooling system, the cycle variations are not large. At the very beginning of the molding, an increasing trend for packing time and a decreasing trend for part weight can be observed. The increase in the gate freezing-off time is due to fact that the gate becomes warmer from cycle to cycle, which in turn slows down the gate freezing; the decrease in the part weight is a result of the increasing mold temperature, which consequently reduces the part shrinkage. After several cycles, the part weight obviously becomes steady, with a variation less than 0.1%.

The packing times and part weights of the last 10 cycles of Fig. 10.7 are listed in Table 10.2. The detected packing time varies from 10.4 s to 11.1 s, but the part weights are very close, with a maximum relative variation less than 0.084%, indicating good performance of the developed system.

**Table 10.2** Online Detection Results of Different Cycles

| Cycle number | Detected packing time (s) | Part weights (g) |
|---|---|---|
| 30 | 10.6 | 47.79 |
| 31 | 10.7 | 47.80 |
| 32 | 10.4 | 47.80 |
| 33 | 10.7 | 47.78 |
| 34 | 10.4 | 47.76 |
| 35 | 10.9 | 47.77 |
| 36 | 10.9 | 47.78 |
| 37 | 10.8 | 47.78 |
| 38 | 11.1 | 47.77 |
| 39 | 10.9 | 47.78 |

**Table 10.3** Part Weights of Different Cycles with a Fixed Packing Time

| Cycle number | Part weights (g) |
|---|---|
| 41 | 47.78 |
| 42 | 47.78 |
| 43 | 47.77 |
| 44 | 47.76 |
| 45 | 47.77 |
| 46 | 47.79 |
| 47 | 47.81 |
| 48 | 47.79 |
| 49 | 47.77 |
| 50 | 47.77 |

In practice, the packing time is normally fixed after the machine becomes stable. The averaged packing time of these cycles, 10.7 s, is then used for further tests. Ten more cycles with this fixed packing time are conducted. Table 10.3 lists their part weights, among which the maximum relative variation increased slightly to 0.1%. This indicates that the packing time can be fixed at a value provided by the detection system after the machine reaches a steady state.

### Online Detection Results of Different Materials

The detection system is tested with both amorphous and crystalline materials. Besides HDPE, materials such as LDPE, PS, and PP are also tested to verify if the developed system can work for different polymers. The result comparison between the online detection system and the offline measurements is given in Table 10.4. The online and offline results are very close, even though the gate freezing-off times are very different for different types of materials. These demonstrate the effectiveness of the developed system for different materials.

**Table 10.4** Results of Gate Freezing-Off Points of Different Materials

| | LDPE | PS | PP |
|---|---|---|---|
| Online results (s) | 12.08 | 2.48 | 1.7 |
| Offline results (s) | 12.0 | 2.5 | 1.5 |

### 10.1.4 Tests of Varying Packing Pressure Profile Cases

The detection system is also designed for cases of varying pressure profiles, which may be required in certain cases. The box mold and material HDPE are used for the test. The system is tested with the closed-loop control cases with step-up, step-down, ramp-up, and ramp-down packing profiles, and the open-loop control case with fixed valve opening, as shown below.

#### 10.1.4.1 Online Detection Results of Step Pressure Profile

The effectiveness of the online detection system is first tested for the step-up change in packing pressure profiles, with the algorithm similar to that of the constant case. The packing is divided into several constant periods according to the step changes, and the detection system makes the decision following the constant case for each individual period.

Experiments are conducted with a step-up packing pressure, as shown in Fig. 10.8. The corresponding screw displacements of both filtered and raw data are shown in Fig. 10.9. It can be found that the displacement signals clearly level off before the end of packing. A comparison of the part weights with those of fixed packing times is given in Table 10.5, indicating a timely prediction of the gate freezing-off point.

Experiments are also conducted with step-down profiles. Again, the leveling off in the signal can be observed before the end of packing. A weight comparison with other fixed packing times is shown in Table 10.6, and the detected result is in good agreement with the gate freezing-off point detected by the offline method. A more detailed description can be found in reference [226].

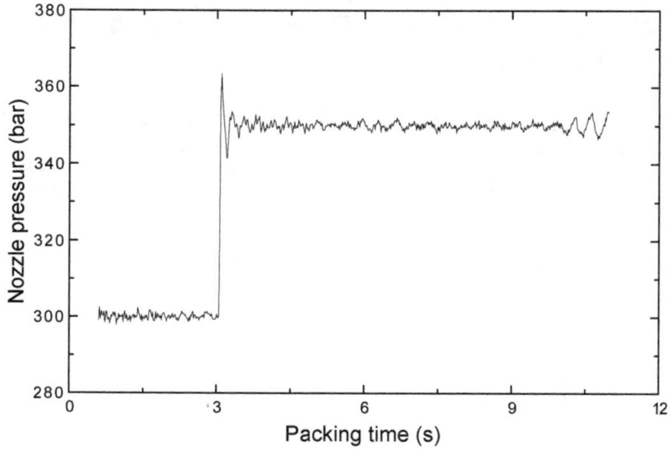


**Figure 10.8** Step-up nozzle pressure

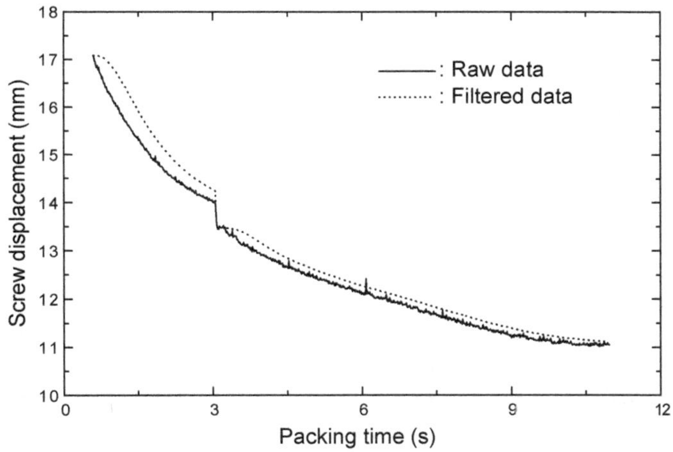


**Figure 10.9** Screw displacement with step-up pressure profile

**Table 10.5** Part Weight Comparison of Step-Up Packing Profile

| | Results with fixed packing time | | | | Results with detection system |
|---|---|---|---|---|---|
| Packing time (s) | 4 | 9 | 10 | 14 | 10.9 |
| Part weight (g) | 46.35 | 47.82 | 48.02 | 48.13 | 48.12 |

**Table 10.6** Part Weight Comparison of Step-Down Packing Profile

| | Results with fixed packing time | | | | | Results with detection system |
|---|---|---|---|---|---|---|
| Packing time (s) | 4 | 6 | 9 | 10 | 14 | 11.0 |
| Part weight (g) | 46.59 | 47.10 | 47.67 | 47.82 | 47.90 | 47.90 |

### 10.1.4.2 Online Detection Results of Ramp Pressure Profile

In this case, the packing pressure is closed-loop controlled with a ramp-up profile, starting from 300 bar and increasing at a rate of 15 bar/s. Following Rules 4, 5, and 6, the packing is stopped at a time of 12.2 s by the autodetection system. Figure 10.10 shows the recorded nozzle pressure during the packing-holding, and the corresponding screw displacement is shown in Fig. 10.11. Compared with the case of constant pressure, no obvious leveling off appears before the end of packing. The detection algorithm for the constant-pressure case based only on the screw displacement measurements is, therefore, not suitable for this case. By running the algorithm for the varying-pressure case, the relative pressure change and screw displacement change are calculated with Eqs. 10.19 and 10.20. The relationship between the two normalized variables is plotted in Fig. 10.12. According to Eq. 10.10, this relationship can be approximated by a series of "linear" functions with increasing intercept and decreasing slope by using least squares estimation. The plot of intercept vs.

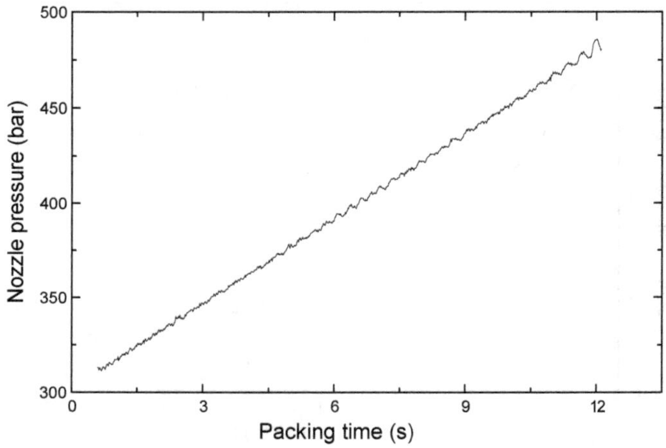


**Figure 10.10** Ramp-up nozzle pressure

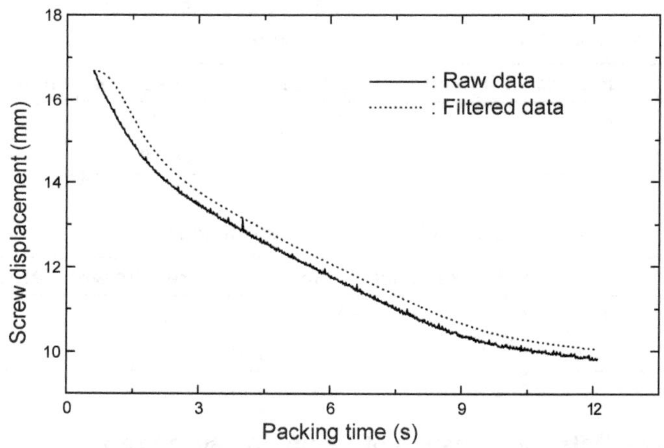


**Figure 10.11** Screw displacement with ramp-up pressure profile

packing time is shown in Fig. 10.13. With the increase in packing time, the intercept increases first and then levels off before the gate is frozen; this is in line with the process analysis.

The part weight comparison between the detection result and offline fixed packing time experiments are listed in Table 10.7. The results indicate that the detection result matches well with the frozen gate detected by the offline experiments.

**Table 10.7** Part Weight Comparison of Ramp-Up Packing Profile

| | Results with fixed packing time | | | | | | Results with detection system |
|---|---|---|---|---|---|---|---|
| Packing time (s) | 2 | 5 | 8 | 10 | 11 | 14 | 12.2 |
| Part weight (g) | 45.41 | 46.78 | 47.93 | 48.40 | 48.52 | 48.58 | 48.58 |

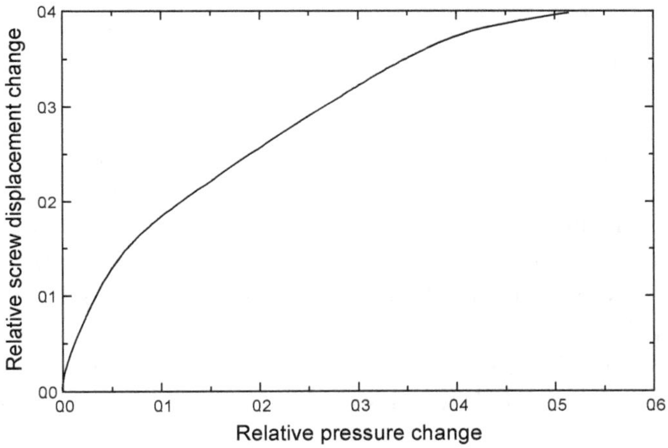


**Figure 10.12** Relative screw displacement change vs. relative pressure change of ramp-up profile case

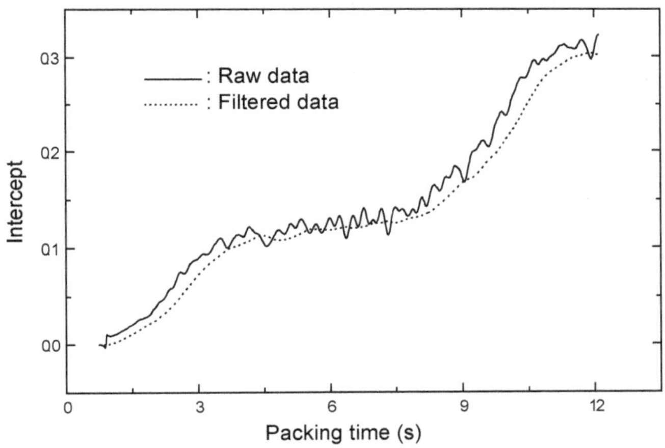


**Figure 10.13** Regressed intercept of ramp-up profile case

The detection system is also tested with a ramp-down pressure profile, starting from 400 bar and decreasing with a rate of 15 bar/s. The comparisons of the part weights are listed in Table 10.8, which again shows the good agreement between the autodetection result and the offline measurements. More detailed descriptions can be found in reference [226].

**Table 10.8** Part Weight Comparison of Ramp-Down Packing Profile

| | Results with fixed packing time | | | | | Results with detection system |
|---|---|---|---|---|---|---|
| Packing time (s) | 3 | 6 | 9 | 10 | 14 | 10.7 |
| Part weight (g) | 46.35 | 47.26 | 47.60 | 47.82 | 47.85 | 47.85 |

## 10.2 Influence of Packing Profile on Part Quality

The packing pressure and its profile throughout the packing-holding stage can influence significantly the final part quality in many aspects. With the accurate control of the packing pressure and other parameters, different types of packing profiles, including constant, ramp, and step-change profiles, have been experimentally studied by Chen and Gao [226, 233]. Four different mold inserts are used in their study. They are referred to as Mold Inserts 1, 2, 3, and 4, with geometry shown in Fig. 10.14. The first one is a rectangular slab, 150 mm × 100 mm × 2 mm, with a fan gate. The second insert has the same mold system as the first one, but with different insert geometry. Its cross-sectional area decreases along the flow path. The third one is also a rectangular slab but with a different mold system. As shown in Fig. 10.14(c), the dimension is reduced to 115 mm × 80 mm × 2 mm. Moreover, this mold has been fitted with four Kistler 6157BA pressure transducers for cavity pressure measurement, as marked by P1, P2, P3, and P4 in the figures.

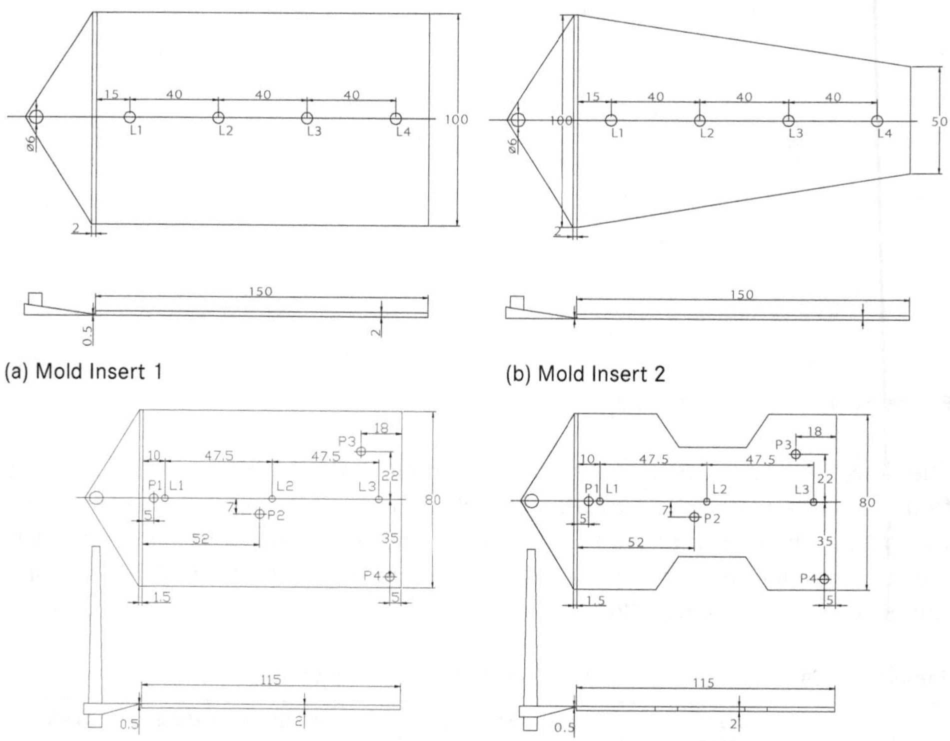


(a) Mold Insert 1

(b) Mold Insert 2

(c) Mold Insert 3

(d) Mold Insert 4

**Figure 10.14** Geometry of mold inserts for packing tests

The fourth insert has the same mold system as the third one, but with a different shape, as shown in the figure. Unless otherwise specified, the experimental results discussed in the following are from the first mold insert. Results of the other mold inserts are also presented for comparison and verification purposes.

Without the packing-holding stage, thermal shrinkage is normally around 10% for an amorphous thermoplastic and 20% for a semicrystalline material [207]. As a typical semicrystalline material, high-density polyethylene (HDPE) (Marlex, HMN6060) is selected for its large dimensional dependence on the packing pressure. In the experiments, the temperature of the first two mold inserts was controlled at 50°C by a mold cooling system. The melt temperature was set at 230°C, and the injection velocity was controlled at 20 mm/s. The experimental conditions of the third and fourth inserts were set differently because the mold system was different. The melt temperature for Molds 3 and 4 was set at 210°C and the injection velocity at 30 mm/s, without mold temperature control. The gate freezing-off time was tested as described before for all inserts via both the proposed online and offline methods. It was found that the two gating systems had close freezing-off time, of about 6 s for all conditions. To simplify the problem, 6 s in packing-holding was used for the profiling study.

The main purpose of packing is to compensate for the part shrinkage due to cooling and at the same time to avoid any part flash caused by overpack. The part shrinkage and flash, therefore, are important quality indexes in the profiling study. In addition to that, part uniformity represented by the thickness distribution is selected in the profiling study. The influence of packing profile on part weight is also studied. To reduce the experimental and measurement errors, five injection molded samples were collected for each molding condition after the machine reached a steady state. Data of the five samples were averaged. The part was weighed by an electronic scale with a resolution of 0.01 g, which is about 0.05% of the weight of the part studied. Shrinkage and evenness were associated with the part thickness, which was measured using a Mitutoyo micrometer with a resolution of 0.01 mm. For the third and fourth inserts, thickness measurement was conducted at three locations, near the gate (L1), close to the end of filling (L3), and a point in between (L2) along the flow path, as shown in Figs. 10.14(c) and 10.14(d). For the first and second inserts, besides the points near the gate and close to the end of filling, two points were selected in the middle flow path as the parts are longer and also to avoid the ejection pin mark in the middle of the part. Therefore, thickness was measured at four locations, L1, L2, L3, and L4, as shown in Figs. 10.14(a) and 10.14(b) for Mold Inserts 1 and 2. Part shrinkage was defined as the normalized difference between the cavity thickness and the mean part thickness, as shown in this equation:

$$\text{Shrinkage} = \frac{2 - \overline{d_i}}{2} \tag{10.27}$$

where $\overline{d_i}$ is the average of the measurements at different locations, and the nominal cavity thickness is 2 mm. The standard deviation of the thickness measurements at different locations was used to represent the part evenness, as shown in this equation:

$$\text{Evenness} = \sqrt{\frac{\sum_{i=1}^{n}\left(d_i - \overline{d_i}\right)^2}{n-1}} \tag{10.28}$$

where $d_i$ is the measured thickness at location $i$, and $n$ is the location number of measurements. Apparently, a smaller value stands for better part evenness. The part flash was simplified as a logic variable with a value of "significant," "slight," or "no" to represent the flash magnitude. This was made according to observations of the appearance of the molded parts by the operator.

### 10.2.1 Constant Packing Profile

Seven constant packing profiles, with each pressure chosen from the pressure level set $\Omega_p$ = {300 bar, 350 bar, 400 bar, 450 bar, 500 bar, 550 bar, 600 bar}, were experimentally studied using Mold Insert 1. An increase in the packing pressure results in the same trend in the part weight as shown in Fig. 10.15. A pressure increase from 300 bar to 600 bar results in the part weight change from 28.19 g to 30.51 g, a greater than 8% change. The tests with other mold inserts reveal a weight trend similar to the packing pressure increase.

A slight flash can be observed with parts molded at the 550 bar packing pressure, and significant flash is found with parts molded at the 600 bar packing pressure.

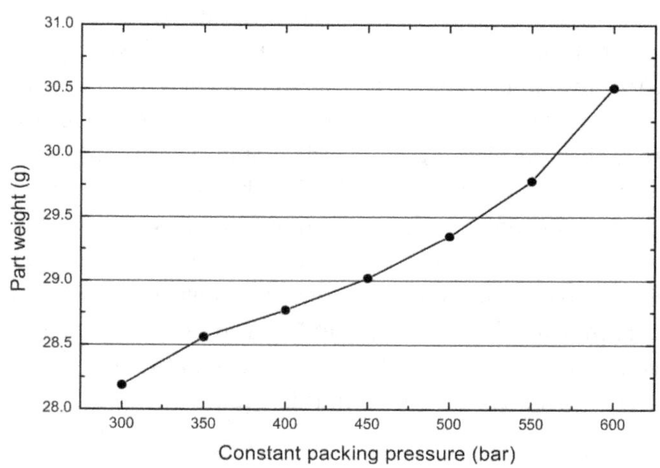


**Figure 10.15** Part weight of constant packing profiles using Mold Insert 1

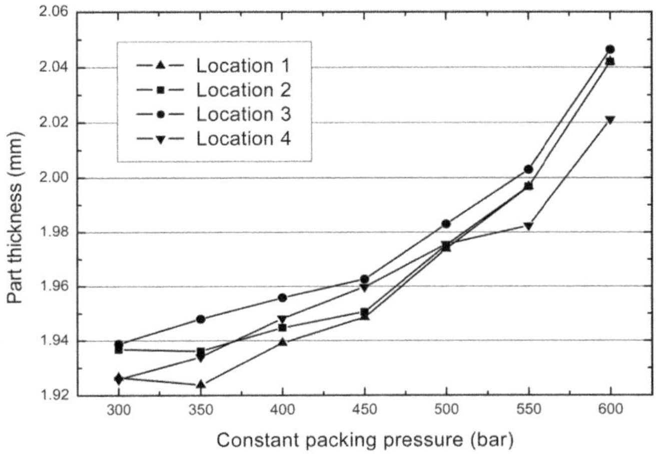


**Figure 10.16** Part thickness distribution of constant packing profiles using Mold Insert 1

This indicates that the packing pressure should be set less than 550 bar for this mold to eliminate flash. The experimental experience suggests that the maximum pressure without flash, which is mold and material dependent, should not exceed the pressure at the end of filling according to the experiences in the tests.

The relation between the packing pressure and the part thickness distribution is more complicated. As shown in Fig. 10.14(a), four thickness measurements are taken at Locations L1, L2, L3, and L4. Location 1 is near the gate, Location 4 is close to the end of filling, and Locations 2 and 3 are in between. Figure 10.16 shows the thickness variation with packing pressure at these four locations. Under all of the packing pressure levels, Location 1 is always thinner than Location 2, and Location 4 is thinner than Location 3, indicating that the middle part is thicker. In addition, when the packing pressure is low, Location 1 is the thinnest; an increase in the pressure results in the most rapid changes in thickness at Location 1. Location 4, on the other hand, is the thinnest point at high packing pressure when flash occurs. It implies that there is a link between flash and the part thickness distribution in which overpack tends to make the part near the gate thicker. This can be seen more clearly with a smaller flat mold, Mold Insert 3. With that mold, part flash appears when the pressure is above 400 bar. The corresponding thickness distribution is shown in Fig. 10.17. The thickness at the near-gate location is also found to be greatly dependent on the packing pressure level.

The part shrinkage calculated according to Eq. 10.27 is shown in Fig. 10.18 for Mold Insert 1. The shrinkage reduces with increases in the packing pressure, resulting in negative shrinkage at excessively high pressures above 550 bar. Experiments indicate that a negative shrinkage is always accompanied by flash, but the flash can occur with a positive shrinkage. In fact, the flash and the shrinkage may not

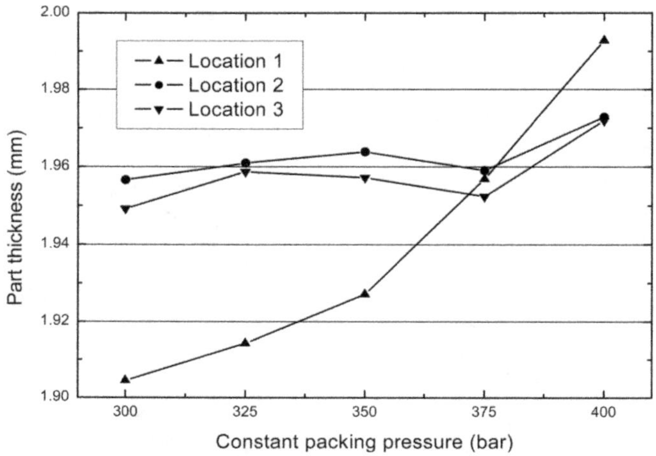


**Figure 10.17** Part thickness distribution of constant profiles using Mold Insert 3

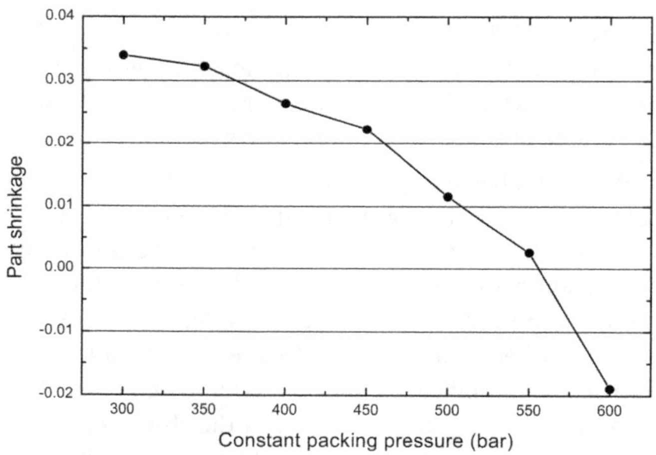


**Figure 10.18** Part shrinkage of constant packing profiles using Mold Insert 1

be eliminated simultaneously in many cases, where a tradeoff has to be made in finding a "proper" packing profile.

To study the part evenness, the standard deviation of the measured thickness at the selected locations is defined by Eq. 10.28. The relation of the part evenness and pressure level is illustrated in Fig. 10.19. An optimal pressure at around 500 bar can be clearly seen to result in the best evenness. In terms of shrinkage, pressure around 500 bar is also good. Besides, a 500 bar pressure setting is the highest pressure that does not result in flash in the experiments. Experimental results with the other mold inserts also prove that a high pressure setting just below the level that results in flash is good for molded parts in terms of evenness and shrinkage.

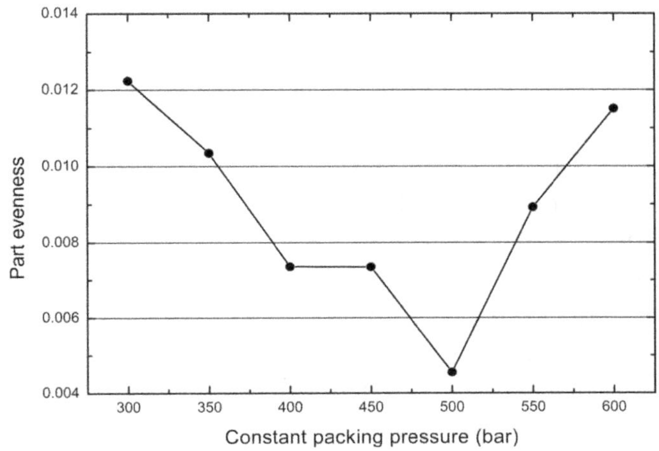


**Figure 10.19** Part evenness of constant packing profiles using Mold Insert 1

For constant packing pressure, based on the results obtained with all of the tested molds, the following statements can be made. When the packing pressure is low, the thickness near the gate is thin; increases in the packing pressure result in a rapid increase in the thickness near the gate, which consequently improves the part evenness, but when the pressure is excessively high, the part near the gate may be overpacked, resulting in poor evenness and possibly flash.

### 10.2.2 Ramp Packing Profile

Unlike a constant packing profile, two parameters, the start pressure, $P_s$, and the end pressure, $P_e$, are required for the determination of a ramp profile once the packing time is given. Hereafter, $(P_s, P_e)$ is used to represent a ramp profile with a start pressure, $P_s$, and an end pressure, $P_e$. When the start pressure is equal to the end one, it turns into a constant profile. The constant profile, therefore, can be treated as a special ramp profile and is also included in the following ramp profile study.

Chen and Gao [226, 233] studied a total of 49 different packing profiles, which are determined by the start and end packing pressures chosen from the pressure set $\Omega_p$ = {300 bar, 350 bar, 400 bar, 450 bar, 500 bar, 550 bar, 600 bar}. Among them, results of seven constant packing profiles with the same start and end pressures were presented in the preceding section. The remaining 42 settings are ramp profiles that can be divided into ramp-up and ramp-down types of profiles. A profile with the end pressure higher than the start one is referred to as a ramp-up, while a profile with the end pressure lower than the start pressure is referred to as a ramp-down. Among the 42 ramp profiles, the ramp-up profile and the ramp-down profile each make up a half. Influences of the ramp profiles on the molded part are analyzed in the following.

### Part Weight

The part weights under the 49 packing profiles are listed in the following table. The list along the diagonal line from the upper left to the lower right with gray shading gives the results under constant packing profiles. Data above this diagonal line are the weights of ramp-down profiles, and those below are data of ramp-up profiles. A contour plot of the part weights is also drawn in Fig. 10.20, in which points on the same color line represent equal weights.

The plot is divided into two halves by a diagonal line, points on which represent the constant packing profiles. Points above the diagonal line represent the ramp-up profiles, and those below are of ramp-down profiles. It can be seen that the part weight of the constant packing profiles increases with pressure, as mentioned previously.

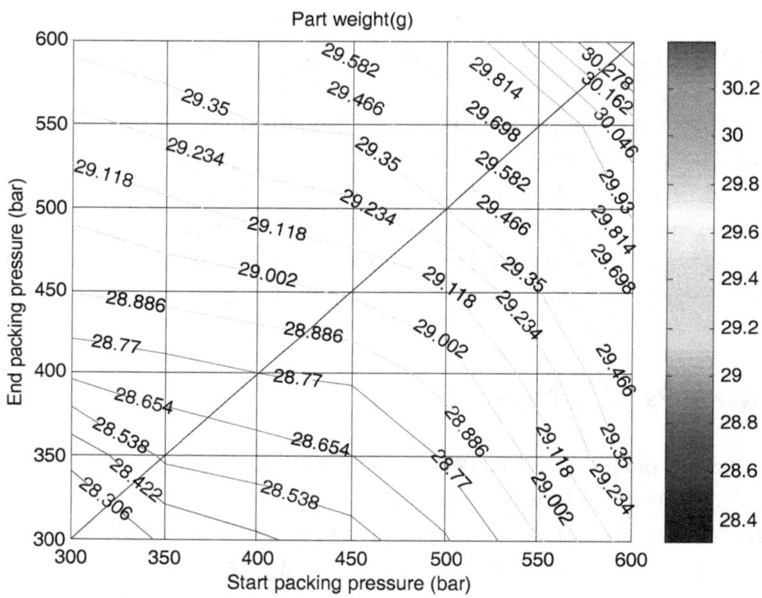


**Figure 10.20** Contour plot of part weight

**Table 10.9** Part Weight of Ramp Profiles Using Mold Insert 1

| Part weight (g) | | Start pressure (bar) | | | | | | |
|---|---|---|---|---|---|---|---|---|
| | | 300 | 350 | 400 | 450 | 500 | 550 | 600 |
| Ending pressure (bar) | 300 | 28.19 | 28.32 | 28.40 | 28.49 | 28.64 | 28.87 | 29.18 |
| | 350 | 28.33 | 28.56 | 28.60 | 28.65 | 28.78 | 29.04 | 29.39 |
| | 400 | 28.68 | 28.72 | 28.77 | 28.79 | 28.94 | 29.17 | 29.51 |
| | 450 | 28.89 | 28.93 | 28.96 | 29.02 | 29.09 | 29.34 | 29.72 |
| | 500 | 29.03 | 29.09 | 29.16 | 29.21 | 29.35 | 29.56 | 29.96 |
| | 550 | 29.20 | 29.27 | 29.35 | 29.37 | 29.60 | 29.78 | 30.12 |
| | 600 | 29.39 | 29.44 | 29.53 | 29.64 | 29.79 | 30.12 | 30.51 |

The part weight of the ramp profile, however, has a more complex relation with the packing pressure level. If either the start or the end pressure is fixed, the part weight is found to increase with the other parameter. Described in a mathematical form, it means that the partial derivative of part weight with respect to the start pressure ($P_s$) or end pressure ($P_e$) is always positive, as shown below:

$$\frac{\partial W}{\partial P_s} > 0 \tag{10.29}$$

$$\frac{\partial W}{\partial P_e} > 0 \tag{10.30}$$

However, if we simply use $\overline{P}$, the mean of $P_s$ and $P_e$, to represent the pressure level of a ramp packing profile, then $\frac{\partial W}{\partial \overline{P}}$, the partial derivative of part weight with respect to the pressure level, is found to be complex. Hereafter we use $W(P_s, P_e)$ to stand for the part weight molded under a ramp profile $(P_s, P_e)$. The data listed in Table 10.9 show that for a group of $n$ ramp-up packing profiles with the same mean pressure, $\overline{P}$, but different start and end pressures, $(P_{s_i}, P_{e_i})$, and with a slope decreasing with $i$ ($i = 1, 2, \cdots, n$), the corresponding part weight, $W(P_{s_i}, P_{e_i})$, is found with the following relation:

$$W(P_{s_1}, P_{e_1}) > W(P_{s_2}, P_{e_2}) > \cdots > W(P_{s_n}, P_{e_n}) > W(\overline{P}, \overline{P}) \tag{10.31}$$

where $W(\overline{P}, \overline{P})$ is the part weight under a constant packing pressure, $\overline{P}$. Illustrated in Fig. 10.20, the above relation means that in each grid of the upper left half of the figure (the ramp-up profile region), the weight of the upper left corner is always larger than that of the lower right one, even though they are with the same mean pressure. These results imply that, with the same pressure level, the ramp-up profile tends to increase the part weight greater than the constant profiles do; moreover, this trend becomes more obvious with the increase of the ramp slope.

Part weights under ramp-down profiles present an opposite trend. As shown in Table 10.9, $W(500\text{ bar}, 300\text{ bar})$ is smaller than $W(450\text{ bar}, 350\text{ bar})$, which is smaller than the weight of the constant 400 bar pressure setting. Most of the results show that with the same mean pressure, the ramp-down profile tends to produce lighter parts than does the constant packing profile; moreover, a larger decrease in slope results in lighter parts. An exception happens when the start pressure is very high, at which flash appears; under this case, more material is packed into the mold at the beginning packing stage due to the excessively high pressure. But under a normal condition, the ramp-down profile does show the inclination to reduce the part weight. This is more obvious in the tests using the other mold inserts, as shown in the following, in which a very strange phenomenon is observed such that the relation shown in Eq. 10.29 is invalid.

A series of ramp and constant profiles are tested using Mold Insert 2, and the corresponding part weights are shown in Fig. 10.21. There are three constant packing profiles, three ramp-up profiles, and three ramp-down profiles under the tests. The part weight, as shown in the figure, can be graded into three levels. Each level consists of the three types of packing profiles, and the part weight in each level does not change much. But the part weights in different levels vary significantly. Additionally in the figure, the ramp-up profiles, with consistently lower pressure than the constant profiles and ramp-down profiles, result in close part weights, but the ramp-down packing profiles, though with consistently larger pressure than the constant profile in the same level, result in smaller weights. This is contradictory to the relations presented in Eqs. 10.29 and 10.30. The results further verify that the ramp-up profile tends to raise the part weight, but the ramp-down profile tends to reduce it.

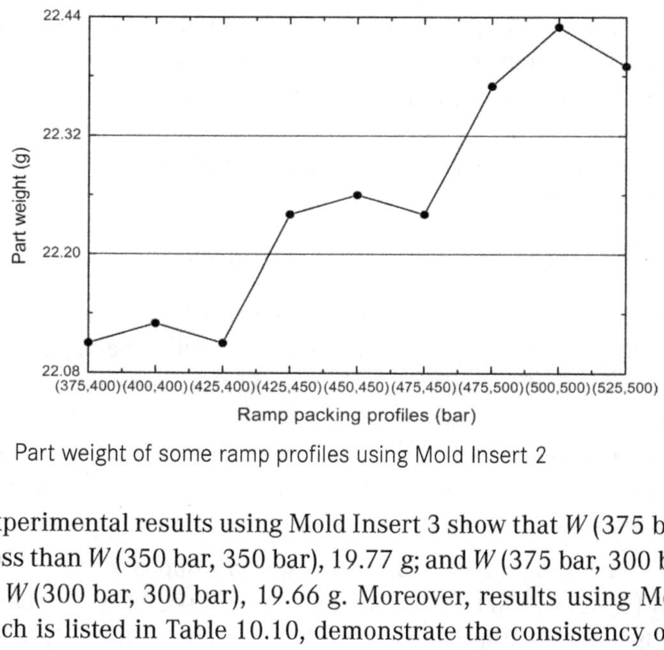


**Figure 10.21** Part weight of some ramp profiles using Mold Insert 2

Similarly, experimental results using Mold Insert 3 show that *W* (375 bar, 350 bar), 19.60 g, is less than *W* (350 bar, 350 bar), 19.77 g; and *W* (375 bar, 300 bar), 19.40 g, is less than *W* (300 bar, 300 bar), 19.66 g. Moreover, results using Mold Insert 4, some of which is listed in Table 10.10, demonstrate the consistency of the weight phenomenon with the ramp-down profile.

In conclusion, the part weight under ramp packing profiles is not a monotonous function of packing pressure level. The ramp-up profiles tend to increase the part weight, whereas the ramp-down profiles tend to decrease it. A possible explanation is that the part weight is determined by the difference between nozzle pressure and cavity pressure, rather than the nozzle pressure alone. The nature of polymer solidification in a mold cavity may result in a larger pressure difference between the nozzle and cavity pressures with a ramp-up nozzle pressure profile, and a smaller pressure difference with a ramp-down profile.

**Table 10.10** Part Weight Comparison Using Mold Insert 4

| Profile type | Start pressure (bar) | End pressure (bar) | Part weight (g) |
|---|---|---|---|
| Ramp-down | 500 | 450 | 15.68 |
| Constant | 450 | 450 | 15.76 |
| Ramp-down | 500 | 400 | 15.45 |
| Constant | 400 | 400 | 15.67 |
| Ramp-down | 500 | 350 | 15.31 |
| Constant | 350 | 350 | 15.58 |
| Ramp-down | 500 | 300 | 15.24 |
| Constant | 300 | 300 | 15.51 |
| Ramp-down | 500 | 250 | 15.18 |
| Constant | 250 | 250 | 15.37 |

### Part Flash

Part flash appears when the start pressure is excessively high, for instance, 550 bar for Mold Insert 1. When the packing pressure is not greater than 500 bar, part flash does not appear, even with the end packing pressure set as high as 600 bar. The influence of the end pressure on the part flash is found to be limited. It affects only the degree of flash when flash has been already caused by a high start pressure. The experimental results have shown that the part flash is mainly determined by the start packing pressure. This is because at the start of the packing stage, the polymer in the mold cavity has not solidified. An excessively large packing pressure can result in a high cavity pressure, which makes the mold open slightly; as a result, flash happens. Toward the end of the packing stage, a high packing pressure cannot effectively transfer the fresh and hot, and consequently less viscous, material toward the part line because the melt front has been effectively frozen. The large end pressure, therefore, is less prone to result in part flash.

### Part Shrinkage

Part shrinkage of the 49 packing profiles is calculated using Eq. 10.27. A contour plot of the part shrinkage is illustrated in Fig. 10.22.

Basically, the part shrinkage is reduced with the increase in packing pressure. The contour lines in a high-pressure region (approaching the upper right corner of the figure) are obviously denser than those in a low-pressure region (approaching the lower left corner of the figure). Besides, negative shrinkage appears in the high-pressure region; this indicates that the part molded at the pressure setting can result in a dimension larger than the mold dimension. The dotted diagonal line in the figure represents the constant packing profiles. Observation of the grids along the diagonal line shows that the part shrinkage of the ramp-up profile (represented

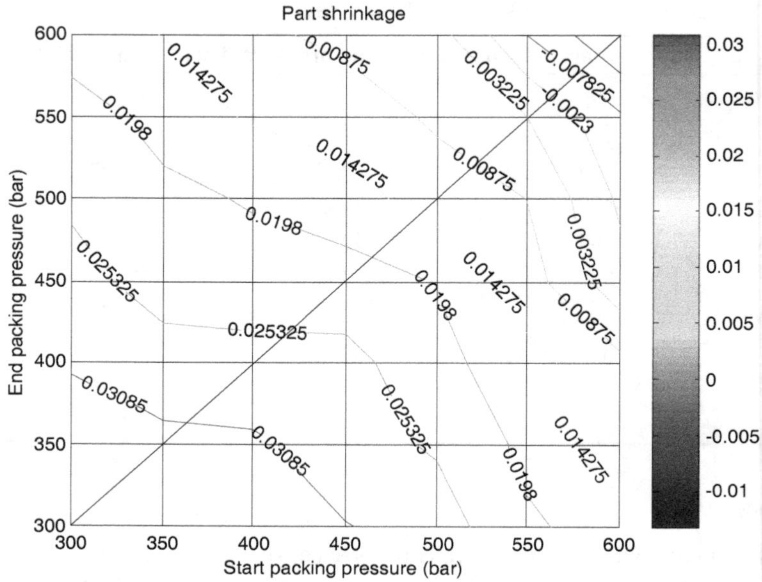


**Figure 10.22** Contour plot of part shrinkage using Mold Insert 1

by the upper-left cross point) is always lower than that of the ramp-down packing profile (represented by the lower-right cross point), though they have the same mean packing pressure. Similar to the results of part weight, this indicates that the ramp-up packing profile tends to reduce the part shrinkage, while the ramp-down profile tends to increase it.

### Part Thickness Distribution and Evenness

The influence of the ramp packing profiles on the part thickness distribution is also found to relate to the packing profile type, as well as the packing pressure level. The part evenness calculated based on Eq. 10.28 varies from 0.004 to 0.036 for different ramp profiles. The experimental results show that the part thickness distribution is influenced by the ramp type and the pressure level. The ramp-up profile tends to make the location near the gate much thinner than the others, unless the pressure is excessively high enough to result in an overpack. The ramp-down profile, on the contrary, tends to make the location near the gate thicker than the others. The detailed results can be referred to in [226]. Figure 10.23 is a plot of the part thickness distribution of a group of ramp-up profiles with the same end packing pressure, 550 bar, using Mold Insert 2. The near-gate location is obviously thinner compared with the other points. The result of a group of ramp-down profiles with the same start packing pressure, 550 bar, however, presents a different phenomenon, as shown in Fig. 10.24. The near-gate location becomes the thickest point, except for the last constant packing profile case.

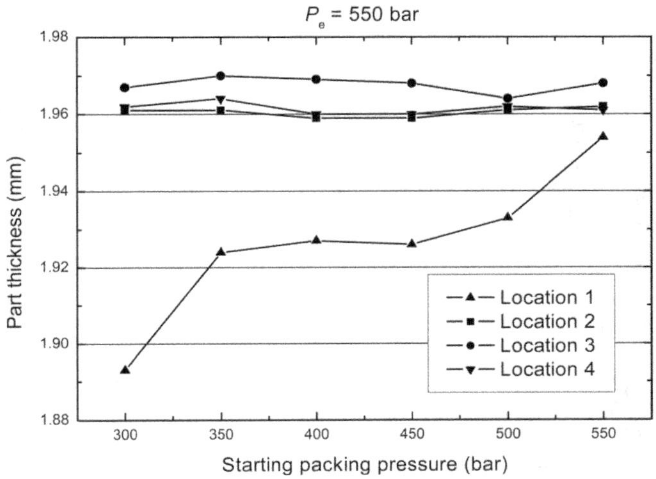


**Figure 10.23** Part thickness distribution of ramp-up profiles with the same 550 bar end packing pressure using Mold Insert 2

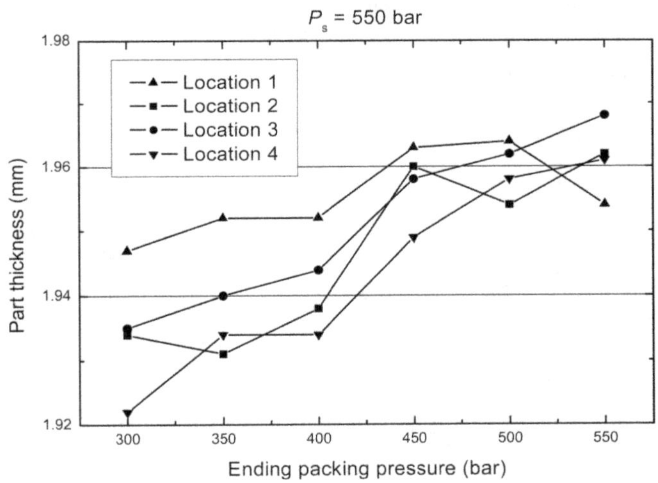


**Figure 10.24** Part thickness distribution of ramp-down profiles with the same 550 bar start packing pressure using Mold Insert 2

To have an overall view of the ramp profile influence on the part evenness, a contour plot of the part evenness of the 49 ramp profiles is illustrated in Fig. 10.25. Generally, the ramp-up profiles, shown in the upper left half of the figure, do not bring about good part evenness. In addition, a larger slope increase results in worse evenness. This can also be observed by experiments using the other mold inserts, as seen in the contour plot of Fig. 10.26, which consists of the results of 36 ramp profiles tested in Mold Insert 2. The part evenness of the ramp-down profiles, on the contrary, is normally much better than that of the ramp-up profiles.

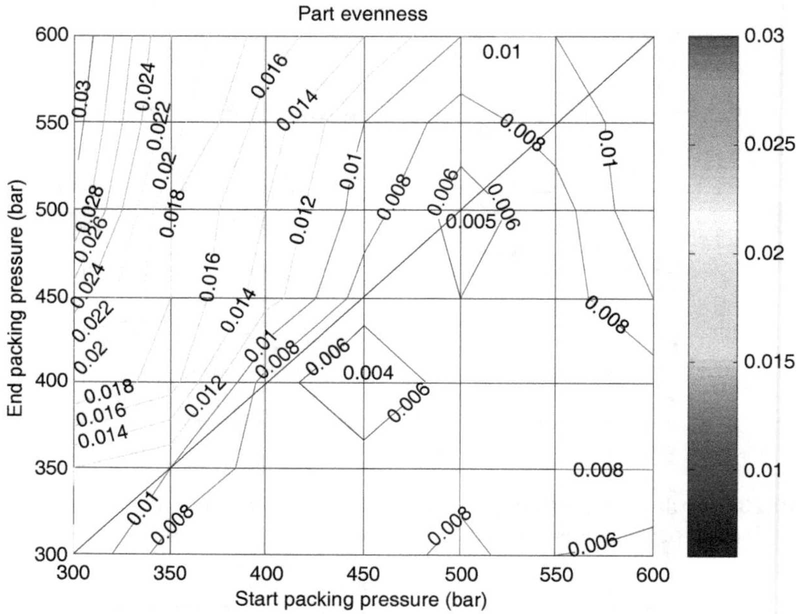


**Figure 10.25** Contour plot of part evenness using Mold Insert 1

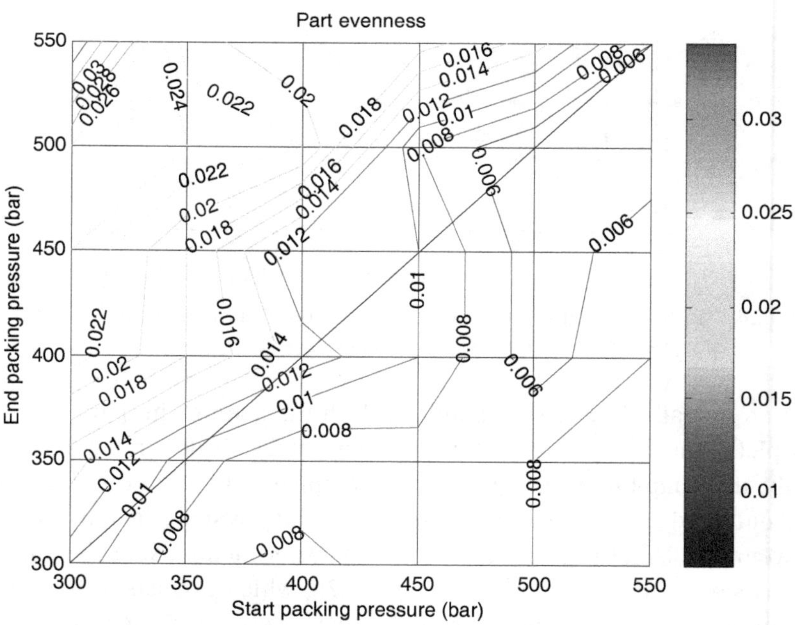


**Figure 10.26** Contour plot of part evenness using Mold Insert 2

Two local minimum points are observed in Fig. 10.25, one with the constant 500 bar setting, and the other with the ramp-down profile (450 bar, 400 bar). The evenness of the ramp-down profile (450 bar, 400 bar) is the best among all the 49 profiles tested in Mold Insert 1. The result of Mold Insert 2 shown in Fig. 10.26 also indicates that the ramp-down profiles result in much better evenness than do the increasing profiles. But the best evenness happens with a constant packing profile, 500 bar. This also indicates that, though most the ramp-down profiles are with good part evenness, a suitable constant packing pressure setting may also bring about good part evenness.

### 10.2.3 Step-Change Packing Profile

The step-change packing profiles are also studied [226, 233]. A group of step-change packing profiles, including both the step-up and step-down types, are experimented with Mold Insert 2. Figure 10.27 shows three step-up profiles with the packing pressure varying from 400 bar to 500 bar. There are one, two, and three step changes, respectively, in the packing profiles in correspondence with labels "incsp1," "incsp2," and "incsp3." The three step-down profiles, labeled as "decsp1," "decsp2," and "decsp3," in correspondence with one, two, and three step changes, respectively, are shown in Fig. 10.28. Since the ramp profile can be viewed as a step-change profile with an infinite number of step changes, two ramp profiles, (400 bar, 500 bar) and (500 bar, 400 bar), are also included for comparison. In addition, the result of a constant 450 bar profile, labeled as "c450," is also compared.

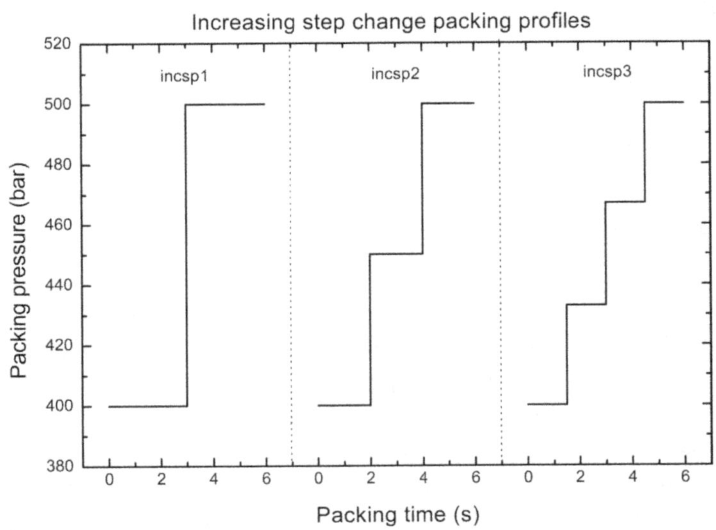


**Figure 10.27** Step-up packing profiles for Mold Insert 2

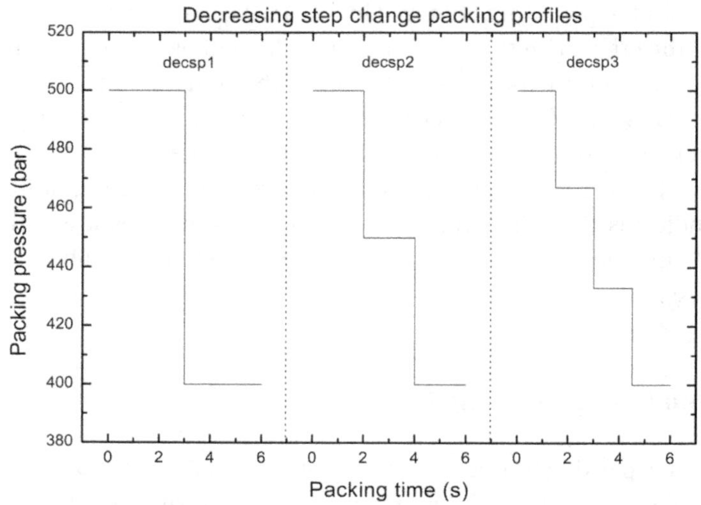


**Figure 10.28** Step-down packing profiles for Mold Insert 2

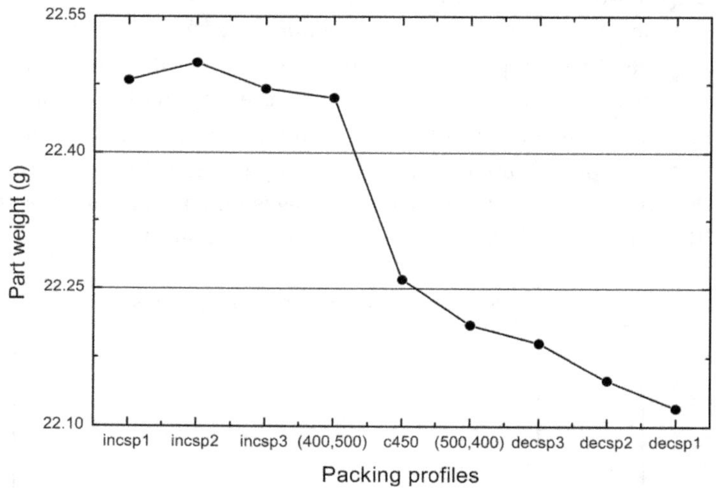


**Figure 10.29** Part weight of step-change profiles using Mold Insert 2

Intuitively, the material packed into a mold cavity during the first half of the packing stage is more than that during the second half because the gate is gradually freezing. Thus the three step-down packing profiles will result in heavier parts than will the step-up profiles because the packing pressure is larger during the first half of the packing stage. But contrary to intuition, the part weights of the step-up profiles are obviously heavier than those of the decreasing type, as shown in Fig. 10.29. This agrees with the result of the ramp profiles in that the increasing profile tends to result in a heavier part, and the decreasing profile tends to reduce the weight.

In addition, the part weights of the constant and ramp profiles are in between. The weight variation among the step-up profiles is small, and the weights of the step-up profiles are slightly larger than that of the ramp-up profiles. As the step number increases, the weight of the step-down profiles is increasing and approaching the weight of the ramp-down profile, which may be viewed as a special type of step-change profile with an infinite number of step changes.

The corresponding part thickness distribution is illustrated in Fig. 10.30. Under the step-up profile case, the thickness of the near-gate location, Location 1, is obviously smaller than the others. When the start pressure is raised to 500 bar, Location 1 gradually becomes the thickest, with the exception of "decsp3," in which backflow may result due to the sharp step change, as seen later.

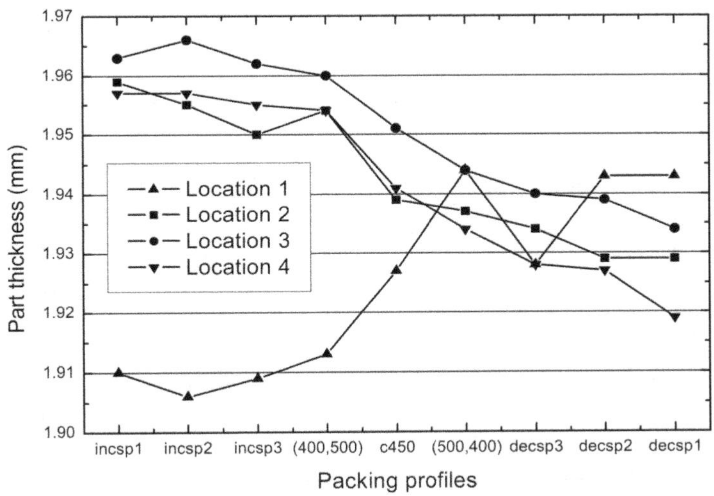


**Figure 10.30** Part thickness of step-change profiles using Mold Insert 2

Figure 10.31 presents a comparison of the part evenness among those profiles. The step-up profiles, like the ramp-up profile, cannot bring about good evenness for the molded parts. The part evenness of the step-down profiles is much improved compared with that of the step-up type. Furthermore, an increase in the number of step changes in a step-down profile improves the part evenness. A similar phenomenon can be found in tests using Mold Insert 3, as shown in Fig. 10.32. Three step-down profiles have the same start pressure, 375 bar, and the same end pressure, 200 bar. But the step-change number varies from 1 to 3, labeled "decsp1," "decsp2," and "decsp3," respectively. A ramp-down profile (375 bar, 200 bar) is also included in the figure for comparison. Though the part evenness has been improved with the increase of step changes, the step-down packing profiles do not show any advantages over the ramp profiles.

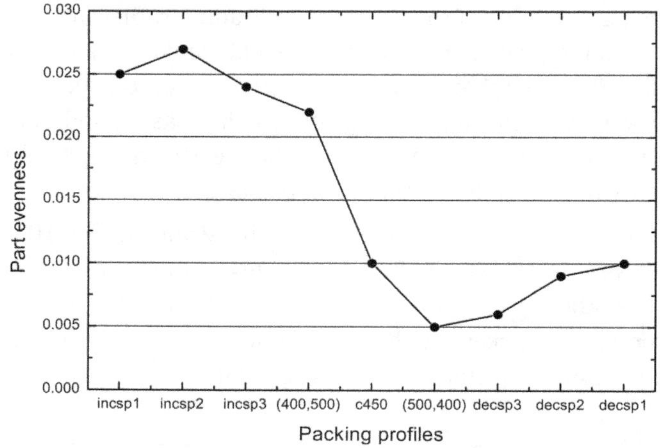


**Figure 10.31** Part evenness of step-change profiles using Mold Insert 2

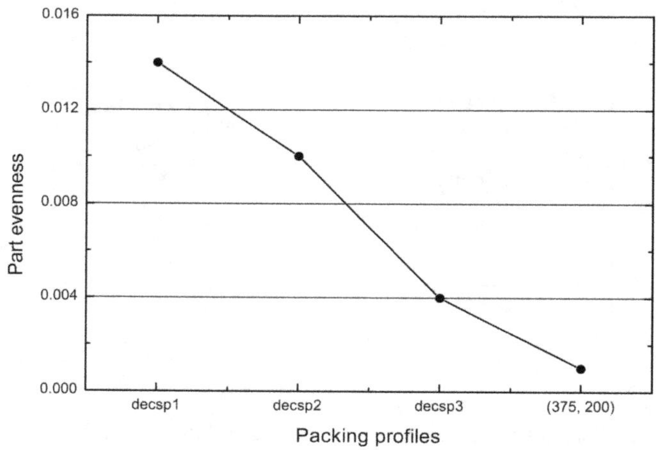


**Figure 10.32** Part evenness of step-change profiles using Mold Insert 3

According to the experimental results, the step-change profile is not recommended for the packing stage, as it may be associated with backflow. This can be revealed by the tests using Mold Insert 3, which has been fitted with four cavity pressure transducers. The recorded cavity pressure data under a step-down packing profile are shown in Fig. 10.33. It takes time for the cavity pressure to respond to the sudden nozzle pressure change. Therefore, there is a period after the step change during which the cavity pressure is larger than the nozzle pressure, as shown in the figure. A similar phenomenon happens also with the two and three step change profiles. This sudden change in the pressure profile may result in backflow, which will influence the final part quality. The ramp packing profile, however, may avoid the problem as shown in Fig. 10.34 because the sudden change is avoided.

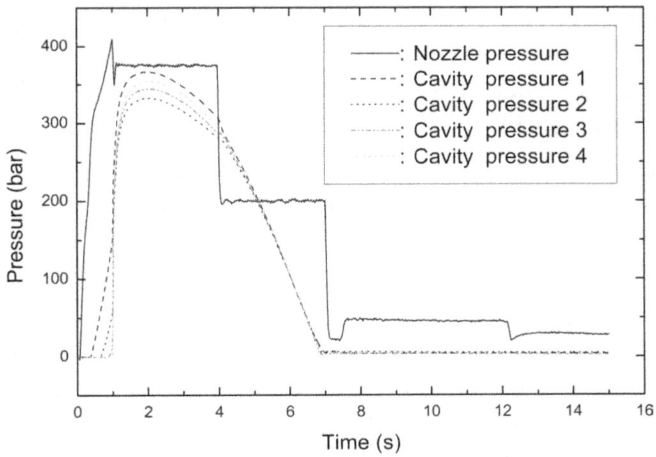


**Figure 10.33** Nozzle and cavity pressures under step-change profile using Mold Insert 3

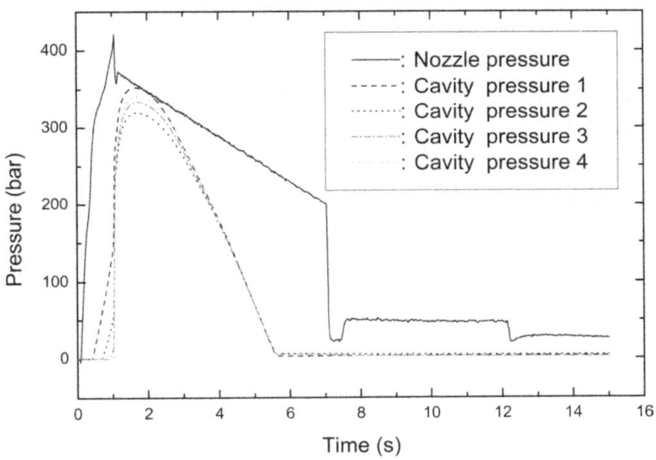


**Figure 10.34** Nozzle and cavity pressures under ramp profile using Mold Insert 3

### 10.2.4 Summary

The part weight is found to be a nonmonotonous function of pressure level. It is greatly affected by the profile type. The increasing packing profiles, of both the ramp and step-change types, tend to increase the part weight more than the constant packing profiles with the same mean pressure. The decreasing packing profiles, on the other hand, tend to reduce the weight.

The part flash is mainly caused by the overpack at the start of the packing stage. Therefore, the start packing pressure should not be set excessively high, to avoid flash.

The part shrinkage has a similar relation with the packing profile type as the part weight. The decreasing type of packing profiles tends to result in large shrinkage. The shrinkage can be eliminated by increasing the pressure level. Overpack may bring about negative shrinkage.

Part flash and part shrinkage may in conflict. They may not be eliminated simultaneously. The flash is found to have happened even with a positive shrinkage. Therefore, an optimal profile is based on demand.

The part evenness, associated with the part thickness distribution, is greatly dependent on the type of the packing profiles as well as the pressure level. When the packing pressure is low, the near-gate location is found to be the thinnest. With the increase of the packing pressure, the near-gate location becomes thicker and gradually becomes the thickest when overpack happens. Both the ramp-up and step-up profiles normally make the near-gate location obviously thinner than locations in the middle and far from the gate. The part evenness, as a consequence, is not good. The ramp-down and step-down packing profiles can normally bring about good evenness. The step-down profile does not show any advantages in improving part quality compared with the ramp-down profile. Moreover, backflow may result, due to the step change in packing pressure. Thus, the step-down profiles suggested by the researchers in [234] are not recommended according to this study. The part evenness of the constant packing profile depends greatly on the pressure level. It is found that there is always an optimal point or a region among the constant-pressure settings that can result in good evenness and reasonable shrinkage.

## 10.3 Profiling of Packing Pressure

### 10.3.1 Profiling Rules

The “optimal” packing profile depends on the requirements of part quality. The influence of packing profiles on part weight, flash, shrinkage, thickness distribution, and evenness have been presented in the preceding section. Different types of packing profiles, including constant, ramp, and step change, are compared, and the following rules can be obtained from the experimental studies:

- Ramp-up profiles tend to make the near-gate location thinner, unless overpack happens.
- Ramp-down profiles tend to make the near-gate location thicker.
- Increasing profiles are not good for part evenness, and a larger increasing rate results in worse part evenness.

- Decreasing profiles tend to result in even parts.
- Decreasing profiles may bring about large part shrinkage.
- The step change may affect part quality.
- In terms of good evenness and small shrinkage, there exists an optimal constant-pressure profile. When the pressure is set lower than this optimal pressure level, the thickness of the near-gate location is thinner than the other locations; when the pressure is higher than it, the near-gate location becomes thicker, and the location far from the gate becomes thinner. In many cases, this optimal point is found to be close to the maximal pressure level that does not result in any flash.

Considering the above factors as a whole, the constant type of packing profile may be the "optimal" one in many cases, and the optimal pressure level can be found by limited runs of experiments. For cases when a specific part thickness distribution is required, ramp profiles are suggested.

### 10.3.2 Online Profiling of Constant Packing Pressure

As concluded in the previous section, the constant packing profile is suitable to most cases. In terms of good evenness and small shrinkage, an optimal constant-pressure profile exists, and this optimal point is close to the maximal pressure level that does not result in any flash. As analyzed in the preceding section, the near-gate location is the thinnest if the packing pressure is set at a low level; it becomes thicker with the increase of the packing pressure, and the thickness increment is larger than the other locations. When the pressure setting is increased to be larger than the optimal point, the near-gate location becomes thicker, and the location far from the gate becomes thinner. Considering the pressure influence on the thickness distribution, an index related to the thickness distribution, $R_{\text{thk}}$, defined below, is introduced:

$$R_{\text{thk}} = \frac{thk_{\text{near}} - thk_{\text{mid}}}{thk_{\text{mid}} - thk_{\text{far}}} \tag{10.32}$$

where $thk_{\text{near}}$, $thk_{\text{mid}}$, and $thk_{\text{far}}$ represent the thickness of locations near the gate, in the middle of the part, and far from the gate, respectively. For measurements at four locations, the following definition is used:

$$R_{\text{thk}} = \frac{thk_{\text{near}} - thk_{\text{mid1}}}{thk_{\text{mid2}} - thk_{\text{far}}} \tag{10.33}$$

where $thk_{\text{mid1}}$ and $thk_{\text{mid2}}$ are the measurements of the two middle locations, and the former is relatively near the gate.

When the pressure is increased from a lower level toward the optimal setting, the near-gate location becomes thicker than the location far from the gate, and its thickness is closer to that of the middle part. Correspondingly, $R_{thk}$ increases from a negative and approaches zero; when it is around the optimal setting, $R_{thk}$ should be in the range of $(\delta, 0)$, where $\delta$ is a negative number in the range $(-1, 0)$, and it is set at $-1/2$ in the following. If the pressure is further increased, it may become even larger and turn into a positive number, which may indicate flash. A profiling method based on the change of the thickness distribution is proposed to seek the optimal constant-pressure setting, as shown in the flowchart below.

In the procedure, an initial setting, $P_0$, is first given as the packing pressure for the first cycle, and $\Delta P_0$ is given as the pressure adjustment for the next cycle. The nozzle pressure at the end of filling, $P_{fe}$, can be used as a reference for the initial pressure setting. This ending pressure is mold-dependent. But normally the packing pressure should be set less than $P_{fe}$; otherwise flash may occur. At the same time, flash is also associated with the clamping force, which depends on the machine setting. Another reference pressure, $P_{ref}$, to ensure flash-free production based on experience, is also used for the initial setting. The initial pressure is, therefore, set as the minimum of $\alpha \cdot P_{fe}$ and $P_{ref}$, where $\alpha$ is a factor less than 1 to ensure flash-free production. An experimental cycle is conducted with the first pressure setting, $P_0$. Then a judgment is made based on the measurements of the part thickness. If the thickness distribution index, $R_{thk}$, is not in the range of $(-1/2, 0)$, indicating the pressure setting is still away from the optimal point, the pressure needs to be updated. A factor, $\beta$, with varying value, is used in updating the pressure. The updated magnitude and direction are also dependent on the thickness distribution index, $R_{thk}$. If it is greater than zero, indicating the near-gate location becomes the thickest and flash may occur, the pressure is decreased with half of the previous magnitude; otherwise, it is increased. When $R_{thk}$ is less than $-1/2$, indicating the pressure has not approached the optimal setting, the pressure is kept increasing with the same magnitude; when it is greater than $-1/2$, implying that the current pressure setting is close to the optimal level, the increasing pressure step is reduced half to avoid flash. The program stops when the pressure setting converges, evaluated by a preset threshold, $\varepsilon$.

According to experience with the machine, a packing pressure under 400 bar is safe for flash-free molding. Therefore, $P_{ref}$ is set at 400 bar. Choosing $\Delta P_0 = 50$ bar, $\alpha = 75\%$, $\varepsilon = 5\%$ and first taking Mold Insert 1, for example, it is found that the nozzle pressure at the end of filling, $P_{fe}$, is about 790 bar. Then initialize $P_0 = \min(400, 790 \times 0.75) = 400$ bar. With this initial pressure setting, a program to seek the "optimal" pressure level starts with the results shown in Table 10.11. The part shrinkage and evenness calculated according to Eqs. 10.27 and 10.28, respectively, are also listed in the table. The final result, 525 bar packing pressure, is a proper setting with good evenness, small shrinkage, and without flash.

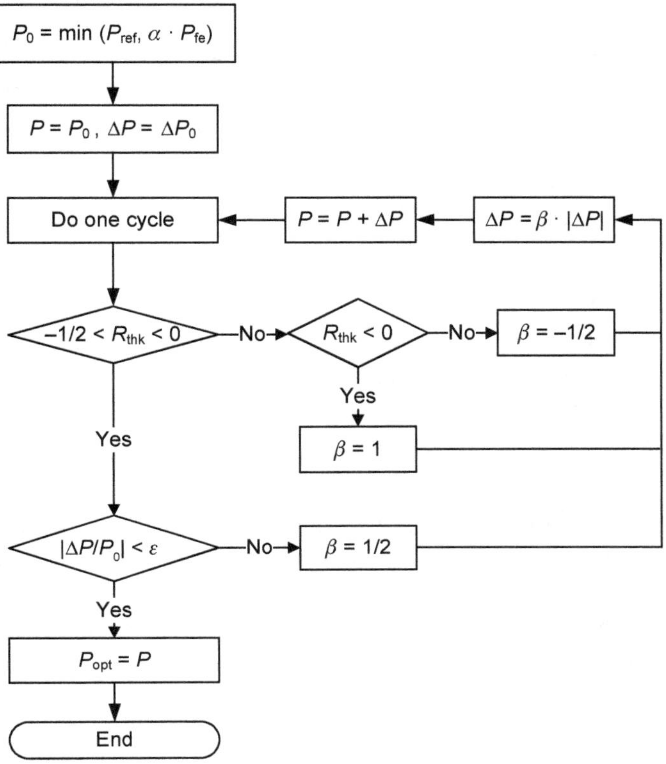


**Figure 10.35** Flowchart of optimal constant packing profiling

**Table 10.11** Profiling Procedure of Mold Insert 1

| Cycle No. | 1 | 2 | 3 | 4 |
|---|---|---|---|---|
| $P$ (bar) | 400 | 450 | 500 | 525 |
| $thk_{near}$ | 1.939 | 1.949 | 1.974 | 1.983 |
| $thk_{mid1}$ | 1.945 | 1.951 | 1.975 | 1.984 |
| $thk_{mid2}$ | 1.956 | 1.963 | 1.983 | 1.990 |
| $thk_{far}$ | 1.948 | 1.960 | 1.976 | 1.977 |
| $R_{thk}$ | -0.75 | 0.67 | -0.14 | -0.08 |
| $\beta$ | 1 | 1 | 1/2 | 1/2 |
| $\Delta P$ (bar) | 50 | 50 | 25 | 12.5 |
| $\Delta P/P_0 < \varepsilon$? | No | No | No | Yes |
| Shrinkage | 0.027 | 0.022 | 0.012 | 0.008 |
| Evenness | 0.007 | 0.007 | 0.005 | 0.005 |
| Flash | No | No | No | No |

The pressure setting for Mold Insert 3 is also experimentally verified. The nozzle pressure at the end of filling is about 450 bar for this mold. The initial pressure is

**Table 10.12** Profiling Procedure of Mold Insert 3

| Cycle No. | 1 | 2 | 3 | 4 |
|---|---|---|---|---|
| $P$ (bar) | 300 | 350 | 400 | 375 |
| $thk_{near}$ | 1.905 | 1.927 | 1.9928 | 1.957 |
| $thk_{mid}$ | 1.957 | 1.964 | 1.9729 | 1.959 |
| $thk_{far}$ | 1.949 | 1.957 | 1.9719 | 1.952 |
| $R_{thk}$ | -6.5 | -5.29 | 19.9 | -0.29 |
| $\beta$ | 1 | 1 | -1/2 | 1/2 |
| $\Delta P$ (bar) | 50 | 50 | -25 | 12.5 |
| $\Delta P/P_0 < \varepsilon$? | No | No | No | Yes |
| Shrinkage | 0.032 | 0.025 | 0.010 | 0.022 |
| Evenness | 0.028 | 0.020 | 0.012 | 0.004 |
| Flash | No | No | Slight | No |

thus set at $P_0 = \min(400, 450 \times 0.75) = 300$ bar. Starting from this initial pressure setting, the program is running with the results shown in Table 10.12. After four runs, the pressure converges to 375 bar, with evenness as low as 0.004 and shrinkage 0.022. The tests show that the increase in pressure can further reduce the shrinkage, but this deteriorates part evenness and brings about flash.

### 10.3.3 Ramp Profile for Specific Thickness Distribution

As stated previously, the thickness distribution can be changed via profiling the packing pressure. For a required thickness distribution, which may not be necessarily a uniform thickness for certain cases, it may be implemented by tuning the ramp packing profile.

As we know from experience, with a constant packing profile, the middle part is always thicker than the locations near the gate and far from the gate, unless flash occurs. If a part is required to be thicker in the near-gate location, the constant packing profile is not effective; a ramp-down profile, however, may be effective to realize it. Taking Mold Insert 1 as an example, if the part is required to be thick at the near-gate location, the following steps can be executed:

1. First determine a high constant-pressure level that does not bring about flash. This can be obtained following the procedure stated in the preceding section with relatively loose termination conditions. For example, 500 bar packing pressure is a satisfied level for Mold Insert 1. At this constant pressure, the thickness measurements at the four locations along the melt flow path are 1.974 mm, 1.975 mm, 1.983 mm, and 1.976 mm, with the middle locations thicker than the start and end locations.

2. Next keep the start packing pressure at 500 bar, and reduce the end pressure with a fixed step, 50 bar, for example, in this case. An experiment with a ramp-down profile (500 bar, 450 bar) is conducted next. The thickness measurements are found to be 1.964 mm, 1.953 mm, 1.964 mm, and 1.963 mm. The near-gate location has been one of the thickest.
3. Then further reduce the end packing pressure to 400 bar. An experiment is conducted with the ramp-down profile (500 bar, 400 bar). The part thickness is found to be 1.967 mm, 1.950 mm, 1.957 mm, and 1.953 mm, in which the near-gate location is obviously thicker than the other points.

Similarly, we can use a ramp profile to mold the part with the near-gate location significantly thinner than the other locations. As shown in Fig. 10.15 of the constant packing pressure results, when the packing pressure is low, the near-gate location can be the thinnest, but the difference is not significant. Moreover, large shrinkage also results. A ramp-up packing profile can achieve such a goal with less shrinkage, as shown by the following steps using Mold Insert 1 as an example.

1. First also set the packing pressure constantly at 500 bar. The thickness measurements at the four locations along the melt flow path are 1.974 mm, 1.975 mm, 1.983 mm, and 1.976 mm. The thickness difference is small.
2. Next increase the end packing pressure to 550 bar, and the experiment is conducted with the ramp-up packing profile (500 bar, 550 bar). The thickness measurements become 1.982 mm, 1.986 mm, 1.992 mm, and 1.977 mm. The thickness difference between the near-gate location and the location far from the gate is not significant.
3. Further increase the end packing pressure to 600 bar. The thickness measurements are 1.981 mm, 1.992 mm, 2.001 mm, and 1.983 mm. The middle part is obviously thicker, but the difference between the near-gate location and the far-from-gate location is not significant. Moreover, slight flash occurs.
4. Then reduce the start packing pressure to 450 bar, and keep the end packing pressure at 600 bar. An experiment is conducted with the ramp-up packing profile (450 bar, 600 bar). The part thickness is found to be 1.966 mm, 1.991 mm, 1.998 mm, and 1.985 mm, in which the near-gate location is obviously thinner than the other points.

In addition, the ramp-down packing profile is normally good for part evenness. In the case where evenness is most important and part shrinkage is not important, a ramp-down profile may be used to improve the part evenness. Taking Mold Insert 3, for example, the "best" constant packing pressure in terms of evenness is 375 bar, as shown in Table 10.12, with an evenness of 0.004. Tests show that the part molded with a ramp-down packing profile (375 bar, 200 bar) can improve greatly the part evenness to 0.002, with the thickness measurements 1.917 mm, 1.915 mm, and 1.913 mm.

## 10.4 Conclusions

The two key elements to properly profile the packing pressure, the gate freezing-off point and the packing profile, have been studied in this chapter.

The existing methods to detect the gate freezing-off point are time-consuming, case-dependent, and determined offline. To avoid the shortcomings, a robust autodetection system using online-measurable parameters to predict the gate freezing-off point is developed. Experimental results under a wide range of molding conditions prove the good performance of the developed system.

Influences of the packing profiles on part weight, evenness, shrinkage, and flash have been experimentally studied. Different types of packing profiles, including constant, ramp, and step-change profiles, are compared with different mold inserts. A constant packing profile with a suitable pressure level is recommended after an all-around analysis based on the extensive experimental results. A procedure to seek the "optimal" constant pressure setting has been proposed and experimentally demonstrated. In the case where evenness is most important and part shrinkage is not important, a ramp-down profile with a reasonable ratio may also be used. Furthermore, ramp packing profiles have been demonstrated to produce parts with a specific thickness distribution.

## References

[207] M. Vincent, in J.A. Covas et al., (Ed.), *Rheological Fundamentals of Polymer Processing*, NATO ASI Series, Kluwer Academic Publishers, Dordrecht, 1995

[225] F. Gao and C. Zhao, *Method and apparatus for the control of injection molding*, US Patent 6,309,571[P], 2001

[226] X. Chen, Ph.D, Thesis, Hong Kong Univ. of Sci. and Technol., 2002

[227] Xi Chen and Furong Gao, *Polym. Eng. Sci.*, **45**(1), 33-41, 2005

[228] B. Hartmannn and M.A. Haque, *J. Appl. Polym. Sci.*, **30**, 1553, 1985

[229] M.R. Kamal and N.T. Levan, *Polym. Eng. Sci.*, **13**, 131, 1973

[230] R.A. Orwoll, in J.E. Mark, (Ed.), *Physical Properties of Polymers Handbook*, AIP Press, New York, 1996

[231] T.W. Parks and C.S. Burrus, *Digital Filter Design*, Wiley, Hoboken, 1987

[232] V.K. Ingle and J.G. Proakis, *Digital Signal Processing Using MATLAB*, Brooks/Cole, Pacific Grove, CA, 2000

[233] Xi Chen and Furong Gao, *Mat. Sci. Eng. A.*, **358**(1), 205-213, 2003

[234] M. Bandreddi, R. Nunn, and R. Malloy, *SPE Annu. Tech. Conf. Papers*, SPE, 40, 348, 1994

# 11 Parameter Setting for the Plastication Stage

Besides the filling and packing stages, plastication is another stage that plays an important role in final part quality. But unlike the other two stages, which focus on profiling of the injection velocity or the packing pressure only, more than one variable has to be considered. The key variable includes barrel and nozzle temperatures, back pressure, and screw rotation speed. A thorough study on parameter setting for the plastication stage is described in this chapter.

Unlike extrusion, injection molding plastication is a non-steady-state process. The relationship among the plastication variables is difficult to predict from models of extrusion. To dynamically observe the special melting behavior of injection plastication, a visual barrel system is first presented in Section 11.1. Experiments are conducted to assist the understanding of polymer behaviors within the screw barrel system with plenty of data and photos, as illustrated in Section 11.2. The plastication influences the part quality mainly through the melt temperature. The temperature of the melt entering the mold, however, is affected by several factors during plastication, including screw rotation speed, back pressure, barrel and nozzle temperature, and dwell time. In addition, the injection velocity also influences the final melt temperature. As described in Section 11.3, a neural network model is developed to correlate the relation between melt temperature and those parameters. For a given injection velocity, how the parameters should be set to achieve a required melt temperature entering the mold cavity is presented in Section 11.4, in which an optimization procedure is established to seek the settings of the plastication parameters.

## 11.1 Visual Barrel System Development

An instrumented visual injection barrel system [226, 235] is developed to assist the modeling and optimization of the injection molding plastication process. This visual system makes it possible to analyze the process by observing the polymer status in the barrel via photography and video recording. The corresponding material conditions can also be measured via a data-acquisition system.

A new barrel is fabricated to be fitted with glass window inserts and temperature and pressure transducers. The detailed locations of the glass windows and the temperature and pressure transducers are shown in Fig. 11.1. The glass window inserts are installed at both the front and rear sides of the barrel, with the end of one window aligned with the beginning of the next, so that a complete melting history can be observed. The glass insert used in this project has a design specification to withstand a pressure of 1000 bar and a temperature of 300 °C. Each glass window is 80 mm in length and 14 mm in width. The introduction of the glass windows will affect, to a degree, the barrel heating system. In the barrel design, the overall volume of the glass windows is kept small, compared to the overall size of the barrel. The observation errors can thus be minimized.

Four Kistler transducers (Model 4091B), each of which allows simultaneous measurements of temperature and pressure at a single location, are also mounted flush with the barrel inner surface to measure the temperature and pressure at the barrel-polymer interface. The barrel and nozzle are surrounded by seven heaters, whose temperatures are closed-loop controlled to be within 1 °C from their set points. All of the other necessary machine conditions can be set or recorded via a data-acquisition and computer-controlled system.

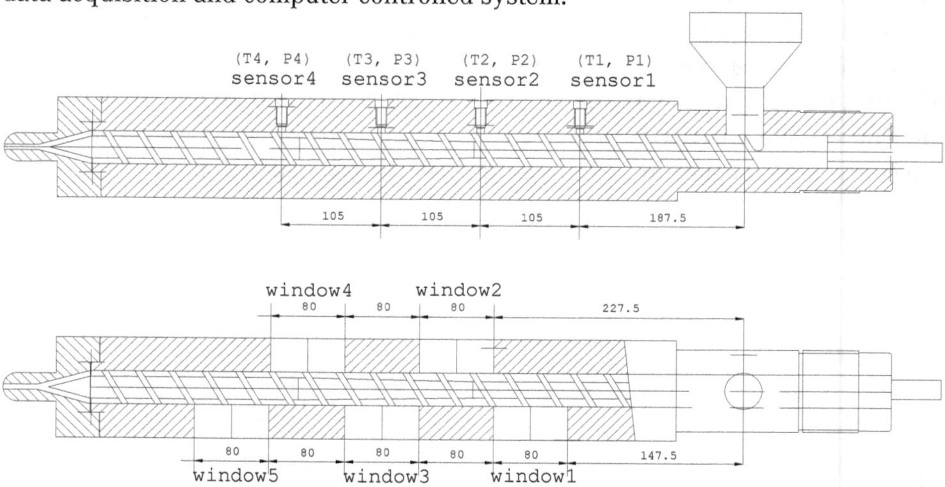


**Figure 11.1** Visual plastication unit design

## 11.2 Plastication Behavior

With the facility described in the previous section, experiments are conducted to observe the polymer behavior inside the injection unit. The material used in the experiments is high-density polyethylene (HDPE) (Marlex, HMN6060) because it has distinctively different appearances before and after melting. The images are recorded via both a digital camera and a video camera during the experiments. The experimental conditions are recorded via the data-acquisition system.

Air-shot experiments, with the nozzle being securely blocked to prevent drooling, are performed with different processing conditions to study their effects on the melting behavior. The experimental results show that the polymer is completely melted after the fourth window. Thus, only pictures from Windows 1 to 4 are illustrated in the following discussion. The injection stroke, rotation speed, back pressure, and barrel temperature are closed-loop controlled during the plastication stage. The nominal operating conditions shown in Table 11.1 are used, unless otherwise specified.

**Table 11.1** Nominal Operating Conditions

| Nominal Operating Conditions | Value |
|---|---|
| Injection Stroke (mm) | 30 |
| Screw Rotation Speed (rpm) | 30 |
| Back Pressure (bar) | 0 |
| Temperature Setting: T1–T7 (°C) | 100, 160, 180, 190, 200, 200, 200 |

### 11.2.1 Melting Behavior

#### Polymer Status in the First Window

The first window is located in the solid conveying zone, within which the solid granules gradually become denser and form a "solid plug" in the screw channel. Figure 11.2 shows photos taken at this window immediately after plastication, at the end of dwell, and immediately after injection.

The right end of each photo is closer to the feeding hopper. The barrel temperature in this zone is set to be lower than the polymer melting point, so no obvious melting can be observed at this window. All photos seem to show that the polymer granules are less compacted close to the screw flights, which are highlighted by two parallel white lines for easy observation. There is no clear difference between Figs. 11.2(a) and 11.2(b), indicating that heat conduction during the dwell phase does not significantly contribute to melting in this zone. A comparison between Figs. 11.2(b) and 11.2(c) clearly indicates that the polymer granules shown in the former are denser than those in the latter. The screw pitch is 36 mm. For this experiment, the

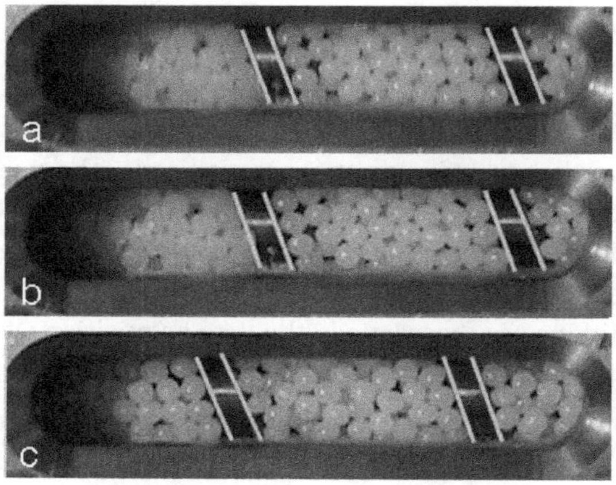


**Figure 11.2** Photos of polymer state at the first window
(a) immediately after plastication,
(b) at the end of dwell, and
(c) immediately after injection

plastication stroke is 30 mm, which suggests that the screw moves back about one screw pitch after plastication. The polymer shown in Fig. 11.2(c), therefore, is about one screw pitch closer to the screw rear end than that in Fig. 11.2(a). This confirms the common belief that the polymer granules become denser along the down-screw direction. The same conclusion can also be supported by the photos shown in Fig. 11.2, as they clearly show that the granule compaction is getting higher from the right to the left end of the photos.

#### Polymer Status in the Second Window

The polymer begins to melt in this window at the barrel-screw interface, due to the heat conduction and the shearing between the polymer and the barrel. A thin layer of melt film, which is transparent compared to the opaque solid, can be clearly observed in Fig. 11.3. The melt pool and the solid granules are clearly separated. The separation lines are nearly parallel to the screw flights. The melt pool is formed at about the eighth screw pitch from the screw rear end. As the screw moves forward 30 mm (about one screw pitch) after injection, the melting pool width observed from this window after injection as shown in Fig. 11.3(c) is smaller in comparison to that in Fig. 11.3(b). It should be pointed out that Figs. 11.3(b) and 11.3(c) show the same barrel position, but different screw positions. There is about one screw pitch difference in terms of the screw position between Figs. 11.3(b) and 11.3(c). All photos indicate that there are melt pools at both the active and passive screw flank sides. Moreover, larger melt pools are found at the active screw flank side. A comparison of Figs. 11.3(a) and 11.3(b) indicates that the heat conduction during the dwell stage contributes significantly to the melting in this zone. The width increments

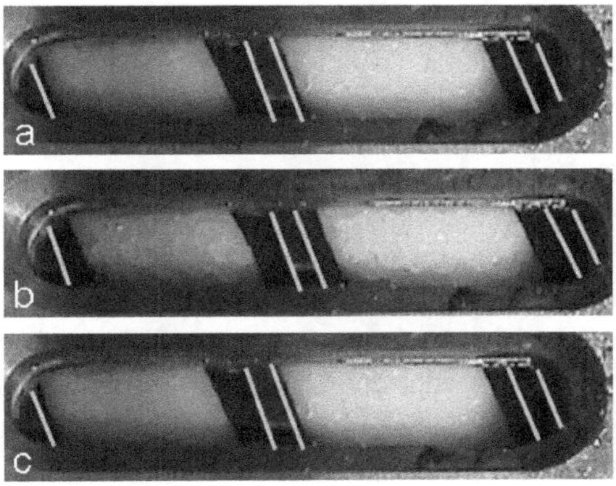

**Figure 11.3** Photos of polymer state at the second window
(a) immediately after plastication,
(b) at the end of dwell, and
(c) immediately after injection

of the melting pools at both ends of the screw channel are about the same during the dwell phase. It is difficult to see if there is any thickness difference in the melt film at the interface of the solid-bed and barrel surface. Most of the melting seems to happen at the solid-bed and melt-pool interface.

### Polymer Status in the Third Window

From this window, we can clearly see that there exists a melt film on top of the solid bed, a solid bed, and two melt pools in a screw channel, as shown in Fig. 11.4.

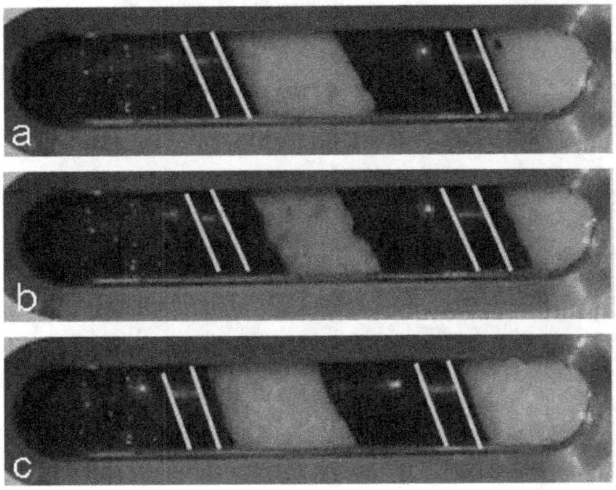

**Figure 11.4** Photos of polymer state at the third window
(a) immediately after plastication,
(b) at the end of dwell, and
(c) immediately after injection

All these figures show that the melt pool at the passive screw flank side (left-hand side) is relatively smaller and that the melt is largely located close to the active screw flank (right-hand side). By comparing Figs. 11.4(a) and 11.4(b), it can be observed that the polymer continues to melt during the dwell phase, indicating that heat conduction during the dwell period makes an important contribution to melting in injection molding. Photos of different dwelling times will be discussed later. A comparison of Figs. 11.4(b) and 11.4(c), just before and after injection, suggests that the melt-pool size at the passive flank side is reduced after injection, even after considering the screw shifting during injection. This suggests that the solid bed has moved toward the passive flank side due to the axial direction shear during injection.

#### Polymer Status in the Fourth Window

The solid bed is nearly melted at the fourth window, as shown in Fig. 11.5(a), which is taken at the end of plastication. Due to the heat conduction, the solid bed is completely melted at the end of dwell, as shown in Fig. 11.5(b). Figure 11.5(c) shows a photo of the material status after injection. The injection moves the screw one pitch closer to the rear end; in theory, there should exist some unmelted solid bed, but that is not observed in Fig. 11.5(c). This suggests that the solid at this screw position is melted during the dwelling phase or injection phase, indicating that the heat conduction during dwell and the shear heating during injection can be significant in injection molding.

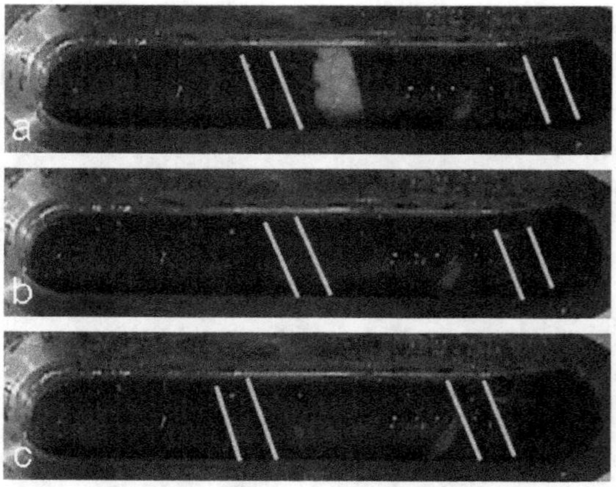


**Figure 11.5** Photos of polymer state at the fourth window
(a) immediately after plastication,
(b) at the end of dwell, and
(c) immediately after injection

**Polymer Status along the Down-Screw Channel Direction**

Figures 11.2(a), 11.3(a), 11.4(a), and 11.5(a) show the polymer status at different screw positions immediately upon the completion of plastication. The entire melting process, including the formulation of the solid bed, the melting of the solid, and the formulation of melt pools, can be clearly observed in these photos. As expected, the solid bed reduces gradually along the screw down-channel direction, while the melt pool, which is largely located at the active screw flank side, increases gradually along the screw down-channel direction. A similar phenomenon can be observed in Figs. 11.2(b), 11.3(b), 11.4(b), and 11.5(b), which show the polymer status after the dwell phase, and in Figs. 11.2(c), 11.3(c), 11.4(c), and 11.5(c), which show the polymer status after injection.

The photos and videos show that three distinctive zones exist: a melting film zone at the interface between the solid bed and the inner barrel surface, a solid-bed zone, and a melt pool zone. This observation agrees with the three-zone division of the well-known "Tadmor model." Further careful observation indicates that a thin melt pool also exists between the solid bed and the passive screw flank. To consider this, a five-zone model is necessary, resulting in a more complex description than the classic "Tadmor model."

The videos are recorded with the corresponding screw positions and speeds, allowing us to convert the video images into plots of the solid-bed width variations versus time and screw position. Figure 11.6 plots the time variation of the solid-bed width and the melt-pool widths at the 12th pitch during the dwell. The melt film thickness can also be seen to change with the dwell time on the video recording, but this thickness is difficult to quantify.

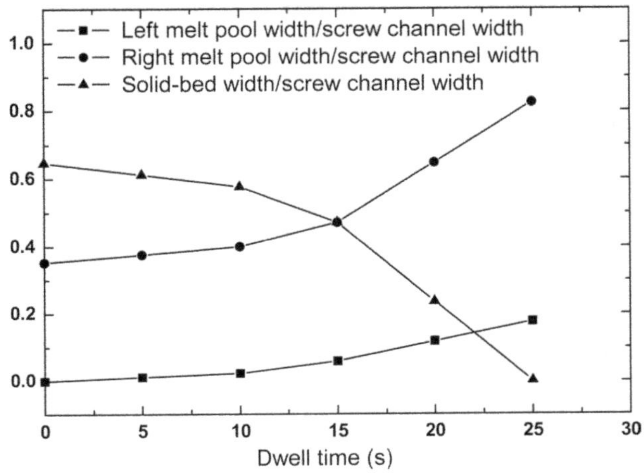


**Figure 11.6** Experimental measurement time-width variations of the solid bed, left-hand, and right-hand melt pools at the 12th screw pitch during dwell

Figure 11.7 is a plot of the solid-bed variation at the 10th pitch during plastication, indicating that the solid-bed width at a given screw location increases with plastication time. Figure 11.8 shows the solid-bed width variations along the down-screw channel direction after the plastication and dwell phases. These figures demonstrate that, unlike in the extrusion process, plastication in injection molding is not a steady-state process.

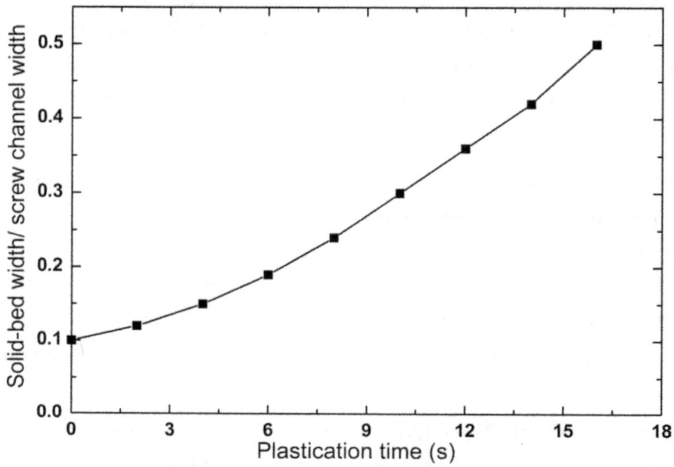


**Figure 11.7** Solid-bed variation at 10th screw pitch during plastication

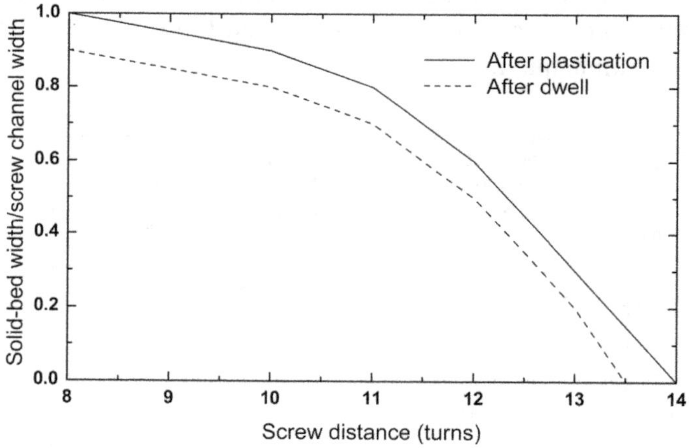


**Figure 11.8** Schematic of melting along the injection screw

**Process Measurement**

Because the screw has an axial motion during plastication, the temperature and pressure measurements at a fixed barrel position give temperature and pressure profiles over a portion of the screw. The temperature profiles along the screw at different plastication times are presented in Fig. 11.9. It is under the nominal operating conditions of Table 11.1, except that the plastication stroke is extended to 75 mm for a longer period of observation. Due to the screw retraction during plastication, the temperature decreases at the same screw position during plastication, which is different from steady-state extrusion.

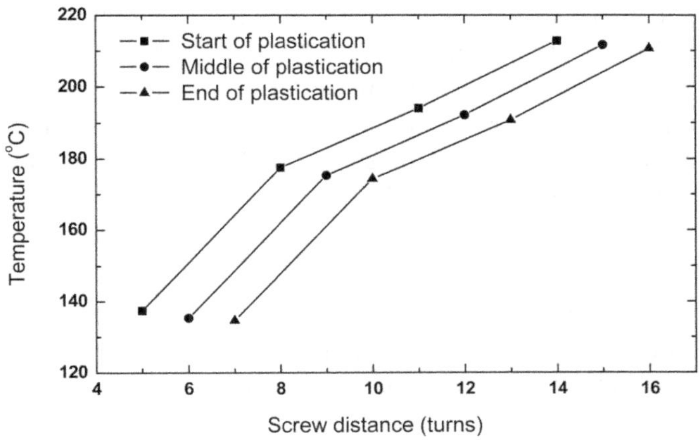


**Figure 11.9** Temperature along the screw during plastication

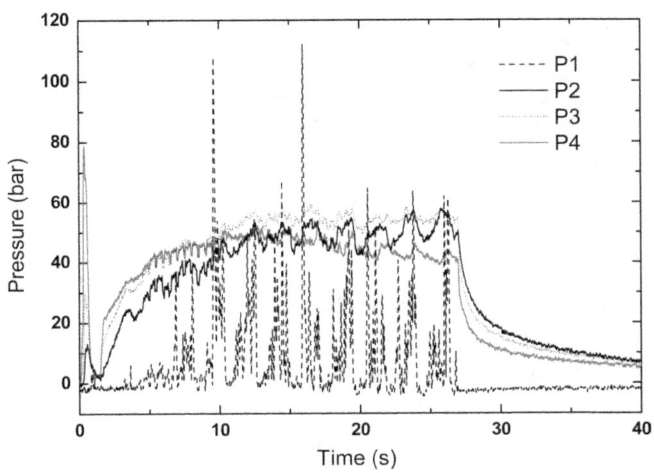


**Figure 11.10** Pressure variations throughout an injection cycle

The pressures measured at the four different barrel positions throughout the injection cycle are presented in Fig. 11.10. The variable P$i$ ($i$ = 1 to 4) represents the pressure measurement at the $i$th window. The first peak represents the pressures experienced during the injection, followed by the pressure variations during plastication from the 2$^{nd}$ to the 26$^{th}$ second, and finally the relatively smooth pressure decreases during the dwell phase. The first sensor (Sensor 1) located opposite Window 1 is within the solid conveying zone over which the polymer has not been melted. As indicated by P1 in Fig. 11.10, the pressure at the beginning of plastication is low, as the polymer granules have not yet formed into a solid block. This pressure becomes oscillatory with some sharp peaks during later plastication, after the granules are formed into a solid bed. This pattern of pressure variation is similar to the pressure of solid-to-solid contact. The pressures at the other three measurement points during the plastication, where a melting polymer film has formed between the solid bed and the sensors, are relatively regular. As the screw is rotated at 30 rpm, the sensor meets the screw flight every 2 seconds. A pressure oscillation period of about 2 seconds can be clearly seen in Fig. 11.10 for P2, P3, and P4.

#### Solid-Bed Breakup

During the operation, we can occasionally observe the phenomenon of solid-bed "breakup" as shown in Fig. 11.11. In this particular case, the injection stroke is 90 mm. We have found that the breakup is generally associated with a high injection stroke, a high rotation speed, or a high back pressure. The breakup may be attributed to the high pressure in the screw channel that forces the melt into the voids within the solid bed.

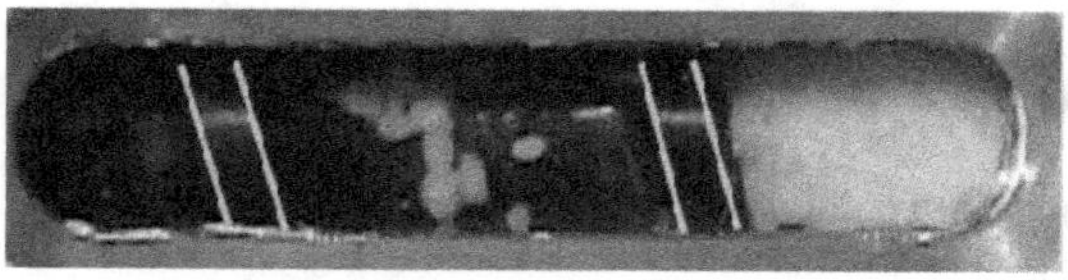

**Figure 11.11** Example photo of solid-bed breakup

### 11.2.2 Processing Condition Effects

Different barrel temperatures, plastication strokes, screw rotation speeds, and back pressures have been used to observe their effects on the melting behavior. Every time, only one processing condition is changed, and all other processing conditions are kept at their nominal conditions of Table 11.1.

### Plastication Stroke

Figure 11.12 shows the polymer status at the third window after plastication, with a stroke of 30 mm and 90 mm. There is about a 1.7 screw pitch difference in terms of the screw position, due to the stroke difference. Obviously, the solid-bed width in Fig. 11.12(b) is larger than that in Fig. 11.12(a).

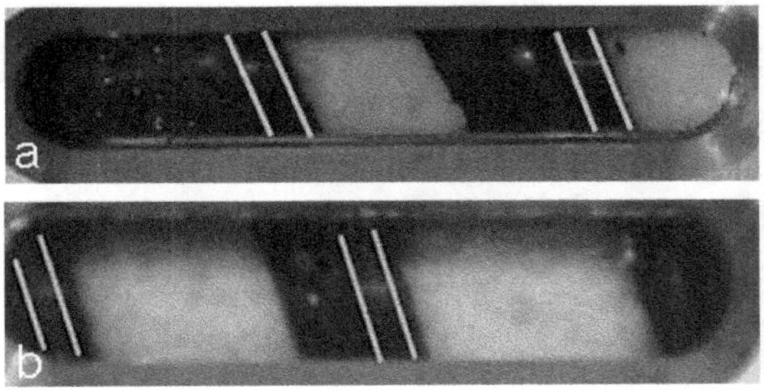


**Figure 11.12** Photos of the polymer state at the third window after plastication: (a) plastication stroke is 30 mm; (b) plastication stroke is 90 mm

Based on the data obtained from video recording, the corresponding solid-bed width variation along the screw channel is plotted in Fig. 11.13. The polymer is melted at the 12th screw pitch for the case of the plastication stroke of 30 mm. For the case of the plastication stroke of 90 mm, the polymer does not completely melt until the 15th screw pitch. With a small injection stroke, the injection volume is small in comparison to the amount of material enclosed in the screw channel. As a result, the residence time will be comparatively long, and hence melting is primarily due to heat conduction. Conversely, with a long injection stroke, the injection volume is large in comparison with the screw channel volume, and the melting is primarily due to the shearing.

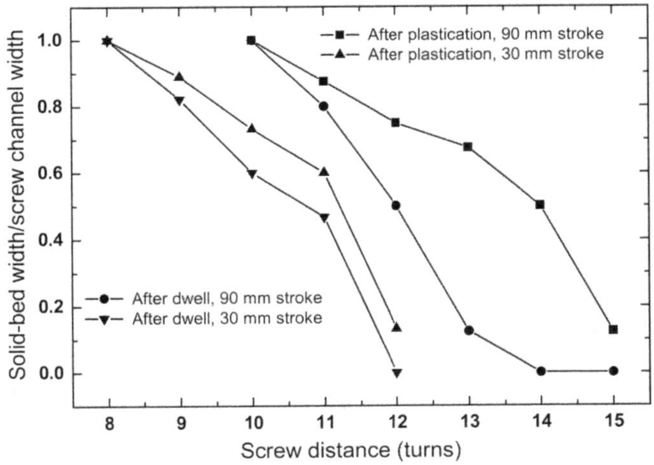


**Figure 11.13** Experimental measurement of melting along the screw

### Screw Rotation Speed

Figure 11.14 shows the material status for a rotation speed of 20 rpm and 40 rpm. There is more polymer melted at a lower rotation speed than at a higher speed. As the plastication stroke is the same, a lower speed results in longer plastication time, while a higher speed has shorter plastication time.

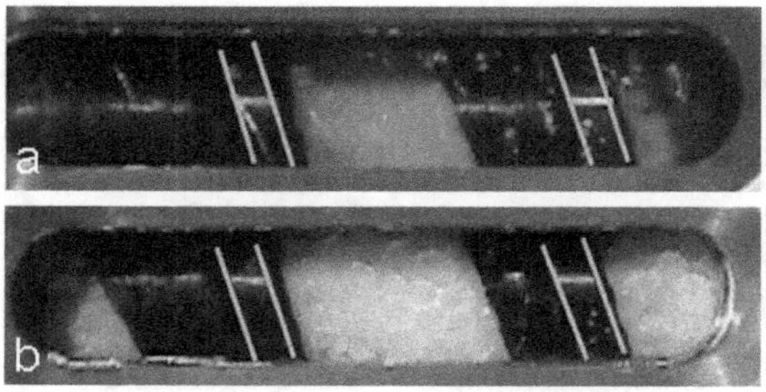


**Figure 11.14** Photos of polymer state at the third window after plastication: (a) screw rotation speed is 20 rpm; (b) screw rotation speed is 40 rpm

### Back Pressure

Back pressure affects the melting, temperature distribution, and pressure distribution in the screw channel. Different back pressures result in different pressure profiles in the screw channel and different plastication times, as shown in Fig. 11.15, which is a plot of the pressure variation at the third window (P3) with a plastication stroke of 30 mm for two different back pressures. Figure 11.16 illustrates the material status at a back pressure of 0 and 5 bar. A comparison of the figures suggests that an increase in back pressure is advantageous to the melting. A higher back pressure also leads to a larger left-hand melt pool size than does a lower back pressure.

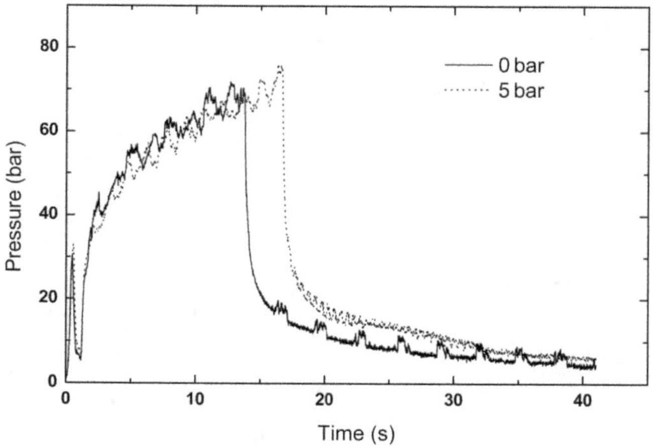


**Figure 11.15** Pressure profiles at the third window

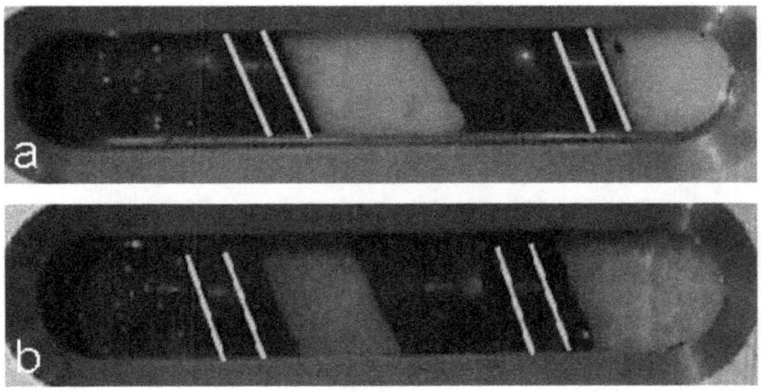


**Figure 11.16** Photos of polymer state at the third window after plastication: (a) back pressure is 0 bar; (b) back pressure is 5 bar

### Barrel Temperature

Figure 11.17 shows a comparison between the profiles of the barrel temperatures and the temperatures at the interface of the polymer and barrel for two different barrel temperature settings. The dotted lines show the barrel metal temperature profiles, with the circles representing the actual barrel heater temperatures. The solid lines indicate the barrel inner surface temperature profiles, with the squares representing the actual temperature measurements at the polymer-barrel interface. The squares and circles are not aligned because the heater thermocouples and the interface temperature transducers are not installed at the same barrel position. Figure 11.17 shows that the polymer-barrel interface temperatures are higher than the barrel temperatures for both barrel temperature settings. The differences seem to increase from the hopper to the nozzle end. The interface temperature can be more than 20 °C higher than the barrel temperature. This may be due to the combination of effects of shear heating at the polymer-barrel interface and the barrel heat loss to the environment. In injection molding modeling, the polymer temperature at the

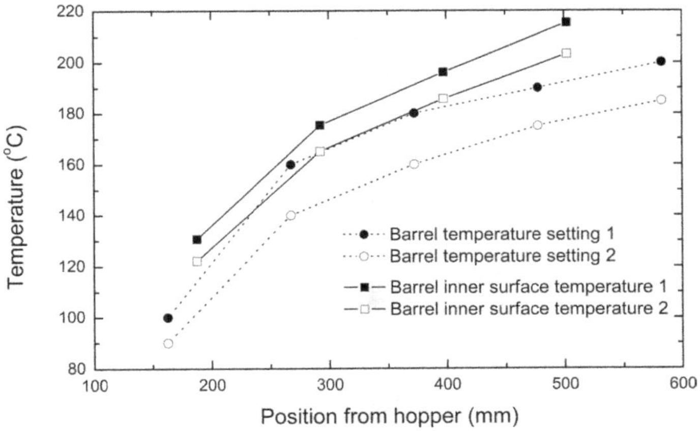


**Figure 11.17** Comparison of barrel and barrel-polymer interface temperature

wall is commonly assumed to be the same as the barrel temperature, which is not an accurate reflection of reality.

### Dwell Time

A long dwelling time conducts more heat and consequently melts more polymer. The size variations with dwell times for the solid-bed width and the melt pools have been previously shown in Fig. 11.6.

## 11.3 Neural Network Modeling of Melt Temperature

The temperature of the melt entering the mold is an important variable that can influence the part quality. In the injection molding process, the melt temperature at the nozzle is mainly determined by three factors: (1) the melt temperature distribution in the reservoir of the injection molding machine at the end of the plastication phase, (2) the heat conduction between the melt and the surrounding during the dwell phase, and (3) the shear heating effect during the injection phase. The experimental results of the visual barrel system have shown that injection plastication is different from that of extrusion. Therefore, the models of extrusion cannot be directly employed. For a given injection velocity, how the plastication parameters shall be set to achieve a required melt temperature will be discussed in this section.

A series of neural network approaches for the predictions of the melt temperature at the end of plastication and during injection phases have been developed [236–238]. A feed-forward neural network was eventually employed to correlate the melt temperature with the plastication parameters and injection parameter. A typical two-layer architecture as shown in Fig. 11.18 was used. Fourteen log-sigmoid neurons in the hidden layer of the neural network receive inputs directly and then broadcast their

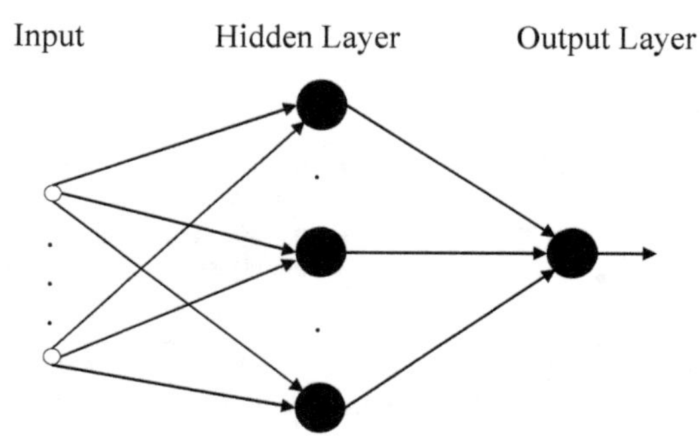


**Figure 11.18** Typical two-layer neural network architecture

outputs to the linear output layer, which calculates the network output. The inputs to the neural network are nozzle heater temperature ($T_n$), barrel heater temperatures ($T_{Z1}$, $T_{Z2}$, and $T_{Z3}$), the screw rotation speed (SRS), back pressure (BP), the required stroke length (SL), and injection velocity (IV). The only output of the neural network is the bulk melt temperature after injection ($T_m$).

The proposed neural network was trained with 74 sets of data and tested with eight sets of untrained experimental data shown in Table 11.2. The variable $T_{exp}$ represents the bulk melt temperature measured experimentally, and $T_{ann}$ represents the bulk melt temperature predicted by the artificial neural network. Among the eight sets of testing data, there is only one testing sample that has a slightly larger prediction error, 2.58°C; for the other testing samples, the prediction errors are small. It shows that the proposed neural network works well and can be used for online melt temperature prediction in the injection molding process.

**Table 11.2** Test Results of Neural Network Model

| Test No. | $T_n$ (°C) | $T_{Z1}$ (°C) | $T_{Z2}$ (°C) | $T_{Z3}$ (°C) | BP (bar) | SL (mm) | SRS (mm) | IV (mm/s) | $T_{exp}$ (°C) | $T_{ann}$ (°C) | Error |
|---|---|---|---|---|---|---|---|---|---|---|---|
| 1 | 200 | 190 | 175 | 165 | 4 | 60 | 90 | 30 | 221 | 222.06 | -1.06 |
| 2 | 200 | 194 | 194 | 189 | 5 | 60 | 100 | 20 | 227.5 | 224.92 | **2.58** |
| 3 | 200 | 194 | 194 | 189 | 5 | 60 | 100 | 40 | 229.8 | 228.72 | 1.08 |
| 4 | 210 | 207 | 207 | 192 | 3.5 | 60 | 110 | 30 | 240.7 | 239.91 | 0.79 |
| 5 | 220 | 210 | 210 | 190 | 4.5 | 60 | 80 | 20 | 246.2 | 245.47 | 0.73 |
| 6 | 220 | 210 | 210 | 190 | 4.5 | 60 | 80 | 40 | 247.5 | 248.40 | -0.90 |
| 7 | 220 | 220 | 210 | 205 | 5 | 60 | 100 | 30 | 249.3 | 248.59 | 0.71 |
| 8 | 230 | 230 | 230 | 220 | 4 | 60 | 90 | 20 | 263 | 261.86 | 1.14 |

## 11.4 Optimal Parameter Setting for the Plastication Stage

Given the plastication and injection parameters, the neural network can predict accurately the melt temperature at the nozzle exit after injection. But how the plastication parameters should be set for a given melt temperature requirement has not yet been studied. A method based on optimization has been developed [226].

Use function *ANN*(●) to represent the relationship between the inputs and output of the artificial neural network model. Then the parameter setting problem is converted to an optimal problem to meet the melt temperature requirement, as shown in the following:

$$\min_{X} f(X) = \left[ ANN(X, \mathrm{SL}, \mathrm{IV}) - T_{\mathrm{m}} \right]^2 \tag{11.1}$$

where $X$ is a vector describing the plastication parameters, including the nozzle heater temperature ($T_n$), the barrel heater temperatures ($T_{Z1}$, $T_{Z2}$, and $T_{Z3}$), the screw rotation speed (SRS), and the back pressure (BP). The required stroke length (SL) and the injection velocity (IV) are given by the injection parameter settings. The variable $T_{\mathrm{m}}$ is the required melt temperature, which is given by the customer.

The solution cannot be sought in an unlimited range. A set of lower and upper bounds is defined for the design variables, $X$, so that a solution is sought in the practical range ($X_{\mathrm{min}}$, $X_{\mathrm{max}}$). This results in the following constraints:

$$T_{n,\mathrm{min}} < T_n < T_{n,\mathrm{max}} \tag{11.2}$$

$$T_{Z1,\mathrm{min}} < T_{Z1} < T_{Z1,\mathrm{max}} \tag{11.3}$$

$$T_{Z2,\mathrm{min}} < T_{Z2} < T_{Z2,\mathrm{max}} \tag{11.4}$$

$$T_{Z3,\mathrm{min}} < T_{Z3} < T_{Z3,\mathrm{max}} \tag{11.5}$$

$$\mathrm{SRS}_{\mathrm{min}} < \mathrm{SRS} < \mathrm{SRS}_{n,\mathrm{max}} \tag{11.6}$$

$$\mathrm{BP}_{\mathrm{min}} < \mathrm{BP} < \mathrm{BP}_{\mathrm{max}} \tag{11.7}$$

Additional constraints may be set based on the interrelations among the plastication parameters. For example, the temperature difference between different zones may have the following relations to prevent excessive temperature change between the adjacent heater zones:

$$\Delta T_1 < T_n - T_{Z1} < \Delta T_2 \tag{11.8}$$

$$\Delta T_3 < T_{Z2} - T_{Z1} < \Delta T_4 \tag{11.9}$$

$$\Delta T_5 < T_{Z3} - T_{Z2} < \Delta T_6 \tag{11.10}$$

And the screw rotation speed may relate to the back pressure as

$$f_{\mathrm{min}}(\mathrm{BP}) < \mathrm{SRS} < f_{\mathrm{max}}(\mathrm{BP}) \tag{11.11}$$

This type of constraint is obtained from the molding experience, typically provided by the experienced operation personnel. They are not indispensable, the presentation of those constraints can improve the solution search, and they may be expressed in a fuzzy form if accurate mathematical descriptions are difficult to obtain.

The parameter-setting problem is converted to a constrained nonlinear optimization problem, which can be described in a general form as the following:

$$\begin{aligned}
&\min_{X} f(X) = \left[ANN(X, \mathrm{SL}, \mathrm{IV}) - T_{\mathrm{m}}\right]^2 \\
&\text{st:} \quad G_i(X) \leq 0 \qquad i = 1, \cdots, n \qquad (11.12) \\
&X_{\min} < X < X_{\max}
\end{aligned}$$

where $G_i(X)$ is a set of constraints transformed from Eqs. 11.2 to 11.11.

A program based on an optimization technique is developed for the plastication parameter setting. It is developed in Matlab software based on the sequential quadratic programming (SQP) method. After inputting the required melt temperature, the stroke length for filling, and the injection velocity, the program can output the "optimal" settings of the plastication parameters to meet the required melt temperature and the corresponding error of the predicted melt temperature. The material under test is HDPE, and the following constraints are set for the parameters according to experience:

$$200°\mathrm{C} \leq T_n \leq 230°\mathrm{C} \qquad (11.13)$$

$$190°\mathrm{C} \leq T_{Z1} \leq 230°\mathrm{C} \qquad (11.14)$$

$$170°\mathrm{C} \leq T_{Z2} \leq 220°\mathrm{C} \qquad (11.15)$$

$$150°\mathrm{C} \leq T_{Z3} \leq 210°\mathrm{C} \qquad (11.16)$$

$$T_n \geq T_{Z1} \geq T_{Z2} \geq T_{Z3} \qquad (11.17)$$

$$50 \leq \mathrm{SRS} \leq 110 \qquad (11.18)$$

$$3 \leq P_b \leq 6 \qquad (11.19)$$

Two case studies are conducted, as presented in the following.

**Case 1**

Please input required melt temperature [210, 270]°C: 230

Please input stroke length [50, 80] mm: 55

Please input injection velocity [20, 40] mm/s: 30

The optimal settings of plastication parameters are derived as $T_n$ = 208.0°C, $T_{Z1}$ = 200.2°C, $T_{Z2}$ = 185.2°C, $T_{Z3}$ = 172.8°C, SRS = 82.6 rpm, and BP = 4.9 bar. A validation experiment with the optimized parameter settings is then conducted.

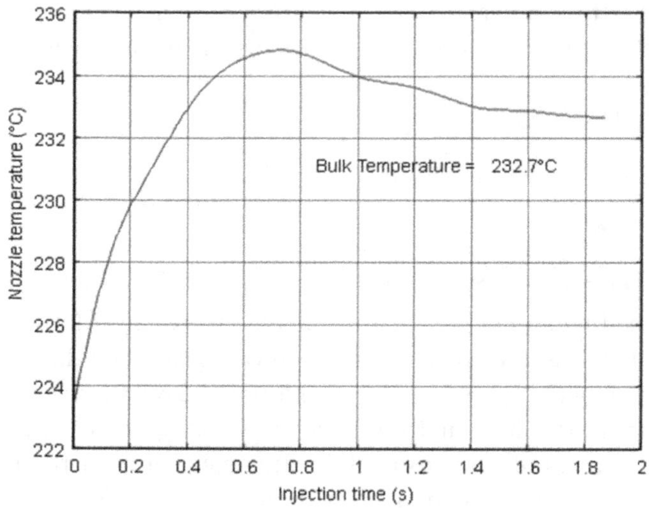


**Figure 11.19** Melt temperature of Case 1

The melt temperature measured at the nozzle is shown in Fig. 11.19. It can be seen that at the beginning of injection, the melt temperature is lower than the bulk temperature, due to the heat loss of the melt at the nozzle exit during the dwell phase. The experimental measured bulk melt temperature is 232.7 °C, slightly higher than the required melt temperature, but the error is within the modeling accuracy.

**Case 2**

Please input required melt temperature [210, 270] °C: 245

Please input stroke length [50, 80] mm: 60

Please input injection velocity [20, 40] mm/s: 25

The optimal settings of plastication parameters are derived as $T_n$ = 222.9 °C, $T_{Z1}$ = 210.5 °C, $T_{Z2}$ = 199.3 °C, $T_{Z3}$ = 181.8 °C, SRS = 75.9 rpm, and BP = 4.3 bar. Similarly, a validation experiment with the optimized parameter settings is then conducted. The measured melt temperature at the nozzle during filling is shown in Figure 11.20. The bulk temperature is about 244.1 °C, 0.9 °C lower than the required setting.

The above case studies show that the parameter settings can be effective, despite the small discrepancy from the desired melt temperature setting. Compared with Case 1, the melt temperature requirement in Case 2 is higher, but with a lower injection velocity. The nozzle and barrel heater temperatures are consequently increased in response in the calculated optimal setting for plastication, as expected. The proposed optimization approach can become more effective with further improvement of the constraint rules and refinement and extension of the neural network model to cover more materials.

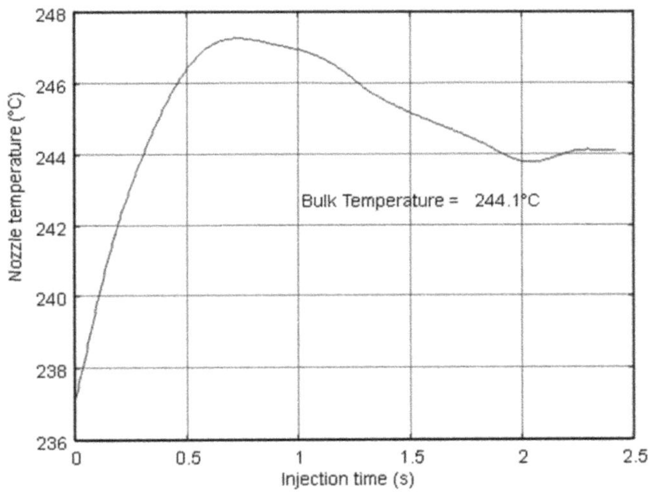


**Figure 11.20** Melt temperature of Case 2

## References

[226] X. Chen, Ph.D. Thesis, Hong Kong Univ. of Sci. and Technol., 2002

[235] F. Gao, Zhiming Jin, and Xi Chen, *Polym. Eng. Sci.*, **40**(6), 1334–1343, 2000

[236] C. Zhao, M.Phil. Thesis, Hong Kong Univ. of Sci. and Technol., 1998

[237] F. Gao, B.G. Wang, and P.L. Yue, *Polym. Process. Soc. 15th Annual Meeting*, 301, 1999

[238] B.G. Wang, F. Gao, and P.L. Yue, *Chinese J. Chem. Eng.*, **8**, 326, 2000

# Subject Index


**Symbol**

2D-DMC 121
2D prediction model 123
$\chi^2$ distribution 157

**A**

acrylonitrile butadiene styrene (ABS) 5
activation energy 26
adaptive control 44
adaptive feed-forward control 52
additional polymerizations 5
addition plastics 5
AFL controller 301
aircraft windshield 2
air-shot experiments 373
Akaike's final predictive errors (FPE) 65
alternating copolymers 7
amorphous plastics 10
Andrade's law 26
anisotropy 11
anomaly detection 149
anti-model-burst 52
approximation 25
arithmetic mean 8
artificial neural networks (ANN) 155, 385
augmented $T^2$ statistic 181
autocorrelation structure 156
automatic control 39
automatic V/P switch-over 169
automobile 2
autoregressive with external input (ARX) model 45
average-flow-length 266, 269

**B**

backflow 362
back pressure 371
back-propagation (BP) algorithm 305
backward elimination 237
barrel heater temperatures 385
barrel-screw interface 374
barrel temperatures 36
batch process 83
batch-to-batch dynamics 169
batch-to-batch time scale 168
batch-to-batch variances 177
batch-wise unfolding 162
biaxial 11
bimodal HDPE 9
binomial distribution 165
block copolymers 7
blow molding 31
bootstrapping-based MPLS model (B-MPLS) 228
bootstrapping improved MPLS 214
branched structure 4
building and construction industry 2
Butterworth filter 334

**C**

capacitance 252
capacitive sensors 252
capacitive transducer 251, 255
cavity pressure 36
cavity thickness 347
center average defuzzifier 74
ceramic aluminum oxide 263
charge-and-discharge frequency 264
charge-and-discharge principle 264
Chebyshev type I filter 73

check-ring failure 187
chemical properties 1
China Railway High-speed 2
closed-loop control 40
C-mold 298
common cause variations 155
compression molding 5
condensation plastics 6
condensation reactions 6
conditional updating 51
cone-plate viscometer 16
confidence interval 156
consistency index 23
constant-trace algorithm 51
continuous process 29
contour plot 355
contribution plot 157, 161
control actuator 40
control chart 85
control horizons 63
controlled autoregressive integrated moving-average (CARIMA) model 112
control limits 151
control system 33
converge procedure 87
copolymer 7
correlation contribution rate 236
cost function 62
Couette flow 16
covariance matrix 158
critical-to-dimension phases 225
critical-to-quality phase 214, 217
critical-to-surface phase 224
cross-correlation structure 156
crossed orientations 11
cross-linked polyethylene 4
cross-linked structure 4
cross-validation 160
crystallinity 10
cumulative sum (CUSUM) 150
curing 5
cycle-to-cycle adaptation 58

D
data-based methods 149
data mining 150
data normalization 176
data quality 154
data reconstruction 159
deformation 18
defuzzifier 71
degrees of freedom 161
density measurement 11
derivative gain 42
detection rule 334
detections of start of filling 266
dielectric constant 252
dielectric properties 251, 255
die swell 16
differential scanning calorimetry (DSC) 11
dimensional stability 35
Diophantine eq. 47
disk extruder 29
distributed control systems (DCS) 150
disturbances 39
dominant poles 65
drag flow 12
dual adaptive control 44
dwell time 371
dynamic matrix control (DMC) 61
dynamic network 298
dynamic PCA/PLS 153
dynamic time warping (DTW) 173

E
effect of pressure on viscosity 28
eigenvalue 133, 158
electrical and electronic devices 2
electrical insulators 260
electrode 260
elongation 12
elongation flow 21
end-of-phase prediction 219
end pressure 351
end-product quality 213
estimator windup 51
Euclidean distance 178
excess ethylene glycol 6
excitation 50
expert knowledge 149
exponential convergence 90
exponentially weighted moving average (EWMA) filter 85, 150
extended Euclidean algorithm 47
extruder 29
extrusion 5, 29
extrusion blow molding 31

F
false alarms 152, 153, 180
fault detection and diagnosis (FDD) 149, 157
F distribution 157, 160
feed-forward control 52
feed-forward neural network 303, 384
feed-forward proportional gain 53
final-phase prediction 220
first-principle models 149
flash 348
forgetting factor 46
forward selection 237
fuzzifier 70
fuzzy IF-THEN rules 70
fuzzy inference engine 71
fuzzy inference system 71
fuzzy rule base 70
fuzzy system 70
fuzzy V/P transfer technique 329

G
gain scheduling 44
gate freezing-off 251, 329
generalized model predictive control (GPC) 62
glass transition temperature 11
glass windows 372
goodness of fit 217
graft copolymers 7
growth of the crystalline regions 10

H
hard-division method 180
heat conduction 374
high-density polyethylene (HDPE) 3
high shear limiting Newtonian viscosity 23
homopolymer 7
Hotelling's $T^2$ statistic 156, 160
hypothesis tests 156

I
in-control 152
in-control process window 153
independent component analysis (ICA) 155
indicator variable 174
infrared spectroscopy 11
initial parameters 57
injection blow molding 31
injection molding 32
injection molding machine 33
injection stretching blow molding 31
injection stroke 274
in-mold polymer melt status 279
integral gain 42
iterative learning control (ILC) 84

J
jackknife 160
John Wesley Hyatt 33

K
Karl Weissenberg 15
Kernel PLS 160
*k*-means algorithm 179
*k*-means clustering 175, 180
knowledge-based methods 149
knowledge-based system 70
knowledge extraction 235

L
landmarks 174
latent variable 158
layer-by-layer solidification model 285
learning rate 90
least mean squares (LMS) 44
least squares regression (LSR) 220
Levenberg-Marquardt (LM) algorithm 305
linear regression analysis 217
linear response 17
linear structure 3
loading matrix 158
loading vector 158
loss function 45
low-density polyethylene (LDPE) 4
lower triangular matrix 130
low shear limiting Newtonian viscosity 23
Lyapunov function 133

M


machine parameters 34
manual control 39
material compression coefficient 333
material constant 27
material disturbance 187
material properties 34
matrix decomposition 159
matrix representation 155
matrix unfolding 162
Maxwell model 19
mean 155
mean prediction squared error 237
measurement unit 40
mechanical properties 35
melt-front area 269
melt-front position 251
melt-front-velocity 295
melt pools 374
melt temperature 371
membership function 70
minimum inference engine 74
missed alarm 152, 153, 180
model algorithmic control (MAC) 61
model-based methods 149
modeling phase 173
model mismatch 67
model order 64
model parameter 46
model predictive control (MPC) 61
model reference 44
mold temperature 36
molecular chain 5
molecular polarization 256
molecular weight 8
molecular weight distribution 9
monomer 3, 7
moving-window PCA 155
multiblock modeling 173
multiblock PCA/PLS 153
multidimensional normal distribution 165
multilayer structure 2
multimode processes 153
multiphase batch processes 170, 173, 174
multiphase nature 109
multiple coefficients of determination 217, 243
multivariate analysis 151
multivariate control chart 152
multivariate projection methods 153
multiway partial least squares (MPLS) 162, 213
multiway PCA (MPCA) 162


N


nanoplastic 3
neural network model 298
Newtonian fluid 14
NM normalization 194
non-Gaussian processes 153
nonlinear 17
nonlinear iterative partial least squares (NIPALS) 159
nonlinear mapping 71
nonlinear optimization problem 318
nonlinear PCA/PLS 153, 155
non-Newtonian fluid 14
nonpolar polymers 256
nonrepeatable uncertainties 121
nonstationary processes 155
normal distribution 157
normalization 155
normal stress difference 16
nozzle heater temperature 385
nozzle temperature 371
nuclear magnetic resonance (NMR) 11
nucleation 10
number-average molecular weight 8


O


offline profiling method 319
online quality prediction 213
opaque 374
open-loop system 40
operation phases 166, 173
orientation 11
output prediction 63
output tracking performance 146


P


packaging industry 1
packing pressure 279
packing profiles 364

packing time 279
parabolic profile 319
parallel plate capacitor 252, 260
parameter setting 371
part evenness 348, 356
part flash 355
partial least squares (PLS) 153, 158, 220
part shrinkage 347, 355
part thickness 347
part thickness distribution 356
part weight prediction and monitoring 287
part weights 352
pathway multiblock PLS 214
PC subspace 159
phase-based PCA modeling 175
phase-based process analysis 218
phase division 173, 174
PID-type ILC 86
piece-wise ramp profiles 323
plastication 371
plasticizing screw 30
plug flow 12
polar polymers 256
pole-placement design 46
polyacrylamide 4
polyamides (nylon) 3
polycarbonate (PC) 5
polyester 5
polyethylene (PE) 3
polyethylene terephthalate (PET) 6
polyimide (PI) 5
polymer age 1
polymerization 5
polymer rheology 12
polypropylene (PP) 5
polystyrene (PS) 5
polyurethanes (PU) 5
poly(vinyl chloride) (PVC) 3
positon form of PID 42
power law index 23
power law relationship 23
prediction error 385
predictive functional control (PFC) 62
pressure flow 12
pressure sensitivity coefficient 28, 29
principal component analysis (PCA) 153, 158
principal component (PC) 158
principal component regression (PCR) 220
process dynamics 150
process monitoring 149
process nonlinearity 44
process safety 149
process startup 169
process temperature 27
process variables 34, 150
product quality 30, 149
profile shift 56
projection algorithm (PA) 44
projection structures 159
proportional gain 42
proportional-integral-derivative (PID) controller 41
proportional valve 84
pseudoplastic fluids 22
P-type learning 86
P-V-T relationship of polymers 332

Q

Q statistic 156
quality consistency 213
quality improvement 167
quality prediction and analysis 150, 157
quality variables 150

R

ram extruder 29
ramp-down 272, 342
ramp-up 272, 342
ramp velocity profile 272
range 155
reactive injection molding 5
reciprocating-screw machine 33
recurrent neural network 304
recursive least-squares (RLS) estimator 44
reference pressure 28
reference temperature 26
reference trajectories 61
regularity condition 91
relative dielectric constant 255
repetitive control (RC) 84
repetitive nature 109
representative loading matrix 180
representative regression parameter matrix 216
representative singular-value diagonal matrix 181

residual matrix 159
residual subspace 159
response 17
ridge regression 220
robust bounded-input-bounded-output (BIBO) stability 94
run-to-run control (R2R) 84

S

score matrix 158
score vector 158
screw channel 375
screw extruder 29
screw rotation speed 371
self-tuning regulators (STR) 44
semicrystalline materials 10
sensor failure 187
sequence manipulation 39
sequential or parallel polymerization 10
sequential quadratic programming (SQP) method 318
servo-valve 49, 83
shape components 264
shape-memory plastic 2
shear 12, 18
shear flow 22
shearing 374
shear rate 12
shear strain 13
shear stress 13
shear viscosity 13
Shewhart chart 150
shift factor 26
shrinkage 11
significance level 161
single-screw extruder 30
single-stage plunger machine 33
singleton fuzzifier 74
singular value decomposition (SVD) 159
SISO (single-input-single-output) 92
soft-sensor 296
solid-bed 375
solid conveying zone 373
solidification rate 284
solid plug 373
special cause variation 155
SPE statistic 161
spherulites 10
spin casting 5
SPM 153
squared simple correlation coefficient 233
stacked modeling methods 219
stage-dependent variables 35
stage division 325
stage-independent variables 35
standard deviation 76, 155
start pressure 351
state-space matrices 91
state-space model 97
state variable 98
static network 298
statistical analysis 150
statistical copolymers 7
statistical distributions 157
statistical features 155
statistical process control 85
statistical process monitoring (SPM) 150, 153
statistical quality control (SQC) 150
step-change 44
step-change profiles 270
step-down 270, 342
step response 63
step-up 270, 342
step-wise regression 237
stochastic approximation (SA) 44
strangulation stresses 15
stress and strain relationship 18
structured data 155
sum of squared error (SSE) 87

T

temperature sensitivities 27
terephthalic acid 6
thermoplastic 5
thermoset 5
three-dimensional array 155
time constant 98
time-dependent process 11
time-invariant 91
time-slice matrices 177
time-slice PLS model 215
time-specific effects 214
time-varying characteristics 44
Toeplitz matrix 91

torsional flow 16
trail-and-error tests 89
training mold 307
transfer molding 5
twin-screw extruder 30
two-dimensional (2D) dynamic 109
two-dimensional data matrix 155, 158
two-dimensional system 109

U

ultrasonic pulse-echo technique 251
unequal-length batch processes 194
unequal-length phenomena 194
uniaxial 11
unimodal HDPE 9
unity variance 155
unity-variance scaling 176
univariate control charts 151
universal gas constant 26

V

validation molds 307
valve sticking 207
variable regulation 39
variable selection 235
variable-wise unfolding 162
varying pressure 342
velocity form of PID 42
velocity/pressure (V/P) transfer 70, 251, 267
video recording 372
viscoelastic 17
viscosity-shear rate relationship 22
viscosity temperature-dependence effect 26
visual injection barrel system 372
visualization 251
Voigt model 20
voltage-capacitance relation 264

W

warm-up stage 57
warpage 11
weight-average molecular weight 9
weighted / distribution 182
weighted loading matrices 178
weighting factor 63
Weissenberg effect 15
Weissenberg extruder 16
Williams-Landel-Ferry (WLF) equation 27
wire coating 17
within-batch dynamics 168
within-batch time scale 168
WKFH normalization 195

X

X-ray diffraction (XRD) 11

Z

zero mean 155
zero-mean centering 176
Ziegler-Nichols (Z-N) method 43

www.ingramcontent.com/pod-product-compliance
Lightning Source LLC
LaVergne TN
LVHW081928200926
848484LV00009B/2391